Lecture Notes in Computer Scie

Edited by G. Goos, J. Hartmanis, and J. van

T0238291

Springer
Berlin
Heidelberg
New York
Barcelona
Hong Kong
London
Milan
Paris
Tokyo

Burkhard Stiller   Michael Smirnow
Martin Karsten   Peter Reichl (Eds.)

From QoS Provisioning to QoS Charging

Third COST 263 International Workshop on
Quality of Future Internet Services, QofIS 2002
and Second International Workshop on
Internet Charging and QoS Technologies, ICQT 2002
Zurich, Switzerland, October 16-18, 2002
Proceedings

Springer

Volume Editors

Burkhard Stiller
Information Systems Laboratory, IIS
University of Federal Armed Forces Munich, UniBwM
Werner-Heisenberg-Weg 39, 85577 Neubiberg, Germany
Computer Engineering and Networks Laboratory, TIK, Swiss Federal Institute of
Technology Zurich, ETH Zurich, Gloriastr. 35, 8092 Zurich, Switzerland
E-mail: stiller@tik.ee.ethz.ch

Michael Smirnow
Fraunhofer Institute FOKUS, Kaiserin-Augusta-Allee 31, 19589 Berlin, Germany
E-mail: smirnow@fokus.gmd.de

Martin Karsten
Multimedia Communications Lab, KOM, Darmstadt University of Technology
Merckstr. 25, 64283 Darmstadt, Germany
E-mail: Martin.Karsten@kom.tu-darmstadt.de

Peter Reichl
Forschungszentrum Wien, FTW, Donau-City-Str. 1, 1220 Wien, Austria
E-mail: reichl@ftw.at

Cataloging-in-Publication Data

Bibliograhpic information published by Die Deutsche Bibliothek
Die Deutsche Bibliothek lists this publication in the Deutsche Nationalbibliografie;
detailed bibliographic data is available in the Internet at http://dnb.ddb.de

CR Subject Classification (1998): C.2, H.4, H.3, J.1

ISSN 0302-9743
ISBN 3-540-44356-8 Springer-Verlag Berlin Heidelberg New York

Springer-Verlag Berlin Heidelberg New York
a member of BertelsmannSpringer Science+Business Media GmbH

http://www.springer.de

© Springer-Verlag Berlin Heidelberg 2002
Printed in Germany

Typesetting: Camera-ready by author, data conversion by Boller Mediendesign
Printed on acid-free paper   SPIN: 10870910   06/3142   5 4 3 2 1 0

Preface

This volume of the Lecture Notes in Computer Science series contains the set of papers accepted for publication at the colocated QofIS/ICQT 2002 workshops, i.e. the 3rd COST Action 263 International Workshop on Quality of future Internet Services (QofIS) and the 2nd International Workshop on Internet Charging and QoS Technology (ICQT), both of which took place at the ETH Zürich, Switzerland, hosted by the Computer Engineering and Networking Laboratory, TIK.

QofIS 2002 was the third in a series of highly successful technical workshops and meetings on Internet services within the framework of the COST Action 263 "Quality of future Internet Services", following previous events in Berlin, Germany in 2000 and in Coimbra, Portugal in 2001. ICQT 2002 was the follow-up to a vivid and extremely well-attended workshop on Internet economics and charging technology that took place within the framework of the Annual Meeting of the German Society for Computer Science (GI) and the Austrian Computer Society in 2001 in Vienna, Austria.

Combining QofIS and ICQT in the QofIS/ICQT 2002 event reflects a natural complement between Internet services technologies and their economics. The specific focus title "From QoS Provisioning to QoS Charging" was chosen deliberately in order to reflect current developments in this rapidly growing research area. As the emphasis lies on the technology for end-to-end and top-down provisioning of QoS, covering Internet as well as end-systems, focusing additionally on the economics of QoS and the required technology for charging support, both workshops taken together target the identification of solutions, investigations of their feasibility, and consolidation of technical and economic mechanisms to enable fast, guaranteed, and efficient provisioning of QoS-based services in the Internet. This is also reflected in the session titles, i.e., End-to-End QoS, Traffic Management, Traffic Marking and Queuing, Signaling, Multi-path Routing, QoS and Congestion Control, Charging Technologies, Pricing and Business Models, and Security.

These QofIS and ICQT workshops are aimed at bringing together researchers from the area of Internet technology and economy in both industry and academia to discuss recent and leading advances and to support further progress in these fields. By their nature, these workshops follow a single-track three-day program, in order to stimulate interaction and active participation. Altogether, the technical sessions of the QofIS and ICQT workshops contained 20 and 10 papers, respectively, which were selected out of 53 and 21 submissions, respectively, via a thorough reviewing process. Showing a truly international scope, the final program of both workshops included 20 European, 7 North and South American, and 3 Asian papers. To complete the technical program, keynote speakers were invited and a panel on "Premium IP: On the Road to Ambient Networking" was added.

QofIS/ICQT 2002 could not have taken place without the enthusiastic and never-abating support of a number of different organizations and people. Firstly, following the event series established during the last three years, COST Action 263 forms the steering lead of these workshops. Secondly, we would like to acknowledge the support of our QofIS/ICQT 2002 patrons, especially IBM, Siemens, and Swisscom for their financial contributions. With their support, these companies demonstrate their interest in the international forum that QofIS/ICQT 2002 provides and the results it will disseminate.

Both workshops owe their technical and research success to all members of the two distinct technical program committees, who devoted their excellent expertise and much of their time to provide this year's QofIS/ICQT with an excellent technical program. Furthermore, we would like to express our thanks to Jan Gerke, Hasan, David Hausheer, Jan Mischke, and Pascal Kurtansky, who performed brilliantly in maintaining the QofIS/ICQT 2002 Web server, managing the electronic system ConfMan for paper submission and review, and dealing with camera-ready papers. In addition, all of them assisted us unceasingly in all phases of the workshops' preparations with technical and administrative help. Thanks go also to Annette Schicker, who ran the QofIS/ ICQT 2002 registration and on-site offices and provided our participants with a first-rate service. Finally, we would like to thank the ETH Zürich and the TIK for hosting the QofiS/ICQT 2002 workshop in a convenient and stimulating environment.

August 2002

Burkhard Stiller
Michael Smirnow
Martin Karsten
Peter Reichl

ETH Zürich, Switzerland

Organization

General Chair

Burkhard Stiller   *University of Armed Forces Munich, Germany and ETH Zürich, Switzerland*

Steering Committee

Jon Crowcroft   *Univ. of Cambridge, UK*
James Roberts   *France Telecom R&D, France*
Michael Smirnow   *FhG FOKUS, Germany*
Burkhard Stiller   *UniBwM, Germany and ETH Zürich, Switzerland*
Fernando Boavida   *Univ. of Coimbra, Portugal*

Program Co-chairs QofIS 2002

Burkhard Stiller   *UniBwM, Germany and ETH Zürich, Switzerland*
Michael Smirnow   *FhG FOKUS, Germany*

Program Co-chairs ICQT 2002

Martin Karsten   *Darmstadt University of Technology, Germany*
Peter Reichl   *FTW Vienna, Austria*

Program Committee QofIS 2002

A. Azcorra   *UC3M, Spain*
D. Bauer   *IBM Research Laboratory, Switzerland*
H. v.d. Berg   *KPN Research, The Netherlands*
C. Blondia   *Univ. Antwerp, Belgium*
F. Boavida   *Univ. of Coimbra, Portugal*
O. Bonaventure   *FUNDP, Belgium*
G. Carle   *FhG FOKUS, Germany*
O. Casals   *UPC, Spain*
J. Crowcroft   *Univ. of Cambridge, UK*
M. Diaz   *LAAS, France*
J. Domingo-Pascual   *UPC, Spain*
H. Esaki   *Tokyo Univ., Japan*
D. Hutchison   *Lancaster Univ., UK*
G. Karlsson   *KTH, Sweden*
R. Kantola   *HUT, Finland*
K. Kilkki   *Nokia Research Center, Finland*
Y. Koucheriavy   *TUT, Finland*
M. Luoma   *HUT, Finland*
H. de Meer   *Univ. College London, UK*

E. Monteiro	*Univ. Coimbra, Portugal*
R. v.d. Mei	*KPN Research, The Netherlands*
G. Pavlou	*Univ. Surrey, UK*
M. Popa	*Procetel, Romania*
J. Roberts	*France Telecom R&D, France*
D. Serpanos	*Univ. of Patras, Greece*
M. Smirnow	*FhG FOKUS, Germany*
J. Solé-Pareta	*UPC, Spain*
I. Stavrakakis	*Univ. of Athens, Greece*
B. Stiller	*UniBwM, Germany and ETH Zürich, Switzerland*
H. Stüttgen	*NEC, Germany*
P. Van Mieghem	*Delft Univ. of Tech., The Netherlands*
G. Ventre	*Univ. Napoli, Italy*
J. Virtamo	*HUT, Finland*
L. Wolf	*Univ. Braunschweig, Germany*
A. Wolisz	*TU Berlin, Germany*
J. Wroclawski	*MIT, U.S.A.*
A. Zehl	*T-Systems, Germany*
M. Zitterbart	*Univ. Karlsruhe, Germany*

Program Committee ICQT 2002

J. Altmann	*HP Palo Alto, U.S.A.*
R. Andreassen	*Telenor, Norway*
T. Braun	*Univ. of Bern, Switzerland*
C. Courcoubetis	*AUEB, Greece*
C. Edwards	*Lancaster Univ., UK*
L. Heusler	*IBM Research Laboratory, Switzerland*
M. Karsten	*Darmstadt University of Technology, Germany*
C. Linnhoff-Popien	*LMU München, Germany*
S. Leinen	*SWITCH, Switzerland*
R. Mason	*Univ. of Southampton, UK*
A. Odlyzko	*Univ. of Minnesota, U.S.A.*
H. Oliver	*HP European Labs, UK*
H. Orlamünder	*Alcatel SEL, Germany*
M. Ott	*Semandex Networks, U.S.A.*
K. Park	*Purdue Univ., U.S.A.*
D. Reeves	*North Carolina State Univ., U.S.A.*
P. Reichl	*FTW Vienna, Austria*
B. Rupp	*Arthur D. Little, Germany*
V. Siris	*ICS Forth, Greece*
D. Songhurst	*BT, UK*
O. Spaniol	*RWTH Aachen, Germany*
B. Stiller	*UniBwM, Germany and ETH Zürich, Switzerland*
L. Wolf	*Univ. of Braunschweig, Germany*

Local Organization

Jan Gerke	*ETH Zürich, Switzerland*
Hasan	*ETH Zürich, Switzerland*
David Hausheer	*ETH Zürich, Switzerland*
Pascal Kurtansky	*ETH Zürich, Switzerland*
Annette Schicker	*Registration Office, Ringwil-Hinwil, Switzerland*

Reviewers

The task of a reviewer required serious and detailed commenting on papers submitted to QofIS/ICQT 2002. Therefore, it is of great pleasure to the Program Committee Co-chairs to thank all those reviewers listed below, in addition to the reviewing PC members, for their important work.

S. Aalto	G. Lichtwald
S. Abdellatif	R. Litjens
P. Antoniadis	H. Lundqvist
G. Auriol	I. Más Ivars
H. Balafoutis	X. Masip-Bruin
P. Balaouras	D. Moltchanov
A. Banchs	E. Ossipov
M. Calderón	P. Owezarski
D. Careglio	A. Panagakis
L. Cerdà	C. Pelsser
C. Chasssot	K. Pulakka
G. Cheliotis	J. Quittek
M. Curado	B. Quoitin
G. Dán	P. Reinbold
E. Exposito	E. Rinde
P. Flegkas	A. Shaker
V. Fodor	S. Spadaro
A. Garcia-Martinez	S. Tartarelli
J. Harju	K. Tepe
I. Iliadis	P. Trimintzios
M. Janic	S. Uhlig
Q. Jiang	H. Vatiainen
C. Kenyon	H. Velayos
V. Laatu	U. Walter
N. Laoutaris	N. Wang
P. Lassila	K. Wu
Y. Li	L. Yang

QofIS/ICQT 2002 Supporters

Swiss Federal Institute of Technology, ETH Zürich, Switzerland

IBM Research, Zurich Research Laboratory, Switzerland

Siemens Schweiz AG, Zürich, Switzerland

Swisscom AG, Bern, Switzerland

Computer Engineering and Networks Laboratory TIK, ETH Zürich, Switzerland

Table of Contents

Quality of future Internet Services (QofIS)

End-to-End QoS

DiffServ Traffic Management

Traffic Marking and Queueing

Signaling and Routing

Multi-path Routing

Panel

Service Differentiation and QoS Control

Congestion Control and MPLS

Internet Charging and QoS Technology (ICQT)

Invited Keynote

Charging Technologies

Pricing Models

Economic Models and Security

Quality of future Internet Services
(QofIS)

Implications for QoS Provisioning Based on Traceroute Measurements

Milena Janic, Fernando Kuipers, Xiaoming Zhou, and Piet Van Mieghem

Delft University of Technology
Information Technology and Systems
P.O. Box 5031, 2600 GA Delft, The Netherlands
{M.Janic, F.A.Kuipers, Z.Xiaoming, P.VanMieghem}@its.tudelft.nl

Abstract. Based on RIPE NCC *traceroute* measurement data, we attempt to model the Internet graph with the graph G_1, which is the union of the most frequently occurring paths in the RIPE *traceroute* database. After analyzing the topological properties of this graph, a striking agreement with the properties of the uniform recursive tree has been found. This similarity allows us to compute the efficiency of multicast. After assigning polynomially distributed link weights to the edges of the graph G_1 and comparing the simulated shortest paths with the *traceroutes*, no clear conclusions on the link weight structure could be deduced. Finally, possible implications of the measurement data on the provisioning of Quality of Service are suggested.

1. Introduction

Many researchers have attempted to provide a map of the Internet as well on a router-level as on an Autonomous Systems (AS) level [11][7][3]. In 1995, Pansiot and Grad [11] created a router-level topology based both on *traceroutes* to 5000 geographically distributed destinations from a single node, as well as on *traceroutes* from a subset of 11 nodes chosen from the set of 5000 nodes to the rest of the destinations. Govindan and Tangmunarunkit [7] obtained a snapshot of the Internet topology by using a Mercator program. Mercator is designed to map the network from a single source location without an initial database of target nodes for probing.

Most of the researchers have used the *traceroute* utility [12] for acquiring a map of Internet. We have analyzed and categorized errors occurring in *traceroute* measurements obtained from RIPE NCC, and based on this *traceroute* data, we attempt to create a graph that will approximate a part of Internet. Moreover, by assigning link weights that are polynomially distributed, and comparing simulated paths with real *traceroute* data, we try to gain better knowledge of the Internet link weight distribution. Investigating the possible implications of *traceroute* measurements on the provisioning of Quality of Service (QoS) has, to our knowledge, never been done.

The remainder of the paper is organized as follows: The construction of the approximate Internet graph and its topological properties (hopcount, node degree) are described in detail in Section 2. The same Section presents the results of the analysis of the Internet link weight structure and of the Internet path dominance. The possible

B. Stiller et al. (Eds.): QofIS/ICQT 2002, LNCS 2511, pp. 3-14, 2002.

implications of *traceroute* measurements for multicast deployment and end-to-end QoS are addressed in Section 3. Finally, we conclude in Section 4.

2. Approximating the Internet Graph

In this section we try to model the Internet based on *traceroute* data provided by RIPE NCC (the Network Coordination Centre of the Réseaux IP Européen). At the moment, RIPE NCC performs *traceroute* measurements between 50 measurement boxes scattered over Europe (and few in the US and New Zealand). Every 40 seconds, probe-packets of fixed size have been sent between each pair of measurement boxes, and the data has been collected once a day in a central point in RIPE (for a further detailed description of the measurements and measurement configuration we refer to [13]). The graph G_1 is the union of the most frequently occurring *traceroute* paths, provided from measurements performed by RIPE NCC in the period 1998-2001. Based on G_1 and the *traceroute* database, we will investigate whether the Internet possesses a polynomial link weight structure. We conclude this section, by evaluating how stable Internet interface paths really are.

2.1. Constructing G_1

For each pair of measurement boxes a large number of different paths have been distinguished in the database. Obviously, some of the *traceroute* records suffer from errors. For 14 test-boxes (out of 50) no record in the database could have been found. Moreover, *traceroute* records for 5 test-boxes have all been erroneous (temporary or persistent loops and unresponsive routers). Therefore, these 19 test-boxes have been excluded from further analysis. In the *traceroute* data from the remaining 31 boxes the most dominant path, i.e. the path occurring most frequently has been determined, resulting in totally 465 most dominant paths. We ascertained further that 17% of the those *traceroutes* suffer from errors mentioned above. The graph G_1 has been created by including every link belonging to each of the remaining 386 non-erroneous paths, resulting in a graph consisting of 1888 nodes and 2628 edges. Here we must make one important remark. The *traceroute* utility returns the list of IP addresses of routers along the path from source to destination. One router can have several interfaces, with several different IP addresses. To determine which IP addresses belong to one router is a rather difficult task, due to, among others, the security reasons (port snooping). As a consequence of this, the graph G_1 represents the approximation of the Internet interface map, not of the Internet router map.

In Subsection 2.3 we will show that the most dominant path alone does not cover the majority of paths between a particular source and destination. Intuitively, a much better approximation to the Internet graph would be a graph created as the union of k most frequently occurring paths. In our study, we have also considered the graph G_{10}, constructed as the union of ten most frequently occurring paths, and we have discovered that properties of G_{10} still resemble those of G_1. Therefore, due to the higher complexity of creating G_k, in further analysis we will confine ourselves to G_1. Let us remark that although the graph G_1 differs from the real underlying Internet graph G_{INT}, (because it is an overlay network on G_{INT}) it seems to present a reasonable approximation to G_{INT} (or a part of G_{INT}). Moreover, nearly 85% of the total number of nodes is already spanned by the most dominant *traceroutes* from (any) 18 sources to all destinations. Including the most dominant *traceroutes* of a new source only adds 1.5% of new nodes to the topology.

2.2.   The Topological Properties of G_1

In Figure 1 and Figure 2, the probability density function of the hopcount and node degree in the graph by G_1 has been plotted, respectively. The data of the hopcount has been fitted with a theoretical law (4.1) derived in [15] based on the random graph with uniformly (or exponentially) distributed link weights. The essential observation in the modeling [15] is that the shortest path tree in the random graph with uniformly (or equivalently exponentially) distributed link weights is independent of the link density p. This implies that even the complete graph with uniformly distributed link weights possesses precisely the same type of shortest path tree as for any connected random graph with uniformly distributed link weights. In other words, to obtain a source-based shortest path tree, the link weight structure is more decisive than the underlying random topology. In [15] has been shown that the shortest-path problem in the class of random graph $Gp(N)$ [2] with exponentially distributed link weights can be reformulated into a Markov discovery process with an associated uniform recursive tree [14]. A uniform recursive tree is defined by the following growth rule: given a uniform recursive tree with N nodes, a uniform recursive tree with $N+1$ nodes is deduced by attaching the $N+1$-th node uniformly (thus with probability $1/N$) to each of the N other nodes in the tree. If the number of nodes $N \to \infty$, the probability density function of the hopcount in the uniform recursive tree is shown [15] to obey precisely

$$\Pr[h_N = k] = \frac{(1+o(1))}{N} \sum_{k=1}^{N-1} c_{m+1} \frac{\ln^{k-m} N}{(k-m)!} \tag{1}$$

where c_k are the Taylor coefficients of $\Gamma(x)^{-1}$ listed in [1]. The law (1) can be approximated by a Poisson law,

$$\Pr[h_N = k] \approx \frac{(E[h_N])^k}{k! N} \tag{2}$$

which implies that $E[h_N] \approx var[h_N] \approx \ln N$.

Figure 1 illustrates that a fit with (1) is reasonable, but the quality remains disputable. A more striking agreement with the uniform recursive tree is shown by the degree law in Figure 2. We observe that the pdf of the node degrees in G_1 follows an exponentially decreasing function with rate -0.668 over nearly the entire range. The ratio of the average of the number of nodes with degree k, denoted by D_N^k, over the total number of nodes in the uniform recursive tree obeys for large N [14]

$$\frac{E[D_N^k]}{N} = \frac{1}{2^k} + O\left(\frac{\log^{k-1} N}{N^2}\right) \tag{3}$$

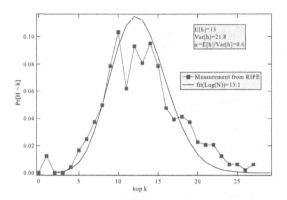

Fig. 1. *Both measurements and theory for pdf of hopcount in G_1 with N = 1888*

which is, for large N, close to *Pr[degree = k]*, the probability that an arbitrary node has degree k. Hence, the decay rate of the pdf of the node degrees in the uniform recursive tree equals $-\ln 2 = 0.693$. In summary, the pdf of the node degrees in G_1 (the union of the most dominant non-erroneous *traceroutes*) follows a same law as that in the uniform recursive tree. This intriguing agreement needs additional comments. It is not difficult to see that, in general, the union of two or more trees (a) is not a tree and (b) has most degrees larger than that appearing in one tree. Hence, the close agreement points to the fact that the intersection of the trees rooted at a measurement box towards the other boxes is small such that the graph G_1 is 'close' to a uniform recursive tree. This conclusion is verified as follows. Starting with the first union $T = T_1 \cup T_2$ of the trees T_1 and T_2, we also have computed the intersection $T_1 \cap T_2$. Subsequently, we have added another tree T_3 to the union $T = T_1 \cup T_2 \cup T$ and again computed the intersection (or overlap) $(T_1 \cup T_2) \cap T_3$. This process was continued until all trees were taken into account. We found that (a) the number of nodes in the intersection was very small and (b) the common nodes predominantly augmented the degree by 1 and in very few cases by more than 1.

It is likely that the overlap of trees would be larger (in terms of common nodes and links) if we had considered the router level map instead the interface map. The similarity in properties of the graph G_1 and the uniform recursive tree might be in that case smaller. This might explain the discrepancy with the results of Faloutsos et al. [5], who have reported a power law for the degrees in the graph of the Internet. This seems to suggest that G_1 is not representing the graph of the Internet well. Or, put differently, the union of *traceroutes* seems to possess a different structure than the underlying graph G_{INT}. Another possible reason for this discrepancy is that the number of test boxes (and path trees) considered here is small compared to the number of nodes. Finally, a source tree in Internet seems thus to be well modeled by imposing uniform or exponential link weight structure on the topology. Both uniform and exponential distributions have the same minimal domain of attraction [15]. Furthermore, this link weight structure models the end-to-end IP paths consisting of concatenations of shortest paths sections [intra-domain routing] "randomized" by BGP's policy-driven path enforcements [inter-domain routing].

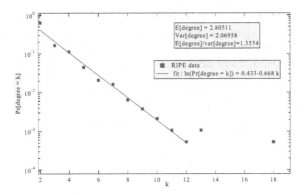

Fig. 2. *The pdf of the node degree (RIPE data) in G1 with N = 1888*

2.3. The Analysis of the Link Weight Structure

In addition to the network topology, knowledge on the link weight distribution is needed in order to solve a particular routing problem. In the Internet topology models proposed earlier, all links weights have unit weight (w=1). However, weights on Internet are assigned according to physical properties, such as link propagation delay or capacity, or in order to enable directing traffic along certain preferred paths, and hence, assigning equal value to all links is not justified. The next step in our analysis was to gain some more insight in the effective Internet link weight structure. In order to perform such an analysis, weights have been assigned to the edges in G_1, and subsequently, for each source-destination pair, the shortest path in G_1 has been compared to the most dominant *traceroute* path. For reasons explained in [15] we have chosen a polynomial distribution for the link weights. Hence, for each link between nodes i and j in G_1, the link weight obeys $P[w \leq x] = x^\alpha 1_{0 \leq x \leq 1} + 1_{x \geq 1}$, for various values of the exponent α (α= 0.05, 0.2, 0.5, 1, 2, 5, ∞). If sufficiently many values of α are applied, we may assume that the value of a that leads to the minimum number of dissimilarities with Internet *traceroutes*, may represent the best polynomially distributed match for the Internet link weight structure. A minor secondary reason for choosing polynomially distributed link weights is motivated by [9]. The asymptotic analysis in [9] reveals that, for random graphs with uniformly (or equivalently exponentially) distributed, independent link weights, the hopcount in r dimensions behaves similarly as in the random graph in r=1 dimension with polynomially distributed link weights, where the polynomial degree a is equal to the dimension r.

For each value of the exponent α, 300 replicas of G_1 were generated, and in each of them, for each source-destination pair, the shortest path was computed using Dijkstra's algorithm and compared to the corresponding traceroute in the database.

Figure 3 shows the difference in hopcount $\Delta = h_{RIPE} - h_{SIM}$ as a function of the exponent α. This figure immediately reveals that a small α is undoubtedly not a good model for link weights on the Internet, since the hopcount of the shortest path can be considerable longer than the traces. With the increase of α, the negative tails decrease in the favor of positive values of Δ_h. Only when $\alpha=\infty$ (Figure 4), corresponding to all link weights being precisely equal to 1, the shortest paths are always shorter than or equal to the *traceroutes*. Even then *traceroutes* are minimal hopcount paths only in about 22% of the cases. These results seem to indicate that either a link weight structure in the Internet is difficult to deduce when assuming that link weights are polynomially distributed or that the latter assumption is not valid. Even the case where all link weights are equal to 1 does not perform convincingly.

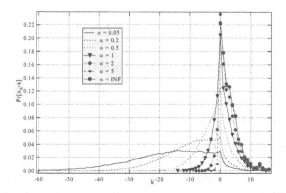

Fig. 3. *The probability density function of the different number of hops in RIPE traces and in simulated shortest paths.*

Inside AS, the intra-domain routing is based on a shortest path criterion, but routing between AS (BGP) does not clearly follow a minimization criterion [8]. Hence, e2e routing may result in paths that obey a different criterion, not necessarily a minimization. This may explain the poor agreement with our simulations.

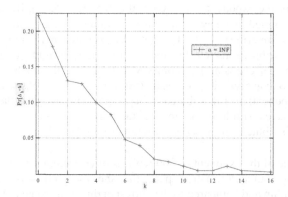

Fig. 4. *The probability density function of the different number of hops in RIPE traces and simulated shortest paths ($\alpha=\infty$)*

2.4. How Dominant Are Internet Paths?

For each source-destination pair (tt_i, tt_j) the number of different interface paths collected in the database is very large, rising up to an amazing number of 9032 different interface paths for some (tt_i, tt_j) pairs, in a period of roughly three years! With the goal of establishing the critical number of different paths that constitute the majority of paths, we evaluated the probability density function and the probability distribution function that a particular non-erroneous path P is the kth dominant path.

We observed that for most source-destination pairs, there is no clear dominant path, which has been illustrated in Figure 5, where we have plotted, for 546 source-destination pairs, the probability that the most dominant path will be chosen. On average, the probability of taking the most dominant path equals 0.13, with a variance of 0.01. This indicates that IP packets regularly follow different paths over the three years time period. This behavior may have several causes:

1.   Growth of the Internet: Over the years, the Internet has grown with respect to the number of nodes and links. This growth and increased link-density has resulted in an increased number of paths between nodes. It is therefore likely that over time better paths emerge, becoming most dominant paths.

2.   Load balancing: Through load balancing, Internet providers try to relieve the shortest most dominant paths in order to avoid overloading them.

3.   Multi-homing: Multi-homing refers to having multiple connections or links between two parties. For instance, we can have multi-homing of hosts, networks and Internet service providers. The reason for doing this is to increase robustness (back-up paths) and for load sharing. A consequence of multi-homing is the increase in the size of the BGP routing tables.

4.   Changing SLAs: Routing on the inter-domain level is subject to policy constraints (Service Level Agreements). These policies may change or new policies may appear, affecting traffic behavior and routing possibilities.

5.   Failures: Our measurements seem to indicate that the Internet has lost in stability and robustness over the past five years. As failures occur more frequently, packets may more often deviate from the most dominant path.

It is debatable whether the behavior in Figure 5 is disturbing or not. If the behavior is mostly due to causes 1-4, then there are no big concerns. However, if the lack of dominant paths is caused by network failures, this would be an alarming situation. Currently the update of router tables only occurs at relatively large intervals of time (15 min.), when some node/link failure has occurred. With QoS routing, we will need to trigger flooding, based on dynamic QoS parameters, like bandwidth and delay. If there are many short-term QoS applications, this will lead to an even greater lack of dominant paths. On the other hand, it is conceivable that for instance a corporate network will request a certain QoS connection for a large period of time (e.g. 1 year) for its users. Such a connection/path will become very dominant.

In [10] it is argued that the lack of interdomain failover due to delayed BGP routing convergence will potentially become one of the key factors contributing to the "gap" between the needs and expectations of today's data networks. In the worst case this delay can grow exponentially with the number of ASs. This can seriously impact the provisioning of QoS, which assumes a stable underlying interdomain forwarding infrastructure and fast IP path restoral. With the emerging of QoS the update frequency will probably increase, leading to an even worse scenario. It is therefore important to have a "smart" update strategy and perhaps desirable to structure the Internet into more hierarchies, reducing the number of domains and updates per hierarchy level.

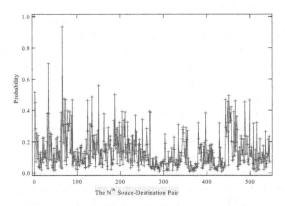

Fig. 5. *For each source-destination pair the probability that the most dominant path will be chosen*

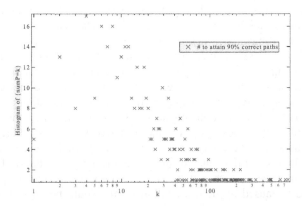

Fig. 6. *The histogram of the number of the non-erroneous paths per source-destination pair over about 3 years time of which k paths are used more than 90% of the time*

Figure 6 represents a histogram of the number of source-destination pairs of which the number of dominant paths, denoted by *numP*, necessary to satisfy the condition $\Pr[P \leq k] = 90\%$ (i.e. which are used more than 90% of the time). We can see that the number of paths is relatively large and varies considerably. The reasons are similar to the 5 reasons listed above. Nonetheless, it would be desirable to evaluate whether those are truly the reasons and if so, which of them is the most dominant.

3.   Implications for Future Internet Services

3.1.   Implications for Multicast Deployment

The indication that the graph G_1 is similar to a uniform recursive tree (see Section 2) has an implication to multicast deployment. The purpose of multicast is to provide savings in bandwidth consumption. However, the network cost for multicast protocols is larger than that of unicast. The computational and administrational overhead of multicast group management increases the deployment cost. Clearly, the deployment of multicast can only be justified if the netto gain defined as savings minus costs is larger than the netto gain for unicast. Therefore, in order to encourage the deployment of multicast on a larger scale, a break-even analysis in the number of users can determine when multicast is more profitable than unicast. To our knowledge, none of the business model proposals to this moment proposes a method for computing the threshold value for the number of users.

In order to define this break-even value, it is necessary to quantify the cost and the gain of multicast. Defining the gain of multicast has been initiated by Chuang and Sirbu [4], but Van Mieghem et al.[17] have presented the general framework, valid for any graph. Van Mieghem et al. derived in [17] the exact mathematical expression for the multicast efficiency over unicast, in terms of bandwidth utilization, in the random graph $Gp(N)$, with independent exponentially distributed link weights w, and m multicast users uniformly chosen out of the total number of nodes N. As already mentioned, the shortest path tree deduced from the Internet measurements was shown to be very close to a uniform recursive tree. For this uniform recursive tree, the corresponding multicast efficiency g_N, defined as the average number of hops in the shortest path tree rooted at a particular source to m randomly chosen destinations, for N large (and all m) is given by

$$g_N(m) \sim \frac{mN}{N-m} \log\left(\frac{N}{m}\right) - 0.5 \tag{4}$$

Since we have shown that the graph G_1 or even better, the shortest path tree from a source to m destinations has properties similar to those of the uniform recursive tree, we were triggered to compare the multicast efficiency in shortest path trees in the graph G_1 with the one computed with the theoretical law (5). The simulations have been conducted as follows: one node, out of 1888 nodes in G_1, has been uniformly chosen to represent a source, and from this source a shortest path tree to m randomly chosen destinations has been constructed. For each number of users m, 105 different combinations of source and destinations have been chosen, and subsequently 105 different shortest path trees have been constructed. The number of users m has been chosen to be small, (the ratio $x=m/N$ is smaller than 2%).

In Figure 7, we plotted the simulation data together with the law (5), which shows a good agreement. It has been shown in [17] that the average hopcount in the r. g. obeys

$$E[h_N] = \log N + \gamma - 1 + o(1) \tag{5}$$

where γ is Euler's constant ($\gamma = 0.5772156...$). Since unicast uses on average $f_N(m)=mE[h_N]$ links, the ratio $g_N(m)/ f_N(m)$ could be used as a good estimate for gain of multicast.

The remaining problem in the break-even analysis of the number of users reduces to quantifying the cost. In other words, all factors that impact the cost of multicast (such as additional cost in construction of multicast trees and in the multicast protocol, maintaining multicast state at routers, etc.) need to be quantified. This not a trivial task is, however, a topic for further research.

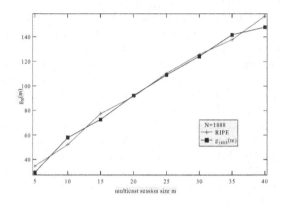

Fig. 7. *The efficiency of multicast*

3.2.   Implications for E2E QOS

The traceroute measurements are end-to-end (e2e) measurements in a part of the current Internet. The fact that there is no QoS architecture implemented yet, prohibits us to measure a "QoS-behavior" in the Internet. Nonetheless, we will reflect on how the traceroute measurements may provide some insight into an important question: Is the current Internet suitable for delivering guaranteed end-to-end QoS?

In order to be able to answer this question, we must first determine the necessary requirements for delivering guaranteed end-to-end QoS. We will restrict ourselves to only enumerate the requirements relevant to this paper:

·   As in current network routing, QoS routing must consist of two identities, namely (1) a QoS routing protocol and (2) a QoS routing algorithm. The QoS protocol should distribute the state of the network along with the relevant QoS parameters (e.g. available bandwidth and delay) as efficient as possible, such that each node in the network will have a consistent and up-to-date view of the network. Based on this view, and the requirements/constraints of a QoS-dependent application, the QoS routing algorithm should find a path through the network that can guarantee the set of constraints. Finding a path based on multiple additive constraints is a NP-complete problem [6][18]. Over the years much research was devoted to finding QoS protocols and algorithms. However, much of this effort was directed to protocols and algorithms on the intradomain level. On this level, it seems possible to guarantee QoS, but on the interdomain level with its many ASs, this will become extremely problematic. In order for end-to-end QoS to become reality, we will need to "QoS-enhance" our routing protocols and algorithms on both intra- and inter-domain level.

·   Current Internet routing is hop-by-hop based. In [16] it has been proven that QoS constraints cannot be guaranteed in a hop-by-hop fashion. We need a connection-oriented approach, reserving the required resources along the path.

·   In order for guarantees to be made, one must understand and be able to control the physical Internet. In other words, routers and switches must be dimensioned properly. The Internet topology must be stable and robust. If link and nodes frequently go down or some other failures occur, this will impact the level of QoS experienced by many users.

Section 2 has indicated that the choice of paths in Internet is currently very volatile. There is hardly ever a path between a source-destination pair that is used more than 50% of the time over the measured period of three years. This implies that routing behavior, due to its volatile nature, is very hard to predict. If predictions about routing or traffic behavior cannot be made, this impacts the ability of a provider to efficiently deliver its services. Predictability of network behavior is essential for QoS negotiation, QoS admission control and maintaining admitted QoS sessions.

Finally we would like to touch on the most difficult part in QoS provisioning, namely QoS routing on the interdomain level. Currently, BGP allows each AS to independently formulate its routing policies and allows these policies to override distance metrics. These policies can conflict, resulting in route oscillations or may not be robust with respect to network failures. Routing on the interdomain level, is therefore not a simple shortest path problem, but could be seen as an instance of the stable path problem [8]. To determine whether such an instance is solvable is a NP-complete problem [8]. This clearly demonstrates that maintaining robustness on the interdomain level is more difficult than on the intradomain level. Moreover, the *traceroute* measurements indicated that the influence of BGP is tremendous. QoS path selection based on policy constraints as well as on other constraints is a task of the QoS routing algorithm, which can be vendor/provider specific. However, the stable paths problem illustrates that there is a need for some kind of control in creating policies, such that the stability and robustness of the Internet is maintained. We believe that the current Internet is still prone to too much failures and therefore, in order to attain QoS in a future Internet, much effort has to be put in making the Internet more robust. We conceive that MPLS based on controlled paths is likely the current best strategy towards providing guaranteed e2e QoS.

4.   Conclusions

The main goal of this paper was to provide some more insight into possible QoS provisioning, based on RIPE NCC *traceroute* measurement data. For this purpose the graph G_1 has been constructed as the union of most frequently occurring paths in *traceroute* measurements. A strong resemblance of the properties of this graph to those of the uniform recursive tree has been found. We showed that this analogy could enable network operators to compute the multicast efficiency over unicast and to determine eventually the break-even point in m to switch from unicast to multicast. After assigning polynomially distributed link weights to the edges of the graph G1 and comparing the simulated shortest paths with the *traceroutes*, no clear conclusions on the link weight structure of the Internet could be deduced. Finally, we analyzed the *traceroute* measurements from a QoS perspective. Two important properties were motivated: (1) QoS provisioning will operate better as the Internet evolves to a more interconnected graph, leading also to a better modeling/understanding of the Internet graph and (2) the Internet has to be made more robust with respect to decreasing failures of routers and links and with respect to controlling interdomain-level policies.

Acknowledgments

We would like to thank dr. H. Uijterwaal of RIPE NCC for providing us the RIPE data.

References

[1] Abramovitz, M. and I.A. Stegun, "Handbook on Mathematical Functions" Dover Publications, Inc., N.Y.,1968.

[2] Bollobas, B., "Random Graphs", Cambrigde University Press, second edition, 2001.

[3] Burch H. and B. Cheswick. "Mapping the Internet". In IEEE Computer, April 1999.

[4] Chuang J. and M. Sirbu, "Pricing multicast communication: A cost-based approach," presented at the INET, 1998.

[5] Faloutsos, M. and P. Faloutsos and C. Faloutsos, "On power-law relationships of the Internet Topology", Proceedings of ACM SIGCOMM'99, Cambridge, Massachusetts, pp. 251-262, 1999.

[6] Garey, M.R., and D.S. Johnson, "Computers and Intractability: A guide to the Theory of NP-completeness", Freeman, San Francisco, 1979.

[7] Govindan R. and H. Tangmunarunkit, "Heuristics for Internet Map Discovery", In proceedings of IEEE INFOCOM'00, Tel Aviv, Israel, 2000.

[8] Griffin T. G., F. B. Shepherd and G. Wilfong, "The stable paths problem and Inter domain Routing", IEEE/ACM Transactions on Networking, vol. 10, No. 2, pp. 232-243, April 2002.

[9] Kuipers F.A. and P. Van Mieghem, "QoS routing: Average Complexity and Hopcount in m Dimensions", Proc. Of 2nd COST 263 International Workshop, QofIS 2001, Coimbra, Portugal, pp. 110-126, September 24-26, 2001.

[10] Labovitz C., A. Ahuja, A. Bose and F. Jahanian, "Delayed Internet Routing Convergence", IEEE/ACM Transactions on Networking, vol. 9, no. 3, June 2001.

[11] Pansiot J. J. and D. Grad. "On routes and multicast trees in the internet". In ACM Computer Communication Review, January 1998.

[12] Richard Stevens W., "TCP/IP Illustrated, volume 1, The Protocols", Addision-Wesley, Reading, Massachusetts, 1994.

[13] RIPE Test Traffic Measurements, http://www.ripe.net/ripencc/mem-services/ttm/

[14] Smythe, R. T. and H. M. Mahmoud, "A Survey of Recursive Trees", Theor. Probability and Math Statist., No. 51, pp. 1-27, 1995.

[15] Van Mieghem P., G. Hooghiemstra and R. W. van der Hofstad, "Scaling Law for the Hopcount", Delft University of Technology, report2000125, 2000.

[16] Van Mieghem P., H. De Neve and F.A. Kuipers, "Hop-by-hop quality of service routing", Computer Networks, vol. 37, pp. 407-423, 2001.

[17] Van Mieghem, P., G. Hooghiemstra and R. W. van der Hofstad, "On the Efficiency of Multicast", IEEE/ACM Transactions On Networking, vol. 9, pp. , 2001.

[18] Wang Z. and J. Crowcroft, "QoS Routing for supporting Multimedia Applications", IEEE J. Select. Areas Commun., 14(7):1188-1234, September 1996.

A Receiver-Driven Adaptive Mechanism Based on the Popularity of Scalable Sessions

Paulo Mendes[1,2], Henning Schulzrinne[1], and Edmundo Monteiro[2]

[1] Department of Computer Science, Columbia University
New York NY 10027-7003. Phone: +1 212 939-7000. Fax: +1 212 666-0140
{mendes,schulzrinne}@cs.columbia.edu
[2] CISUC, Department of Informatics Engineering, University of Coimbra
3030 Coimbra, Portugal,
{pmendes,edmundo}@dei.uc.pt

Abstract. Receiver-driven adaptation allows streaming of multimedia content to different receivers across heterogeneous networks. However, receivers are only encouraged to adapt if network providers guarantee a fair distribution of bandwidth and also the punishment of receivers that do not adjust their rate in case of congestion. We define a receiver-driven adaptive mechanism based on a new fairness protocol that provides the required guarantees for adaptation. We use simulations to evaluate the proposed mechanism and to compare its performance with other receiver-driven mechanisms.[1]

Keywords: quality adaptation, scalable sessions, SAPRA, fairness, TCP

1 Introduction

Source-based rate adaptation performs poorly in a heterogeneous multicast environment because there is no single target rate: the different bandwidth requirements of receivers cannot be simultaneously satisfied with one transmission rate. This problem can be solved if sources use scalable encoding and the adaptation task is performed by receivers. Scalable encoding [21, 7] divides a video stream into cumulative layers with different rates and importance. Layers are then sent to different multicast groups. All layers belonging to the same stream form a *session*, as in the Real-Time Protocol (RTP) [20]. The rate of each session is obtained by adding the rates of all its layers.

When a receiver-driven approach is combined with scalable encoding, receivers can adapt to the best quality the network offers, by selecting subsets of layers for their session. But, to motivate receivers to adapt, the network has to have three fairness properties. The first is *inter-session* fairness, the ability to guarantee a fair distribution of bandwidth between sessions sharing a service. The second is *intra-session* fairness, the ability to respect the importance of

---

[1] This work is supported by POSI-Programa Operacional Sociedade de Informação of Portuguese Fundação para a Ciência e Tecnologia and European Union FEDER

B. Stiller et al. (Eds.): QofIS/ICQT 2002, LNCS 2511, pp. 15–24, 2002.

each layer of a session. The third is the ability to punish *high-rate* sessions, i.e., sessions with a rate higher than their fair share of bandwidth, due to the fact that their receivers do not reduce the reception rate when packets are lost.

The current Differentiated Services (DS) model [2] aggregates traffic into services with different priorities at the boundaries of each network domain. Among the services DS can provide, the Assured Forwarding PHB (AF) [4] is ideal for transporting scalable sessions, since flows are assigned different drop precedences. Although AF services provide intra-session fairness, the DS model lacks the other two properties. Therefore, we proposed [14, 15] a protocol named *Session-Aware Popularity-based Resource Allocation* (SAPRA) that allows a fair allocation of resources in each DS service. SAPRA provides inter-session fairness by assigning more bandwidth to sessions with higher audience size, and intra-session fairness by assigning to each layer a drop precedence that matches its importance. SAPRA has a punishment function and a resource utilization maximization function. The former increases the drop percentage of high-rate sessions during periods of congestion. The latter avoids waste of resources when sessions are not using their whole fair share: the remaining bandwidth is equally distributed among other sessions. SAPRA is implemented in edge routers to handle individual traffic aggregated in each service: interior routers are not changed.

In this paper, we propose a simple receiver-driven adaptive mechanism named *SAPRA Adaptive Mechanism* (SAM). Simulation results show that when network resources are fairly distributed using SAPRA, a simple mechanism such as SAM has a good performance.

The remainder of this paper is organized as follows. In Section 2, we describe related adaptive mechanisms. Section 3 describes SAPRA support to receiver-driven adaptive mechanisms and characterizes SAM operation. In Section 4 we evaluate SAM using simulations. Section 5 presents some conclusions.

2   Related Work

McCanne et al. [12] developed the Receiver-driven Layered Multicast (RLM) mechanism, the first receiver-driven adaptive mechanism for scalable sessions. However, RLM has high instability with bursty traffic such as Variable Bit Rate (VBR), poor fairness and low bandwidth utilization [17]. Vicisano et al. [22] described a protocol called Receiver-driven Layered Congestion control (RLC) that complements RLM with a TCP-friendly functionality. However, RLC does not solve issues such as slow convergence and losses provoked by the adaptive mechanism on other flows.

An analysis of the pathological behavior of RLM and RLC [9] showed that their bandwidth inference mechanism is responsible for transient periods of congestion, instability and periodic losses.

Legout et al. [11] developed a receiver-driven protocol for scalable sessions named packet Pair receiver-driven cumulative Layered Multicast (PLM) that uses Packet-pair Probing (PP) [8] and Packet-level Generalized Processor Sharing (PGPS) scheduling [16] to infer the available bandwidth in the path. However, PLM has four major disadvantages. First, PP measurements can have large oscillations, since PP depends on packet size and the burstiness of traffic. Second, PLM requires all packets from all layers to be sent back-to-back and adds an extra bit to the packet header to identify the start of a PP burst. Third, PLM is not suitable for DS scenarios, since all routers have to implement PGPS. Fourth, PGPS uses the max-min fairness definition [1]. However, this fairness definition cannot be applied to discrete sets of rates [19], does not take into account the audience size of sessions, not increasing the number of receivers with good reception quality, and does not punish high-rate sessions.

3   SAM Description

In this section we describe how receivers use SAM to adapt to different network conditions. We also explain briefly how SAPRA support simple receiver-driven adaptive mechanisms, such as SAM. SAPRA is described and evaluated in [15] and a detailed study of its fairness policy can be found in [14].

3.1   SAPRA Support for Receiver-Driven Adaptation

SAPRA computes the *fair rate* of a session for each outgoing link of an edge router. The fair rate represents a percentage of the link bandwidth, given by the ratio between the session audience size and the total population of the link. Eq. 1 gives the fair rate F_{ui} of a session S_u in a link i, where n_{ui} is the audience size of S_u and C_i is the bandwidth of the service shared by m_i sessions.

$$F_{ui} = \left(\frac{n_{ui}}{\sum_{x=1}^{m_i} n_{xi}}\right) * C_i \tag{1}$$

The *sustainable rate* of a session is also computed for each outgoing link. The sustainable rate is the larger of the fair rate of the session and the sum of the session rate plus the bandwidth not being used in the link. Eq. 2 gives the sustainable rate U_{ui} of a session S_u in a link i, where r_{ui} is the rate of S_u, and b is the bandwidth not being used in that link.

$$U_{ui} = max(F_{ui}, (r_{ui} + b)) \tag{2}$$

Traffic is marked in each edge router using the fair rate of sessions. SAPRA also provides receivers with periodic reports about the minimum sustainable rate in the path from their session source. Reports are updated with a minimum interval of 1 s. In case the sustainable rate does not change significantly (25% or more in our experiments), reports are suppressed.

Receivers use SAM to add layers when notified of an increase in the sustainable rate. The sustainable rate can increase if: the session has a higher audience; other sessions have a lower audience; there is bandwidth not being used in the path of the session.

Receivers that join later reach an optimal quality level quickly, since they always get a report immediately after joining their session, even if the sustainable rate does not change significantly due to their arrival.

3.2   Overview of SAM Operation

Receivers can join sessions by, for instance, listening to Session Announcement Protocol (SAP) [3] messages, which may include information about the address and rate of each layer. Receivers join first the multicast group for the most important (lowest) layer and then the SAM algorithm controls the reception quality by joining and dropping additional (higher) layers.

The decision to join or drop layers depends on the session sustainable rate and on the existence of congestion in the session path. The sustainable rate, provided by SAPRA, gives SAM an indication of the maximum number of layers that receivers can join. Packet loss is a sign of penalization for high-rate sessions in a congested path. Packets start to be dropped from the less important layer, since SAPRA protects the most important ones. When losses happen in any layer, SAM is triggered to leave layers. In this paper we assume a loss limit of 2.5% as the maximum quality degradation allowed by receivers. We chose this value based on the study made by Kimura et al. [5], which shows that in MPEG-2 layering with Signal to Noise Ratio scalability, 5% of losses in the most important layer in addition to 100% of losses in all other layers, lead to a decrease of the quality of sessions from good to bad accordingly to the ITU-500-R rating [6].

SAM operation is divided into three states as shown in Fig. 1: steady state, join state and drop state. Receivers remain in the steady state as long as they do not receive a report and while losses are lower than 2.5%. Upon receiving a report, receivers enter the join state. If the new sustain-

Fig. 1. SAM States

able rate is higher than the previous one, receivers increase their reception quality adding layers. Since receivers know in advance the average rate of each layer, they immediately join as much layers as possible. The number of layers they can join is upper bounded by the sustainable rate of their session. After this, receivers return to the steady state. Receivers react to a lower sustainable rate considering the percentage of lost packets and not the report. If losses rise above 2.5%, receivers enter the drop state. In the drop state, receivers drop a layer every 500 ms, the *non-reacting* period, while losses are higher than 2.5%. The non-reacting period avoids over-reacting to losses. With losses equal or below 2.5%, a receiver enters the join state if it receives a new report while in the drop state. Otherwise, it returns to the steady state.

Since we compare SAM with PLM in section 4, we use the same non-reacting period as PLM. Although we use a loss limit of 2.5% and a non-reacting period of 500 ms, these values can be changed to suit other configurations.

4   SAM Evaluation

In this section we present simulations (using NS) that aim to show that SAM has small convergence time, remains stable in the optimal quality level, is fair towards TCP, and allows receivers to use a rate proportional to the audience size of their session and to the amount of bandwidth not being used. We use the three scenarios shown in Fig. 2. A complete set of results can be found in [13].

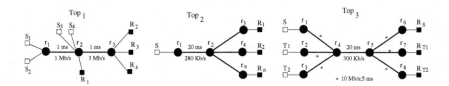

Fig. 2. Topologies

As metrics, we use the rate measured at receivers, presented with a precision of 10 kb/s. As system parameters, we use the type of traffic and the granularity of layers, since these factors affect the adaptation performance. We use layers with exponential rates, which are common in scalable codecs [18], and thin layers as diagnostic tool, since they can identify pathological behaviors.

The first simulation uses the topology *Top1* of Fig. 2. We aim to show how the sustainable rate computed by SAPRA allows SAM to reach an optimal quality level. We use four sessions: S_1 spans seconds 30 through 130, and S_2, S_3 and S_4 span seconds 10 to 240. Sessions S_1, S_2 and S_3 have one receiver each, and S_4 has one receiver until second 170 and five receivers after that. Each session has six layers: the most important layer, l_1, has 32 kb/s and each layer l_i has a rate equal to twice the rate of l_{i-1}. Sessions S_1 and S_2 share the link between routers r_1 and r_2, (r_1, r_2), and S_2, S_3 and S_4 share (r_2, r_3). R_i represents receivers of S_i. We assume that queues have a size of 64 packets, which is the default value in Cisco IOS 12.2, and data packets have 1,000 bytes, a middle value between the 576 bytes MTU of dial-up connections and the 1,500 bytes MTU of ethernet and high speed connections.

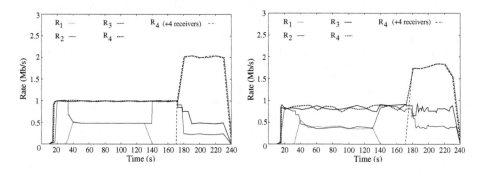

Fig. 3. Rate that receivers get when sessions have CBR (left) and VBR sources (right)

Fig. 3 (left) shows the rate that receivers get when sessions have Constant Bit Rate (CBR) sources. Until $t = 30$ s, S_2 is the only session in (r_1, r_2), so it has a sustainable rate of 1 Mb/s in that link. In (r_2, r_3), there are three sessions, S_2, S_3 and S_4. Since these three sessions have one receiver each, SAPRA distributes the link bandwidth equally between them, giving each a sustainable rate of 1 Mb/s. Therefore, R_2, R_3 and R_4 receive a sustainable rate of 1 Mb/s in the first report. This allows them to join four more layers, reaching a rate of 992 kb/s, as show in Fig. 3 (left). At $t = 30$ s, S_2 starts sharing (r_1, r_2) with S_1. Therefore, the sustainable rate of S_2 is diminished by half, becoming 500 kb/s, the same value of the sustainable rate of S_1. As a consequence, R_1 joins four layers, reaching 480 kb/s, and R_2 leaves one layer, decreasing its rate from 992 kb/s to 480 kb/s. R_3 and R_4 get a new report since the sustainable rate of their sessions increase more than 25% due to the decrease of the sustainable rate of S_2. However, R_3 and R_4 do not join l_6, maintaining their rate of 992 kb/s. This happens because the new sustainable rate is lower than 2.016 Mb/s, the total rate of the six layers. At $t = 130$ s, R_1 leaves and the sustainable rate of S_2 increases from 500 kb/s to 1 Mb/s. Therefore, R_2 gets a new report and grabs the bandwidth not being used by S_1 in (r_1, r_2), reaching again a rate of 992 kb/s. At $t = 170$ s, four more receivers join S_4, increasing the sustainable rate of the session from 1 Mb/s to 2.142 Mb/s. Therefore, the five receivers of S_4 receive a new report, which allows the four new receivers to join six layers and the previous receiver to join one more layer, reaching each of them a rate of 2.016 Mb/s. This shows that SAM allows late-join receivers to get the same quality as previous members of an existing session. Due to the higher audience size of S_4, the sustainable rates of S_2 and S_3 decrease to 428 kb/s each, which is insufficient to maintain their quality level. R_2 is the first to react to losses leaving l_5 with 2.54% of losses and leaving l_4 500 ms after that, with 9.68% of losses. Due to the reaction of R_2, the sustainable rate of S_3 increases to 760 kb/s, which is sufficient to maintain l_4. This shows that with SAM, competition for bandwidth does not increase quality oscillations.

Fig. 3 (right) shows SAM behavior when sessions have a VBR source. Each layer has a mean rate equal to its rate with CBR, a maximum and minimum rate 1.5 times higher and lower than the mean rate, respectively, and a burst time of 2 s with a deviation of 0.5 s. Results show that SAM is able to adjust fairly the reception quality even when sessions have oscillatory rates. We can observe that S_2 and S_3 have a higher rate with VBR than with CBR after $t = 170$ s. This happens because they have a higher sustainable rate, since S_4 has a rate lower with VBR than with CBR.

The simulation with topology *Top1* of Fig. 2 evaluates the operation of SAM. Results shows that reports provided by SAPRA allow a simple adaptive mechanism, such as SAM, to keep receivers rate close to the sustainable rate of their sessions, guaranteeing inter-session fairness and increasing bandwidth utilization. Results also show that a loss threshold of 2.5% does not lead to quality oscillations.

In what concerns the convergence time, stability and fairness with TCP, Legout et al. show that PLM performs better than RLM and RLC. Hence, we compare SAM with the results presented by Legout et al. for PLM. For that, we use the same scenarios used by Legout. et al.. These simulations use the topologies *Top2* and *Top3,* shown in Fig. 2.

We use the topology *Top2* shown in Fig. 2 to evaluate the time SAM takes to convergence to an optimal quality level, its accuracy and stability. We also show the performance that SAM would have, if receivers didn't know in advance the rate of each layer. Links (r_2, r_n) have a bandwidth uniformly chosen between [500,1000] kb/s and a delay uniformly chosen between [5,150] ms. We use one session, S, and layers with a thin granularity of 50 kb/s. At $t = 5$ s, twenty receivers join S. From $t = 30$ s to $t = 50$ s, a receiver joins S every 5 s. At $t = 80$ s, five more receivers join this session. Each receiver is positioned in a different leaf router. We use the packet and queue size used for PLM, i.e., packets with 500 bytes and queues with 20 packets.

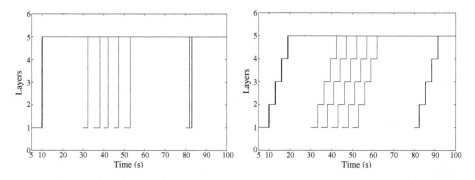

Fig. 4. Convergence time with (left) and without (right) knowledge of layers rates

Fig. 4 (left) shows that the first twenty receivers start to converge to their optimal rate at $t = 10$ s, 5 s after joining S, while late-join receivers wait a little less (3 s) to converge. This happens because receivers only start to converge after receiving the first report. First receivers wait longer, since the fair rate of S has to be computed for the entire path, and so the first report is originated by the node nearest to the source. Nevertheless, all receivers converge immediately to an optimal rate, which is maintained without losses.

Fig. 4 (right) shows that the convergence time would be slower if receivers did not know in advance the average rate of each layer. This happens because receivers would have to wait before joining each layer, in order to estimate the current rate of the session and predict the rate of the next higher layer. In these experiments, receivers use an exponential equation to estimate the average rate of each received layer, and predict that the next layer has a rate equal to the last joined layer.

This simulation shows that neither the audience size or late-joins influence the convergence time and stability of SAM. The results shown in Fig. 4 (left) are similar to the ones presented by Legout et al. for PLM [10], except that with

PLM receivers start to converge 2 s after joining a session. This happens because with PLM receivers are notified only about the available bandwidth in the path and not about the sustainable rate of their session.

We use the topology *Top3* of Fig. 2 to evaluate the behavior of SAM in the presence of TCP flows, which are handled by SAPRA as scalable sessions with one layer and one receiver. This scenario has one scalable session, S, and two TCP flows, T_1 and T_2. S has one receiver, R_S, and layers with granularity of 20 kb/s. T_1 starts at $t = 0$ s, S_1 at $t = 20$ s and T_2 at $t = 60$ s.

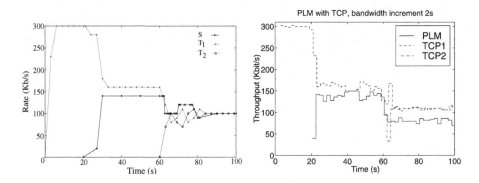

Fig. 5. Fairness of SAM (left) and PLM (right) with TCP

Fig. 5 (left)[2] shows that SAM is fair in the presence of TCP. R_S joins the lowest layer at $t = 20$ s, reaching a rate of 32 kb/s. At $t = 27$ s, it increases its rate after receiving the first report. Since S has only one receiver, the bandwidth of (r_4, r_5) is equally divided between S and T_1. Therefore, after $t = 20$ s, S and T_1 have a fair rate of 150 kb/s, and so R_S joins seven layers, reaching a rate of 140 kb/s. Since S does not use 10 kb/s of its fair share, the rate of T_1 reaches 160 kb/s. When T_2 starts, the fair rates of S, T_1 and T_2 reach 100 kb/s. R_S starts to experience losses and decreases its rate to 100 kb/s, but it maintains seven layers since losses are lower than 2.5%.

Due to T_1 and T_2 oscillations until $t = 84$ s, R_S grabs the bandwidth not being used by the TCP flows, increasing its rate to 120 kb/s. From then until the end of the simulation, the rate of S, T_1 and T_2 stabilizes at 100 kb/s.

	60 s	63 s	66 s	69 s	72 s	75 s	78 s	81 s	84 s
S	140	100	100	120	120	120	120	90	100
T₁	160	120	80	90	110	80	70	110	100
T₂	0	70	100	80	70	100	110	90	100
Total	300	290	280	290	300	300	300	290	300

Fig. 6. Link (r_2, r_4) utilization in kb/s

In the meantime, R_S leaves layer seven with losses of 2.91%, but maintains the rate of 120 kb/s, which is the maximum possible rate with six layers. At $t = 80$ s, R_S leaves layer six, since losses reach 3.28%. Fig. 6 shows that the bandwidth of (r_4, r_5) is completely used, except for the interval from $t = 63$ s to $t = 81$ s, where the utilization rate decreases to 98.1% due to T_1 and T_2 oscillations.

---

[2] Fig. 5 (right) was taken from [10] with the author's permission.

The results shown in Fig. 5 show that SAM and PLM are fair in the presence of TCP, but SAM has smaller quality oscillations and a fairer distribution of bandwidth: after $t = 60\,\text{s}$, S, T_1 and T_2 get the same share of bandwidth, while in the case of PLM, the two TCP flows get a higher share.

5 Conclusion

Receiver-driven adaptive mechanisms can accommodate heterogeneity when combined with scalable encoding. However, receivers are only motivated to adapt if the network guarantees a fair distribution of bandwidth and also punishes receivers that do not adjust their rate in case of congestion.

This paper describes and evaluates SAM, a receiver-driven adaptive mechanism based upon SAPRA. SAPRA is a signaling protocol that has the required punishment and fairness properties. SAM controls the reception quality by joining and dropping layers. The sustainable rate, provided by SAPRA, indicates to SAM the maximum number of layers that receivers can join, while the measured packet losses triggers SAM to drop layers.

Simulation results show that with SAM, receivers get always a rate near the sustainable rate of their session, independently of the number of sessions and their audience size. Results also show that SAM has small convergence time and remains stable even in the presence of bursty traffic, such as VBR. Compared to PLM, SAM has less quality oscillations, is fairer in the presence of TCP flows and requires few changes in the network structure.

As major improvement, SAM motivates receivers to adapt, since SAPRA guarantees a fair distribution of bandwidth and the punishment of high-rate sessions.

References

[1] Dimitri Bertsekas and Robert Gallager. *Data Networks*. Prentice-Hall, Englewood Cliffs, New Jersey, 1987.

[2] S. Blake, D. Black, M. Carlson, E. Davies, Z. Wang, and W. Weiss. An architecture for differentiated service. Request for Comments 2475, Internet Engineering Task Force, December 1998.

[3] M. Handley, C. Perkins, and E. Whelan. Session announcement protocol. Request for Comments 2974, Internet Engineering Task Force, October 2000.

[4] J. Heinanen, F. Baker, W. Weiss, and J. Wroclawski. Assured forwarding PHB group. Request for Comments 2597, Internet Engineering Task Force, June 1999.

[5] Jun ichi Kimura, Fouad A. Tobagi, Jose-Miguel Pulido, and Peder J. Emstad. Perceived quality and bandwidth characterization of layered MPEG-2 video encoding. In *Proc. of SPIE International Symposium*, Boston, Massachussetes, USA, September 1999.

[6] ITU-500-R. "Methodology for the subjective assessment of quality of television pictures". Itu-500-r recommendation bt.500-8, ITU, 1998.

[7] Mathias Johanson. Scalable video conferencing using subband transform coding and layered multicast transmission. In *Proc. of International Conference on Signal Processing Applications and Technology (ICSPAT)*, Orlando, Florida, USA, November 1999.

[8] Srinivasan Keshav. A control-theoretic approach to flow control. In *Proc. of SIGCOMM Symposium on Communications Architectures and Protocols*, pages 3–15, Zürich, Switzerland, September 1991. ACM. also in *Computer Communication Review* 21 (4), Sep. 1991.

[9] Arnaud Legout and Ernst Biersack. Pathological behaviors for RLM and RLC. In *Proc. of the International Workshop on Network and Operating System Support for Digital Audio and Video (NOSSDAV)*, Chapel Hill, North Carolina, USA, June 2000.

[10] Arnaud Legout and Ernst W. Biersack. PLM: Fast convergence for cumulative layered multicast transmission schemes. technical report, Institut Eurécom, Sophia-Antipolis, France, November 1999.

[11] Arnaud Legout and Ernst W. Biersack. PLM: Fast convergence for cumulative layered multicast transmission schemes. In *Proc. of ACM SIGMETRICS performance evaluation review*, Santa Clara, California, USA, June 2000.

[12] Steven McCanne, Van Jacobson, and Martin Vetterli. Receiver-driven layered multicast. In *Proc. of SIGCOMM Symposium on Communications Architectures and Protocols*, pages 117–130, Palo Alto, California, USA, August 1996.

[13] Paulo Mendes. "SAPRA: Session-Aware Popurality-based Resource Allocation fairness protocol". http://www.cs.columbia.edu/~mendes/sapra.html.

[14] Paulo Mendes, Henning Schulzrinne, and Edmundo Monteiro. Session-aware popularity resource allocation for assured differentiated services. In *Proc. of the Second IFIP-TC6 Networking Conference*, Pisa, Italy, May 2002.

[15] Paulo Mendes, Henning Schulzrinne, and Edmundo Monteiro. Signaling protocol for session-aware popularity-based resource allocation. In *To appear in Proc. of the IFIP/IEEE International Conference on Management of Multimedia Networks and Services*, Santa Barbara, California, USA, October 2002.

[16] Abhay K. Parekh and Robert G. Gallager. A generalized processor sharing approach to flow control in integrated services networks: The single node case. *Journal of IEEE/ACM Transactions on Networking*, 1(3):344–357, June 1993.

[17] Gopalakrishnan Raman, James Griffioen, Gisli Hjalmtysson, and Cormac Sreenan. Stability and fairness issues in layered multicast. In *Proc. of the International Workshop on Network and Operating System Support for Digital Audio and Video (NOSSDAV)*, Basking Ridge, New Jersey, USA, June 1999.

[18] Kenneth Rose and Shankar L. Regunathan. Toward optimality in scalable predictive coding. *Journal of IEEE Transactions on Image Processing*, 10(7):965–976, July 2001.

[19] Dan Rubenstein, Jim Kurose, and Don Towsley. The impact of multicast layering on network fairness. In *Proc. of SIGCOMM Symposium on Communications Architectures and Protocols*, Cambridge, Massachussetes, USA, September 1999.

[20] H. Schulzrinne, S. Casner, R. Frederick, and V. Jacobson. RTP: a transport protocol for real-time applications. Request for Comments 1889, Internet Engineering Task Force, January 1996.

[21] David Taubman and Avideh Zakhor. Multirate 3-D subband coding of video. *Journal of IEEE Transactions on Image Processing*, 3(5):572–588, September 1994.

[22] Lorenzo Vicisano, Luigi Rizzo, and Jon Crowcroft. TCP-Like congestion control for layered multicast data transfer. In *Proc. of the Conference on Computer Communications (IEEE Infocom)*, San Francisco, California, USA, March/April 1998.

Large-Scale Behavior of End-to-End Epidemic Message Loss Recovery

Öznur Özkasap

Koç University, Department of Computer Engineering, Istanbul, Turkey
oozkasap@ku.edu.tr

Abstract. An important class of large-scale distributed applications is insensitive to small inconsistencies among participants, as long as these events are temporary and not frequent. An efficient way for propagating information to participants in such cases is referred to as epidemic protocols. Epidemic protocols are simple, scale well and robust again common failures, and provide eventual consistency as well. They combine benefits of efficiency in hierarchical data dissemination with robustness in flooding protocols. These communication mechanisms have been mainly used for resolving inconsistencies in distributed database updates, failure detection, message loss recovery in multicast communication, network news distribution, group membership management, scalable system management, and resource discovery. In this paper, we focus on an end-to-end epidemic loss recovery mechanism for multicasting and give our simulation results discussing the performance of the approach in large-scale network settings.

Keywords: loss recovery, epidemic communication, scalable multicast, end-to-end protocols, Bimodal Multicast.

1 Introduction

End-to-end protocols belonging to transport level of the network architecture support communication between the end application programs. There are two major forces that shape an end-to-end protocol. At one end, from the level above, the application processes that use its services have certain needs. Some of the common properties that a transport protocol can offer are message loss recovery that guarantees message delivery, ordered message delivery, no message duplication, support for arbitrarily large messages transmitted by the application, support for flow control, and provision of multiple application processes at each host. At the other end, from the level below, the underlying network has certain constraints in the level of service it can provide. Typical constraints of the network are dropping messages, reordering messages, delivering duplicate copies of a given message, limiting messages to a finite size, delivering messages after an arbitrarily long delay. Such an underlying network structure is referred to as a best-effort level of service as exemplified by the Internet. The challenge of an end-to-end protocol is, therefore, to provide algorithms that turn these best-effort properties of the underlying network into the high level of service required by application programs [1].

B. Stiller et al. (Eds.): QofIS/ICQT 2002, LNCS 2511, pp. 25–35, 2002.

Increasing popularity of group-based applications in large-scale settings given the varying quality of service requirements stipulates efficient multicast communication mechanisms. Multicast paradigm, which is a type of one-to-many communication, has several practical applications such as multimedia, videoconferencing, distributed computation, data dissemination, database and real-time workgroups. Multicast applications have a much wider range of requirements than their unicast counterparts.

Transport level multicast requirements can be broadly classified as loss-sensitive reliable services and delay-sensitive interactive services. While interactive applications such as multimedia conferencing can tolerate reliability in support of real-time delivery, data dissemination applications such as multicast file transport tolerate longer transfer delays. However, several distributed applications exploiting multicast communication require reliable delivery of messages to all destinations. The degree of reliability guarantees required by such applications differs from one setting to another. Some protocols offer strong reliability guarantees such as atomicity, delivery ordering, virtual synchrony, real-time support, security properties and network-partitioning support. These protocols allow limited scalability, and hence are not suitable for large-scale high-speed network platforms. The main drawback is that in order to obtain strong reliability guarantees, costly protocols are used and the possibility of unstable or unpredictable performance under failure scenarios is accepted. Support for best-effort reliability in large-scale settings becomes important as the widespread availability of IP multicast and the MBone have considerably increased both the geographic extent and the size of the communication groups. Although the other class of protocols, offering support for best-effort reliability in large-scale, overcomes message loss and failures, they do not guarantee end-to-end reliability. Common failure scenarios such as router overload and system-wide noise can cause these protocols to behave pathologically [2] and hence lead to negative protocol effects on network performance. Within this context, example transport level reliable multicast protocols are the Internet Muse protocol [3] for network news distribution and Scalable Reliable Multicast (SRM) [4]. In the spectrum of scalable reliable multicast protocols, novel options such as Bimodal Multicast [5] provide better quality of service and performance.

Loss recovery is an essential part of a multicast service offering reliability. In this study, we investigate an end-to-end epidemic loss recovery mechanism for multicasting, demonstrate that it works efficiently in large-scale network settings, and give our comparative simulation results discussing the performance of the approach.

The paper is organized as follows: Section 2 explains message loss recovery techniques and scalability in reliable multicasting. In section 3, we describe the Bimodal Multicast protocol and its epidemic loss recovery mechanism. Section 4 presents simulation model, analysis and results of our study. Section 5 gives information on epidemic communication and related work. Section 6 concludes the paper.

2 Message Loss Recovery and Scalability in Multicasting

A significant protocol element for ensuring reliability at the transport level is message loss recovery. Traditional reliable unicast (point-to-point) protocols, such as TCP, use feedback (ACKs) to recover from message loss. Since it is the responsibility of the message source to detect losses, this approach is often referred to as sender-initiated. Our focus in this study is on scalable reliable multicast services that attempt to ensure that messages are delivered to all members in a process group. In the case of large-scale multicast applications, the sender-initiated approach may cause feedback implosion, since each delivered message initiates an ACK from every group member. This, in turn, leads to overflow on the sender's buffer and congests the network. Alternatively, in the receiver-initiated approach to loss recovery; upon detecting message losses, receivers request their retransmission by generating negative acknowledgments (NACKs). Performance comparison studies confirm that receiver-initiated multicast transport protocols demonstrate better performance than their sender-initiated counterparts in terms of scalability. However, a problem with returning only NACKs is that the sender would need to keep a message in its buffer for a long time. Most reliable multicast transport protocols are either pure receiver-initiated or a hybrid of sender and receiver-initiated approaches.

Two approaches that are representatives of many existing solutions for providing loss recovery in scalable multicasting are nonhierarchical feedback control and hierarchical feedback control. Overall, the key issue is to reduce the number of feedback messages that are returned to the sender. In the former approach, a model that has been adopted by several wide-area applications is referred to as *feedback suppression*. A well-known example is SRM protocol. In SRM, when a receiver detects that it missed a message, it multicasts its feedback to the rest of the group. Multicasting feedback allows another group to suppress its own feedback. A receiver lacking a message schedules a feedback with some random delay. An improvement to enhance scalability is referred to as local recovery, which is related to restraining the recovery of a message loss to the region where the loss has occurred. In the latter approach, hierarchical approaches are adopted for achieving scalability for very large groups of receivers [6]. Scalable reliable multicasting is an area of active research. In addition to the mentioned approaches, another alternative for ensuring reliability is forward error correction (FEC). The idea behind this approach is predicting losses and transmitting redundant data. On the other hand, recent epidemic or randomized approaches to loss recovery have promising outcomes in terms of robustness and overhead. In this direction, Bimodal Multicast provides an epidemic loss recovery mechanism. Epidemic protocols are simple, scale well and robust again common failures, and provide eventual consistency as well. They combine benefits of efficiency in hierarchical data dissemination with robustness in flooding protocols.

3 Bimodal Multicast and Two-Phase Anti-entropy Protocol

Bimodal Multicast [5] is a novel option in the spectrum of multicast protocols that offers throughput stability, scalability and a bimodal delivery guarantee as the key properties. The protocol is inspired by prior work on epidemic protocols [7], Muse protocol for network news distribution [3], the SRM protocol [4], and the lazy transactional replication method of [8]. Bimodal Multicast is based on an epidemic loss recovery mechanism. It has been shown to exhibit stable throughput under failure scenarios that are common on real large-scale networks [5]. In contrast, this kind of behavior can cause other reliable multicast protocols to exhibit unstable throughput. Bimodal Multicast consists of two sub-protocols, namely an optimistic dissemination protocol, and a two-phase anti-entropy protocol.

The former is a best-effort, hierarchical multicast used to efficiently deliver a multicast message to its destinations. This phase is unreliable and does not attempt to recover a possible message loss. If IP multicast is available in the underlying system, it can be used for this purpose. For instance, the protocol model implemented on ns-2 network simulator [9] in this study uses IP multicast. Otherwise, a randomized dissemination protocol can play this role.

The second stage of the protocol is responsible for message loss recovery. It is based on an anti-entropy protocol that detects and corrects inconsistencies in a system by continuous gossiping. The anti-entropy is an epidemic communication strategy that was previously used for error recovery in wide area database and large-scale direct mail systems [7,10]. Further information and background related to anti-entropy strategy are given in section 5. The two-phase anti-entropy protocol progresses through unsynchronized rounds. In each round:

o   Every group member randomly selects another group member and sends a digest of its message history. This is called a 'gossip message'.
o   The receiving group member compares the digest with its own message history. Then, if it is lacking a message, it requests the message from the gossiping process. This message is called 'solicitation', or retransmission request.
o   Upon receiving the solicitation, the gossiping process retransmits the requested message to the process sending this request.

One of the differences of Bimodal Multicast's anti-entropy protocol from the other gossip protocols is that during message loss recovery, it gives priority to the recent messages. If a process detects that it has lost some messages, it requests retransmissions in reverse order: most recent first. If a message becomes old enough, the protocol gives up and marks the message as lost. By using this mechanism, the protocol avoids failure scenarios where processes suffer transient failures and are unable to catch up with the rest of the system. One of the drawbacks of traditional gossip protocols is that such a failure scenario can slow down the system by causing processes' message buffers to fill.

The duration of each round in the anti-entropy protocol is set to be larger than the typical round-trip time for an RPC over the communication links. The simulations conducted in this study use a round duration of 100msec. Processes keep buffers for storing data messages that have been received from members of the group. Messages from each sender are delivered in FIFO order to the application. After a process receives a message, it continues to gossip about the message for a fixed number of rounds. Then, the message is garbage collected.

4 Analysis and Results

In this section, we describe our simulation settings, and give our simulation results. Simulation methods allow gaining control over the parts of the network and lead to better understanding of protocols. For instance, in a simulation model, link loss probabilities can be set easily, several network topologies can be constructed. Many process group applications and scenarios can be built on top of these settings. Our simulation study uses ns-2 simulator [9] to model network and protocol behavior. Our implementation for Bimodal Multicast developed over ns-2 is used in this study [19].

4.1 Topologies and Simulation Settings

Our simulation scenarios consist of large-scale transit-stub topologies with 1500 nodes where the sender located on a central node and receivers located at randomly selected nodes on the network. Transit-stub topologies approximate the structure of the Internet that can be viewed as a collection of interconnected routing domains where each domain can be classified as either a stub or a transit domain. Stub domains correspond to interconnected LANs and the transit domains model WAN or MANs [11]. We used gt-itm topology generator for producing transit-stub topologies. A sample transit-stub topology consisting of total 20 nodes where there exists one

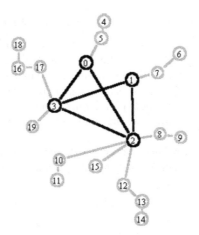

Fig.1. Sample Transit-stub topology

transit domain with 4 nodes (0,1,2, and 3) is shown in Fig.1. In fact, our simulations use 1500-node transit-stub topologies with 6 transit domains and on average 10 nodes per transit domain. A certain link noise probability is set on every link that forms a randomized system-wide noise. We varied three operating parameters, namely group size, multicast message rate of the sender, and system-wide noise rate. This scenario primarily focuses on the impact of randomized message loss over traffic generated by the protocol. We obtain our results from several runs of simulations, each run consisting of a sequence of 32000 multicast data messages transmitted by the sender.

4.2 Recovery Overhead Distribution

In this section, we investigate loss recovery distribution of Bimodal Multicast. During each simulation, the total number of multicast messages distributed by the sender is 32000, and the message rate is either 10 or 100 messages per second. Fig.2 shows request messages received by each receiver in the multicast group where group size is

10 and system-wide noise rate on all links is 0.01. Similar distribution is found for retransmission messages generated by each receiver in the group (because of the page limitation, we do not include this graph). It is observed that the load of overhead traffic is balanced over group members, a desirable property that avoids overloading one member or a portion of members with high overhead traffic. Moreover, increasing multicast transmission rate of the sender by a multiple of 10 (from 10 messages per second to 100) does not change the overhead distribution, only causing a slight increase in the number of requests received by each group receiver. As we will examine next, increasing message rate in fact decreases number of requests triggered by each receiver most of the time. As a future work, we plan to analyze this behavior in detail.

Fig.3 shows request messages received by each receiver in the multicast group in the same settings. If there are n links on the route from the sender to a specific receiver where noise rate on each link is p_i, then the probability of message loss experienced by the receiver can be calculated as follows. Let $P\{loss\}$ be the probability of loosing a message on its route from sender to the receiver.

$$P\{loss\} = 1 - P\{no\ loss\} = 1 - (1-p_i)^n$$

Our simulation results are consistent with the above probability calculation. As an example, consider the receiver 4 in fig.3 that is 3 links away from the sender. According to the above calculation; for all multicast data messages transmitted by the sender, we would expect 951 messages to be lost on the way to receiver 4. This would essentially trigger the same amount of request messages to be generated by the receiver. That is actually what we have measured as our analysis result that is consistent with theoretical findings.

Fig.4 demonstrates our results when we scale the group size to 50 on the network of 1500 nodes. Similar to our previous results, we observe that the load of overhead

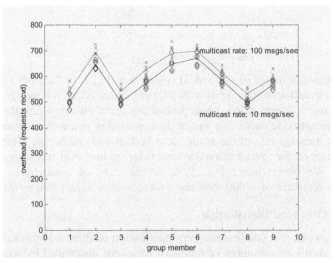

Fig.2. Request messages received by each group member, Bimodal Multicast, 1500-node network, group size: 10, noise rate: 0.01

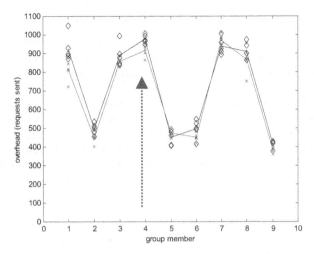

Fig.3. Request messages generated by each group member, Bimodal
Multicast, 1500-node network, group size: 10, noise rate: 0.01

traffic is balanced over group members for both request (Fig.4a), and retransmission
traffic (Fig.4b). Increasing message rate of the sender does not cause a regular
increase on each member, which is an indication demonstrating the scalability of
epidemic loss recovery. For all these simulations, we observe full reliability during
multicast communication where all message losses are recovered and all receivers
successfully deliver multicast data. In the next section, we further examine the impact
of message rate and randomized noise on the scalability of the approach.

4.3 Scalability: Impact of Message Rate and Randomized System-wide Noise

Now, we explore the impact of an increase in message rate and randomized noise over
network links. Table.1 shows average loss recovery overhead (in the form of requests
received and retransmissions generated) observed on all receivers of the multicast
group where three parameters, namely system-wide noise rate, group size and
multicast message rate, vary. For an increase in noise rate, probability calculation of
the previous section suggests a linear increase in recovery traffic where the other
parameters are constant. In practice, our simulation results demonstrate better
behavior with an expected increase in overhead traffic that is below linear an effect
indicating the scalability of the loss recovery. As an example, for a group size of 50
and message rate of 100, overhead changes from 3% to 21% as noise rate increases
from 0.01 to 0.1.

Scaling the group size from 10 to 50 on the identical network settings causes a very
slight change in the overhead load that is another indication of scalability. For
instance, for a noise rate of 0.1 and message rate of 100, overhead changes from 17%
to 21% as group size increases from 10 to 50. Moreover, increasing message rate of
the sender, while keeping the other factors constant, has an insignificant effect on
overhead which makes epidemic loss recovery a good candidate for high-speed data
distribution.

4.4 Comparison with Nonhierarchical Feedback Control

Recent studies [12,13,5] have shown that, for the SRM protocol with its nonhierarchical feedback control mechanism for loss recovery, random packet loss can trigger high rates of overhead messages. In addition, this overhead grows with the size of the system.

Related to this scalability problem, some of our simulations explore the behavior of Bimodal Multicast and SRM protocol versions on a large-scale network with high-speed data transfer. In these simulations, we construct large-scale tree topologies consisting of 1000 nodes. Up to hundred of the 1000 nodes are randomly chosen to be group members. We set the message loss rate to 0.1% on each link with the sender located at the root node injecting 100 210-byte multicast messages per second. The

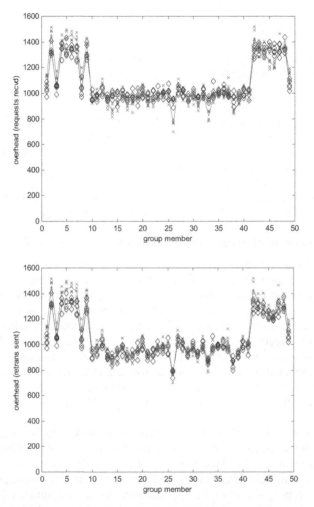

Fig.4. a)Requests received and b) Retransmissions generated by each group member, Bimodal Multicast, 1500-node network, group size: 50, noise rate: 0.01

results for the background overhead of each protocol in the form of request message traffic are shown in Fig.5. We include error bars showing minimum and maximum values recorded over the set of runs, using different seeds for the random number generator. The results demonstrate that, as the network and process group size scale up, the number of control messages received by group members during loss recovery increases linearly for SRM protocols, an effect previously reported in [12,13]. These costs remain almost constant for Bimodal Multicast versions (in graphs these are labeled as Pbcast and Pbcast-ipmc for short). Pbcast-ipmc is the version of Bimodal Multicast that uses IP multicast for message repairs during loss recovery. Compared to the basic Pbcast, Pbcast-ipmc has a slightly lower overhead in the form of request messages. If multiple receivers missed a message, Pbcast-ipmc increases probability of rapid convergence during loss recovery.

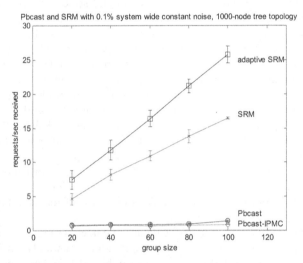

Fig.5. Overhead in the form of requests per second for Bimodal Multicast and SRM, 1000-node tree topologies with 0.1% system-wide noise

Table 1. Impact of message rate and randomized noise

Parameters			Loss recovery overhead		
Noise rate	Group size	Message rate	Requests	Retransmissions	%
0.01	10	10	575.88	575.88	2
		100	614.17	614.17	2
	50	10	1083.33	1044.92	3
		100	1079.98	1079.98	3
0.1	10	10	5204.97	5204.97	16
		100	5577.13	5577.13	17
	50	10	6873.06	6870.45	21
		100	7076.72	7076.72	21

5 Epidemic Communication and Related Work

There exists a substantial class of large-scale distributed applications that are insensitive to small inconsistencies among participants, as long as these events are temporary and not frequent. Epidemic communication is suitable in this case where it allows such inconsistencies in shared data and offers low overhead as a benefit. Information changes are spread throughout the participants without incurring the latency and bursty communication that are typical for systems achieving a strong form of consistency [14]. In fact, this is especially important for large systems, where failure is common, communication latency is high and applications may contain a large number of participants.

Epidemic communication mechanisms were first proposed for spreading updates in a replicated database. Epidemic protocols are simple, scale well and robust again common failures, and provide eventual consistency as well. They combine benefits of efficiency in hierarchical data dissemination with robustness in flooding protocols. *Anti-entropy* is an epidemic communication strategy introduced for achieving and maintaining consistency among the sites of a widely replicated database. Compared to deterministic algorithms for replicated database consistency, this strategy also reduces network traffic [7]. Anti-entropy has been proposed as a mechanism that runs in background for recovering errors of direct mail in large network, as well [10]. Bimodal Multicast protocol utilizes this mechanism for probabilistically reliable multicast communication. Periodically, every site chooses another site at random and exchanges information to see any differences and achieve consistency. This technique is called gossiping. For the case of database maintenance, the information exchanged during gossip rounds may include database contents. For epidemic multicast communication, the information may include some form of message history of the group members.

The anti-entropy method is based on the theory of epidemics [15]. According to the terminology of epidemiology, a site holding information or an update it is willing to share is called 'infective'. A site is called 'susceptible' if it has not yet received an update. In the anti-entropy process, non-faulty sites are always either susceptible or infective. One of the fundamental results of epidemic theory shows that simple epidemics eventually infect the entire population. If there is a single infected process at the beginning, full infection is achieved in expected time proportional to the logarithm of the population size.

Epidemic or gossip style of communication has been used for several purposes. Examples include use of epidemic communication techniques for group membership tracking [14], for support of replicated services [8], for deciding when a message can be garbage collected [16], for failure detection [17], and for error recovery in reliable multicast [18].

6 Conclusions

A significant protocol element for ensuring reliability at the transport level end-to-end multicasting is message loss recovery. This study yields some conclusions about the large-scale behavior of end-to-end epidemic loss recovery in multicast communication. Our results show that epidemic loss recovery produces balanced overhead distribution among group receivers and it is scalable as group size, multicast message rate and system-wide noise rate increase. We also compare the epidemic loss recovery with nonhierarchical feedback control for demonstrating the efficiency of our approach on the identical network settings. An area for further study within our simulation model would be a detailed exploration of hierarchical epidemic mechanisms that would help to increase efficiency of the approach.

References

1. Peterson, L. and Davie, B.S., 1999, Computer Networks: A Systems Approach, Second Edition, Morgan Kaufmann Series in Networking, ISBN 1-55860-514-2.
2. Paxson, V., 1997, End-to-End Internet Packet Dynamics, Proceedings of SIGCOMM '97, 139-154p.
3. Lidl, K., Osborne, J. and Malcome, J., 1994, Drinking from the Firehose: Multicast USENET News, USENIX Winter 1994, 33-45p.
4. Floyd, S., Jacobson, V., Liu, C., McCanne, S. and Zhang, L., 1997, A Reliable Multicast Framework for Light-weight Sessions and Application Level Framing, IEEE/ACM Transactions on Networking, 5(6), 784-803p.
5. Birman, K.P., Hayden, M., Ozkasap, O., Xiao, Z., Budiu, M. and Minsky, Y., 1999, Bimodal Multicast, ACM Transactions on Computer Systems, 17(2), 41-88p.
6. Tanenbaum, A.S. and van Steen, M., 2002, Distributed Systems: Principles and Paradigms, Prentice Hall, ISBN 0-13-088893-1.
7. Demers, A., Greene, D., Hauser, C., Irish, W., Larson, J., Shenker, S., Sturgis, H., Swinehart, D. and Terry, D., 1987, Epidemic Algorithms for Replicated Database Maintenance, Proceedings of the Sixth ACM Symposium on Principles of Distributed Computing, Vancouver, British Columbia, 1-12p.
8. Ladin, R., Lishov, B., Shrira, L. and Ghemawat, S., 1992, Providing Availability using Lazy Replication, ACM Transactions on Computer Systems, 10(4), 360-391p.
9. Bajaj, S., Breslau, L., Estrin, D., et al., Improving Simulation for Network Research, 1999, USC Computer Science Dept. Technical Report 99-702.
10. Birrell, A.D., Levin, R., Needham, R.M. and Schroeder, M.D., 1982, Grapevine, An Exercise in Distributed Computing, Communications of the ACM, 25(4), 260-274p.
11. Ken Calvert, Matt Doar and Ellen W. Zegura. Modeling Internet Topology. IEEE Communications Magazine, June 1997.
12. Liu, C., Error Recovery in Scalable Reliable Multicast, Ph.D. dissertation, University of Southern California, 1997.
13. Lucas, M., Efficient Data Distribution in Large-Scale Multicast Networks, Ph.D. dissertation, Dept. of Computer Science, University of Virginia, 1998.
14. Golding, R.A. and Taylor K., 1992, Group Membership in the Epidemic Style, Technical Report, UCSC-CRL-92-13, University of California at Santa Cruz.
15. Bailey, N.T.J., 1975, The Mathematical Theory of Infectious Diseases and its Applications, second edition, Hafner Press.
16. Guo, K., 1998, Scalable Message Stability Detection Protocols, Ph.D. dissertation, Cornell University Dept. of Computer Science.
17. Van Renesse, R., Minsky, Y. and Hayden, M., 1998, A Gossip-style Failure Detection Service, Proceedings of Middleware'98, 55-70p.
18. Xiao, Z. and Birman, KP., 2001, A Randomized Error Recovery Algorithm for Reliable Multicast, Proceedings, IEEE Infocom 2001.
19. Ozkasap, O., Scalability, Throughput Stability and Efficient Buffering in Reliable Multicast Protocols, Technical Report, TR2000-1827, Dept. of Computer Science, Cornell University.

Evaluation of a Differentiated Services Based Implementation of a Premium and an Olympic Service

Volker Sander[1] and Markus Fidler[2]

[1] Central Institute for Applied Mathematics,
Forschungszentrum Jülich GmbH, 52425 Jülich, Germany
[2] Chair of Computer Science IV, RWTH Aachen,
Ahornstr. 55, 52074 Aachen, Germany

Abstract. In this paper we describe the implementation of a network providing advanced services such as a premium service that aims at providing low loss, low delay, and low delay jitter and an olympic service that allows for a service differentiation in terms of delay within three additional classes. Our implementation of this network is based on the Differentiated Services Architecture, which is the most recent approach of the Internet Engineering Task Force towards Quality of Service. Access to service classes is controlled by a Bandwidth Broker, which can perform Traffic Engineering by the means of Multiprotocol Label Switching. The premium service is implemented as Expedited Forwarding and the olympic service as a group of Assured Forwarding Per-Hop-Behavior. We present a thorough evaluation of the proposed services and their respective implementations.

1 Introduction

We wish to provide Quality of Service (QoS) mechanisms for *high-end network applications* Such applications consist of complex mixes of flows, with a variety of data rates and widely differing latency requirements. Applications with these characteristics arise in areas as remote visualization, analysis of scientific databases, Table 1 lists the various flows of a future teleimmersion application together with their networking demand. Characteristics such as these place substantial demands on networks which cannot be fulfilled by today's Best-Effort (BE) Internet.

Table 1. Networking flows and requirements of future teleimmersion applications [6]

	Latency	Bandwidth	Reliable	Multicast	Security	Streaming	Dyn QoS
Control	< 30 ms	64 Kb/s	Yes	No	High	No	Low
Text	< 100 ms	64 Kb/s	Yes	No	Medium	No	Low
Audio	< 30 ms	128 Kb/s	No	Yes	Medium	Yes	Medium
Video	< 100 ms	5 Mb/s	No	Yes	Low	Yes	Medium
Tracking	< 10 ms	128 Kb/s	No	Yes	Low	Yes	Medium
Database	<100 ms	> 1 Gb/s	Yes	Maybe	Medium	No	High
Simulation	< 30 ms	> 1 Gb/s	Mixed	Maybe	Medium	Maybe	High
Haptic	< 10 ms	> 1 Mb/s	Mixed	Maybe	Low	Maybe	High
Rendering	< 30 ms	> 1 Gb/s	No	Maybe	Low	Maybe	Medium

B. Stiller et al. (Eds.): QofIS/ICQT 2002, LNCS 2511, pp. 36–46, 2002.

The Differentiated Services (DS) [2] framework defines an architecture for implementing scalable service differentiation in the existing Internet by an aggregation of flows to a small number of different traffic classes. DS can be complemented by Traffic Engineering (TE), which is of special interest, if not only a relative differentiation between services shall be implemented, but, if absolute service guarantees need to be given. We apply Multiprotocol Label Switching (MPLS) [13] for this purpose. MPLS is based on a functional decomposition of the network layer into a control component and a forwarding component. This distinction gives a number of options for the implementation of the control component. In [14,15] we presented the General-purpose Architecture for Reservation and Allocation (GARA), which implements an advance reservation framework using heterogeneous resource ensembles. GARA includes a DS reservation manager, which allows to use it as a Bandwidth Broker (BB) for an automated DS network management and to apply DS TE by means of MPLS.

In this paper, we describe an implementation of a premium service based on the Expedited Forwarding (EF) Per-Hop Behavior (PHB) [5] and an olympic service based on the Assured Forwarding (AF) PHB group [10]. We perform a careful evaluation of the services by applying both Transmission Control Protocol (TCP) and User Datagram Protocol (UDP) flows. The remainder of the paper is organized as follows: Sections 2 describes the DS architecure in detail. In section 3 we show results obtained from measurements in the implemented network. Section 4 concludes the paper.

2 Differentiated Services Architecture

The DS architecture [2] addresses the scalability problems of Integrated Services by defining the behavior of aggregates. Packets are identified by simple markings that indicate according to which aggregate behavior they should be treated. In the core of the network, routers need not determine to which flow a packet belongs, only which aggregate behavior should be used. Edge routers mark packets and indicate whether they are within profile or if they are out of profile, in which case they might even be discarded by a dropper at the edge router. A particular marking on a packet indicates a PHB that has to be applied for forwarding of the packet. Currently, the EF PHB [5] and the AF PHB group [10] are specified. While the specification of PHBs is the basic building block for service guarantees, the actual provisioning of QoS depends on the composition of the aggregates. Bandwidth brokers such as GARA are introduced to carefully control the access to aggregates and are thus another building block for advanced network services.

The EF PHB is intended for building a service that offers low loss, low delay, and low delay jitter, namely a premium service. The specification of the EF PHB was recently redefined to allow for a more exact and quantifiable definition. Besides the premium service a so called olympic service [10] is proposed by the IETF to be based on the AF PHB group by extending it by means of a class based over-provisioning. Three of the four currently defined classes of the AF PHB group are used for such an olympic service. The service differentiation between the three classes gold, silver, and bronze is proposed to be performed by the means of admission control, i.e. assigning only a light load to the gold class, a medium load to the silver class and a high load to the bronze class.

3   Experimental Studies

We report on experiments designed to examine a DS implementation based on commodity products. In the following subsections we first give the experimental configuration and then address the implementation and evaluation of the different traffic classes. We show problems observed when using the BE service and address these with similar measurements using the olympic or the premium service.

3.1   Experimental Configuration

Our experimental configuration comprises a laboratory testbed at the Research Centre Jülich, donated by Cisco Systems. The testbed allows controlled experimentation with basic DS mechanisms. Four Cisco Systems 7200 series routers were used for all experiments. These are either connected by OC3 ATM connections, by Fast Ethernet, or by Gigabit Ethernet connections. End-system computers are connected to routers by switched Fast Ethernet connections. Hence, the minimum MTU size of our testbed is that of the end-systems: 1500 B. To create a point of congestion, we configured an ATM Permanent Virtual Circuit (PVC) between an ingress and an interior router to 60 Mbps.

 We performed several experiments demonstrating the performance of high-end TCP applications [15] like a guaranteed rate file transfer with deadline and typical UDP applications like video streaming [8] or videoconferencing. The following tools have been used for traffic generation:

- *gen_send/gen_recv – BE UDP traffic generator.* This traffic generator was applied to generate BE UDP traffic with a mean rate of 50 Mbps and different burst characteristics. These UDP flows do not aim to model any specific applications, but we assume that the applied burst characteristics reflect effects that occur in today's and in the future Internet. TCP streams are initially bursty, UDP based real-time applications are emerging, which create bursts, for example by intra-coded frames in a video sequence. Further on burst sizes increase in the network, due to aggregation and multiplexing [4].
- *rude/crude – Delay-sensitive UDP traffic generator.* This traffic generator allows to measure the one-way delay and delay jitter. In our experiments we used real-time traffic patterns from script files, which we created from publicly available video traces [8]. We applied IP fragmentation for the transmission of frames that exceed the MTU, which we consider as being allowed here, since we configured the DS classes to prevent from dropping fragments. The sequence, which we applied for the experimental results shown in this paper, is a television news sequence produced by the ARD. The sequence is MPEG-4 encoded with a minimum frame size of 123 B, a maximum frame size of 17.055 KB, a mean rate of 0.722 Mbps and a peak rate of 3.411 Mbps. The Hurst parameter is about 0.5 and decays with an increasing aggregation level. Figure 1 illustrates the traffic profile of the sequence.
- *ttcp – TCP stream generator.* We used the widely known TCP benchmark ttcp to generate TCP load. In the experiments reported on in this paper we selected an end-system which was not capable of generating a rate of more than 1.8 MB/s and if not stated otherwise we applied a socket buffer corresponding to a maximum window size of about 15 MTU.

3.2 Implementation and Evaluation of the Best Effort Service

Applying the plain BE service to the two example applications used throughout this paper we generate the baseline for our evaluation. Our configuration allocates the remaining capacity of the ATM bottleneck link, which is not used by any other class, to the BE class. In the following experiments no other class than BE is used, resulting in an assignment of 60 Mbps of the bottleneck ATM link to the BE class. The tx-ring-limit parameter on the ATM interface card that specifies the queue size, which is assigned to the applied ATM PVC, was set to 16 particles each of 512 B allowing to store upto four MTU on the ATM interface. This value is by far smaller than the default value, but it has to be applied to allow for an efficient QoS implementation [7]. The BE layer 3 queue was configured to hold at most 256 packets. We consider this queue size, which is a trade off between delay and loss rate, as being feasible for BE traffic, which is rather sensitiv to packet drops than to queuing delay in a range of a few tens of milliseconds.

In figure 2 the delay measured when transmitting the news sequence in the BE class is shown. Congestion is generated by applying an UDP stream with two bursts, each of ten seconds duration. As can be seen from figure 2, the delay is bounded to about 42 ms, showing some minor effects on the measurements due to tail-drop in the router. The delay corresponds to an effective data rate on the ATM interface of about 48 Mbps after subtracting the ATM induced overhead. While this delay is acceptable for streaming video applications, it can be critical for real-time video applications like video conferencing.

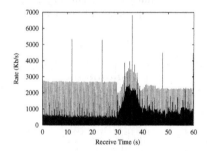 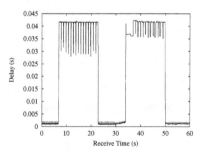

Fig. 1. Data Rate UDP News Sequence.   **Fig. 2.** Delay BE UDP News Sequence.

Next we address the performance of TCP flows in the BE class under either downstream or upstream congestion. Upstream congestion only affects the acknowledgements of the TCP stream, but we can show a major impact of the delay and the loss rate of the acknowledgements on TCP performance, which also has been addressed analytically by Hasegawa et al. [9] and in simulations by Wu and Williamson [16]. Congestion is generated by applying an unresponsive UDP flow with static burst characteristics. Figure 3 illustrates the TCP congestion window adaptation over the time when downstream congestion occurs and packets are dropped. This typical behavior of TCP allows for an efficient congestion contol [1], but it can lead to an effective data rate which makes it hard to predict the remaining transfer time. Several TCP throughput models [11,12,3] have been proposed which characterize the perceived throughput in terms of round-trip delay and packet loss rate. However, their applicability relies on the accuracy of these parameters and – if considered by the model – on the prediction of the number of time outs and their average duration. Figure 4 illustrates the respective plot giving the TCP sequence number over the time. Selective acknowledgements marked with an S indicate packet loss and trigger retransmissions marked by an R. The slope of the curve represents the throughput.

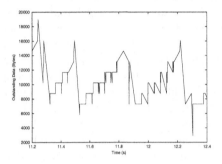
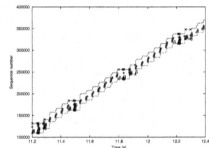

Fig. 3. TCP CWin Downstream Congestion.

Fig. 4. TCP SeqN Downstream Congestion.

The following figures show the effects of upstream congestion. Figure 5 gives the throughput over time. About 10 s after the start of the TCP transmission upstream congestion is generated by an UDP flow, leading to a significantly reduced throughput. The effects on the throughput are generated by an increase of the Round-Trip-Time (RTT) shown in figure 6.

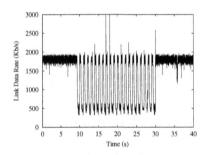
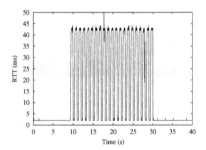

Fig. 5. TCP Tput Upstream Congestion.

Fig. 6. TCP RTT Upstream Congestion.

Besides acknowledgements can get delayed due to congestion they can also get lost. While the throughput impact of a lost acknowledgement is less significant than a lost data packet, it makes the TCP stream more bursty. Figure 7 illustrates the effect of an increasing RTT on the sequence number over the time. At first the TCP transmission is bounded by the application, which writes at a speed of 1.8 MB/s to the socket buffer, leading to a quite smooth transmission without actually using the maximum available window size. When the upstream congestion starts, the application is no longer the limiting factor. The available window slows the transmission down, as can be seen from the utilization of sequence numbers at the upper border of the available window in figure 7. Furthermore, the decreasing gradient denotes a significantly reduced data rate. This reduction of the data rate R can be addressed analytically by equation 1 with RTT denoting the round trip time and W denoting the window size.

$$R = W/RTT \qquad (1)$$

Clearly, equation 1 shows that the problem of an increased RTT because of delayed acknowledgements can be addressed by an increase of the window size, but since the window scale option is only negotiated during connection establishment, any dynamic right-sizing is bounded to the specific window scale value of its connection. Unfortunately, most operating systems do not allow the explicit specification of the window scale option. Instead its value is derived from the socket buffer size during connection establishment. Whenever servers have to maintain a huge number of parallel TCP connections, a trade-off between the desired window size and the dedicated amount of kernel memory has to be made. Additionally, we can observe from figure 7 that lost or delayed acknowledgements make the TCP flow significantly bursty. Especially upstream delay jitter increases this effect. When congestion is resolved, acknowledgements for a complete window can be received by the sender as an acknowledgement burst, thus leading to huge downstream data bursts. This clustering of data segments might result in downstream congestion. Of course, the effect is increased if bigger TCP windows are used and can even lead to downstream packet drops. Hence, the assumption of an upper bound for the window size of a single TCP connection is reasonable. Another negative effect of acknowledgement bursts is shown in figure 8. Here a complete burst of acknowledgements is locked out from queuing space and discarded, due to upstream congestion. Unfortunately TCP in this case has to wait for the retransmission timer to expire before it is allowed to perform a retransmission of one TCP segment. Furthermore, TCP has to perform slow start after the timeout, reducing the transmission rate unnecessarily.

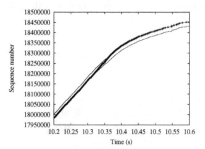 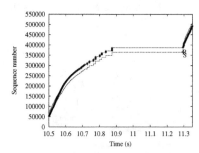

Fig. 7. TCP SeqN Upstream Congestion.   **Fig. 8.** TCP SeqN Upstream Congestion.

Taking the negative effects of the BE class – a not guaranteed transmission rate, possible packet loss and a chance of high delay and delay jitter – we argue that certain applications exist, like a file transfer with deadline or videoconferencing, which require services that are better than BE.

3.3 Implementation and Evaluation of an Olympic Service

A Weighted Fair Queuing (WFQ) environment is used for the implementation of the olympic service based on three AF PHB classes. Within these classes GARA is capable of managing the allocated resources and the relative load in order to allow for a service differentiation in terms of delay. The olympic service [10] proposed by the IETF is realised by admission control and a class based over-provisioning. We carried out experiments with the transmission of the news sequence in each of the olympic classes, with the classes configured according to table 2. Within each of the olympic

Table 2. Core Configuration of the Olympic Classes

Class	Percent	Gross Capacity	Approximate Net Capacity	Over-Provisioning Factor
Bronze	5 %	3 Mb/s	2.4 Mb/s	≥ 1
Silver	10 %	6 Mb/s	4.8 Mb/s	≥ 2
Gold	15 %	9 Mb/s	7.2 Mb/s	≥ 3

classes a differentiation of the drop probability for differently marked excess traffic can be performed by applying Multiple Random Early Detection (M-RED). Nevertheless, we consider excess traffic in an over-provisioned class as harmful for the BE class. Therefore we mark the conforming traffic green and we drop excess traffic in the over-provisioned classes, whereas we allow a marking of excess traffic as red in the Bronze class. The layer 3 queue size of each of the three olympic classes was configured to 128 packets in the WFQ environment. Consequently, the ingress meter and marker is based on a token bucket with a confirmed information rate of 2.4 Mbit/s for all olympic classes thereby leading to the over-provisioning factors given in table 2. A confirmed burst size of 32 MTU is used at the ingress. This value is intentionally smaller than the queue size applied, to avoid packet drops in the olympic classes within the network and also to avoid a high utilization of the queuing space and thus to reduce queuing delays. Besides it has to be noted that the WFQ queue sizes are configured in packets, which can be smaller than the MTU, whereas the confirmed burst size is configured in bytes.

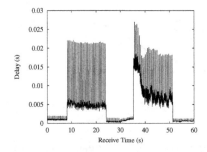

Fig. 9. Delay Bronze UDP News Sequence.

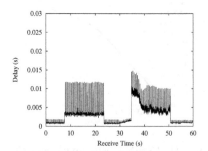

Fig. 10. Delay Silver UDP News Sequence.

Figure 9 shows the measured delay for the news sequence in the Bronze Class and the impacts of congestion in the BE class on the Bronze class. Compared to the transmission of the sequence within the BE class, which is shown in figure 2, the delay is reduced significantly. Furthermore, packet drops did not occur in the Bronze class. Thereby AF based services can be applied as guaranteed rate service without packet loss for conforming traffic. The delay and delay jitter differentiation, which can be achieved in addition by the olympic service, is shown in figure 10 and 11 for the Silver and the Gold class respectively, compared to the Bronze class in figure 9.

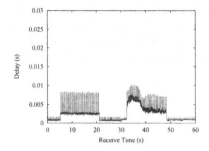
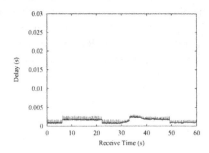

Fig. 11. Delay Gold UDP News Sequence.

Fig. 12. Delay Prem. UDP News Sequence.

Additionally, we present experiments with TCP in the Bronze class and demonstrate how TCP can be configured in a guaranteed rate (GR) environment to achieve the desired throughput. We show that, if the pertaining class is configured properly, packet drops do not occur, which prevents from halving the TCP congestion window. The data rate instead corresponds to the capacity allocated for the flow. To avoid effects on the RTT by upstream congestion, the acknowledgements are also transmitted in the Bronze class. The maximum window size is in our experiments controlled by setting the socket buffer size. The resulting RTT can be computed according to (1) with W denoting the maximum window size and R denoting the configured GR capacity. The RTT adjusts to the available or configured capacity and to the configured maximum window size.

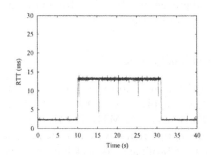
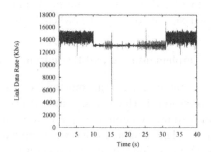

Fig. 13. TCP RTT Bronze 15 MTU Sock-Buf.

Fig. 14. TCP Tput Bronze 15 MTU SockBuf.

For these experiments we configured the Bronze class to 25 % of the bottleneck link capacity, corresponding to a net data rate of about 1.6 MB/s. Figure 13 shows the RTT for a configured socket buffer of 15 MTU. Congestion in the BE class starts after 10 s and leads to an increase in the RTT, which corresponds to the queuing delay added by queuing the data of a complete TCP window. Figure 14 shows the corresponding throughput. At the beginning the application limits the data rate to about 1.8 MB/s and after the BE downstream congestion started, the limitation is given by the configured capacity for the Bronze class at about 1.6 MB/s and from the TCP point of view leads to a limitation of the sending rate by the offered window. The same effect on the throughput can be observed, if the maximum window is increased by configuring a socket buffer of 32 MTU. Figure 15 and figure 16 show the resulting RTT and throughput for this configuration. Again the RTT by increased queuing delay is adjusted to the available capacity and the window size, being about twice as high as in the previous experiment.

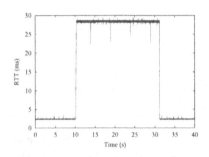

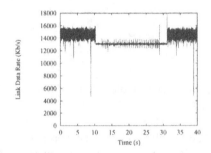

Fig. 15. TCP RTT Bronze 32 MTU Sock-   **Fig. 16.** TCP Tput Bronze 32 MTU
Buf.   SockBuf.

From these TCP experiments it can be seen that during periods of BE congestion WFQ acts as an aggregate traffic shaper with a rate corresponding to the configured WFQ weight. The achieved TCP throughput is independent of the TCP window size, as shown in figure 14 and 16. An explicit per-flow based traffic shaping functionality can be located at the ingress side of a DS source domain by applying a leaky bucket mechanism, while any service differentiation in the core is still strictly class-based. On the other hand such a static traffic shaping would prevent the TCP flow from utilizing unused capacity during times when the BE class is not loaded.

3.4   Implementation and Evaluation of a Premium Service

The premium service was implemented based on EF using Priority Queuing (PQ) and conforms to [5]. The ingress router was configured to apply a meter and marker with a confirmed information rate of 4.8 Mbps and a burst size of 32 MTU. Excess traffic is dropped. The parameters that were applied at the ingress router were reflected by the core configuration. The PQ scheduler was bound to 10% of the bottleneck link capacity, corresponding to about 4.8 Mbps. Bursts of up to 48 KB are permitted in the core.

Since we assume that rather UDP than TCP is applied for extremely time-critical transmissions, we only performed an UDP experiment with the premium class. Of course, any guaranteed bandwidth service for TCP could be build using the EF aggregate when no olympic service is available. Figure 12 shows the results of a transmission of the news sequence. A reduction of the transmission delay and delay jitter especially for big video frames, which lead to packet bursts, becomes obvious for PQ compared to WFQ. In this configuration the tx-ring-limit parameter, which is used to configure the outgoing non-preemptive interface queuing capacity, is of major importance [7].

4   Conclusions and Future Work

We have presented a quantitative evaluation of a DS implementation providing a premium service and an olympic service automatically configured by GARA. Our evaluation addressed the QoS demand of heterogeneous type of flows. The experiments presented used commodity hardware. This demonstrates that real application can actually use DS, especially if access to services is automated by a BB such as GARA.

Our future work will focus on larger scenarios with multiple flows per class and several possible bottleneck links. While we assume to be able to handle multiple flows by an adequate per-flow traffic shaping at ingress nodes, the second problem requires complex TE mechanisms and an advanced resource management, which we aim at addressing with GARA and MPLS.

Acknowledgements

This work was supported in part by the Path Allocation in Backbone networks (PAB) project funded by the German Research Network (DFN) and the Federal Ministry of Education and Research (BMBF).

References

1. Allman, M., Paxson, V., and Stevens, W.: TCP Congestion Control. RFC 2581, (1997)
2. Blake, S. et al.: An Architecture for Differentiated Services. RFC 2475, (1998)
3. Cardwell, N., Savage, S., and Anderson, T.: Modeling TCP Latency. Proceedings of IEEE Infocom, (2000)
4. Charny, A., and Le Boudec, J.-Y.: Delay Bounds in a Network With Aggregate Scheduling. Proceedings of QofIS, (2000)
5. Davie, B. et al.: An Expedited Forwarding PHB. RFC 3246, (2002)
6. DeFanti, T., and Stevens, R.: Teleimmersion. In Foster, I., and Kesselman, C.: The Grid: Blueprint for a Future Computing Infrastructure. Morgan-Kaufmann, (1998)
7. Ferrari, T., Pau, G., and Raffaelli, C.: Priority Queuing Applied to Expedited Forwarding: A Measurement-Based Analysis. Proceedings of QofIS, (2000)
8. Fitzek, F., and Reisslein, M.: MPEG–4 and H.263 Video Traces for Network Performance Evaluation. IEEE Network, (2001), 15(6):40-54
9. Hasegawa, G., Murata, M., and Miyahara, M.: Performance Evaluation of HTTP/TCP on Asymmetric Networks. Int. Journal of Communication Systems, (1999), 12(4):281-96
10. Heinanen, J. et al.: Assured Forwarding PHB Group". RFC 2597, (1999)
11. Mathis, M., Semke, J., Mahdavi, J., and Ott, T.: The Macroscopic Behavior of the TCP Congestion Avoidance Algorithm. Proceedings of ACM SIGCOMM, (1997)
12. Padhye, J., Firoiu, V., Towsley, D., and Kurose, J.: Modeling TCP Troughput: A simple model and its empirical validation. Proceedings of ACM SIGCOMM, (1998)

13. Rosen, E., Viswanathan, A., and Callon, R.: Multiprotocol Label Switching Architecture. RFC 3031, (2001)
14. Sander, V., Adamson, W., Foster, I, and Roy, A: End-to-End Provision of Policy Information for Network QoS. IEEE Symposium on High Performance Distributed Computing, (2001)
15. Sander, V., Foster, I., Roy, A., and Winkler, L.: A Differentiated Services Implementation for High-Performance TCP Flows. Terena Networking Conference, (2000)
16. Wu, Q., and Williamson, C.: Improving Ensemble-TCP Performance on Asymmetric Networks. Proceedings of MASCOTS, (2001)

Unfairness of Assured Service and a Rate Adaptive Marking Strategy

Seung-Joon Seok[1], Sung-Hyuck Lee[2], Seok-Min Hong[1], and Chul-Hee Kang[1]

[1] Department of Electronics Engineering, Korea University,
1, 5-ga, Anam-dong, Sungbuk-gu, 136-701, Seoul, Korea
{ssj, starsu, mickey, chkang}@widecomm.korea.ac.kr
http://widecomm.korea.ac.kr/~ssj
[2] Network Protocol T.G., i-Networking Lab., Samsung Advanced Institute of Technology
San 14-1, Nongseo-ri, Kiheung-eup, Yongin-shi, Kyungki-do, Korea
starsu@sait.samsung.co.kr

Abstract. Assured Service, which is a service model of the Internet Differentiated Services (DiffServ) architecture is not currently well implemented on the Internet, mainly because TCP employs an AIMD (Additive Increase and Multiplicative decrease) mechanism to control congestion. In this paper, we analyze the problem and describe a marking rule, called RAM (Rate Adaptive Marking), which diminishes the problem. The RAM method marks sending packets, at a source node or at an edge router, in inverse proportion to throughput gain and in proportion to the reservation rate and throughput dynamics. We also discuss methods to implement the RAM strategy. Several experiments with the ns-2 simulator are performed to evaluate the RAM scheme.

1 Introduction

Previous work reports that there are two types of unfairness in Assured Service. Several studies [1][2] have found that TCP flows with large profile rate and/or long round trip time (RTT) barely meet their target rate in the presence of numerous flows with small profile rate when these two flows are competed in a common bottleneck link. The reason of this problem is that flows with large profile rate to have longer time to reach to its target rate after a packet loss or not to reach to. For convenience, this problem is herein referred to "*inter-SLA unfairness*". The unfairness is defined as excess bandwidth of a bottleneck link not being distributed equably among flows that go through the link. A SLA may cover a set of flows. In this case, however, there exists unfair sharing reserved bandwidth or profile rate within aggregated flows [3][4]. This unfairness between the aggregated flows is a serious problem in Assured Service model. The unfairness can be caused by differences in RTTs, in link capacities, or in congestion levels experienced by flows within the network, though the total throughput of the aggregation still reaches the target rate. This type of unfairness is herein referred to as "*intra-SLA unfairness*".

B. Stiller et al. (Eds.): QofIS/ICQT 2002, LNCS 2511, pp. 47–56, 2002.
© Springer-Verlag Berlin Heidelberg 2002

When responsive flows with TCP and non-responsive flows with UDP competes the bottleneck link bandwidth each other, responsive flows are seriously affected by the non-responsive flows. The non-responsive flow's sender does not reduce the sending rate in response to the network congestion. As a result, the non-responsive flow worsens the congestion state of the bottleneck queue. Thus, the OUT packet drop probability of the other responsive (i.e., TCP) flows will increase, and consequently, the throughput of the TCP flows will decrease. The impact of the non-responsive flow has been demonstrated by other studies.

Most studies of Assured Service focus on TCP protocol extensions [1][5], shaping [6][7] and dropping policies [1][8]. However, few approaches propose to change the current general marking rule which is that the total IN marking rate is equal to the reservation rate. In this paper, we analyze the inter-SLA unfairness of Assured Service and implement a new marking rule to resolve the two unfairness problems in section 2 and section 3, respectively. Section 4 studies the performance of RAM using an ns-2 network simulator and section 5 concludes this paper.

2   Analysis of Inter-SLA Unfairness

Before delving into the methods reducing the unfairness problem, we first introduce a mathematical analysis of inter-SLA unfairness that is based on [11][12]. The throughput (Th) of a single TCP flow is able to be modeled approximately as equation 1. This analysis model was generated on the assumption that a TCP flow operates mostly in the congestion avoidance phase.

$$Th = \frac{MSS}{RTT}\sqrt{\frac{3}{2bp_{loss}}} = \frac{MSS}{RTT}\sqrt{\frac{c_1}{p_{loss}}} \tag{1}$$

where MSS is the maximum segment size of the TCP flow, RTT is the average round trip time delay, p_{loss} is the average packet loss rate, and b is the delayed acknowledgement parameter. C_1 is $3/2b$.

If the TCP flows are treated as an Assured Service in the DiffServ domain and marked following the general marking rule, the loss rate can be expressed as in equation 2 where $RsvRate$ denotes the flow's reservation rate, p_{out} the probability of marking a packet as OUT, and $p_{out-loss}$ the loss probability of the OUT packets. Replacing the loss rate in equation 1 with the expression of equation 2 and solving the quadratics, we can derive equation 3, in which C_2 is $C_1*(MSS/RTT)^2$, as a solution. Also it is easy to derive equation 4 for the excess bandwidth that a TCP flow gets additionally.

$$p_{loss} = p_{out} \times p_{out-loss} = \frac{Th - RsvRate}{Th} \times p_{out-loss} \tag{2}$$

$$Th = \frac{RsvRate}{2} + \sqrt{\left(\frac{RsvRate}{2}\right)^2 + \frac{c_2}{p_{out-loss}}} \tag{3}$$

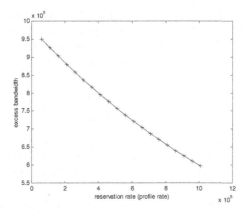

Figure 1. the analysis of inter-SLA unfairness.

$$B_{ex} = \sqrt{\left(\frac{RsvRate}{2}\right)^2 + \frac{c_2}{p_{out\text{-}loss}}} - \frac{RsvRate}{2} \qquad (4)$$

Now consider two TCP flows (flow1 and flow2) that have quite different reserva-tion rates. The *MSS*, *RTT* and *b* values of flow1 are the same as for flow2. Also, it is important to note that the two flows may experience similar loss rates of OUT packets if a multilevel active queue management technique like RIO behaves correctly and the network is not overloaded.

Figure 1 depicts the excess bandwidths (equation 4) for several reservation rates. In this experiment, *RTT*, *MSS*, *b*, and $p_{out\text{-}loss}$ are considered as 1000-byte, 100msec, 1, and 1% respectively. This result proves that the excess bandwidth of a bottleneck link is distributed not equally among the flows that pass through the link. A flow with 0.01Mbps reservation rate comes to have about 0.95Mbps excess bandwidth but an-other flow with 1Mbps about 0.6Mbps. Finally, the result of this analysis is that low profile flows has more excess bandwidth than high profile flows and it is caused mainly by the deviation of reservation rate and general marking rule.

In equation 4, *R* within root term represents marking rate and R within the last term represents reservation rate. Thus, the marking rate of high profile flows should be increased or that of low profile flows should be decreased in order to improve the inter-SLA unfairness.

3   Rate Adaptive Marking Strategy

A RAM marker determines a flow's marking rates periodically, depending on infor-mation about throughput rates collected during a previous period, and marks sending packets according to this rate during the next period. The basic principle of RAM strategy is marking packets as IN in inverse proportion to the throughput gain and in proportion to the reservation rate. Also, RAM makes a flow's IN marking rate to be proportional to the dynamics in that flow's throughput. The RAM marker monitors two throughput levels for all flows; temporal throughput and long term average throughput. The temporal throughput is an average throughput for a short term interval such as an average RTT, while the long-term average throughput is based on a relative long period of time such as the congestion interval. The following equation is the RAM's rule for determining the marking rate.

$$marking\_rate = \frac{(Th\max - Thavg)}{(Th\max - RsvRate)} \times RsvRate \qquad (5)$$

where *Thmax* is the maximum temporal throughput during the immediately preceding congestion interval and *Thavg* is the long term average throughput of the flow during the congestion interval. This equation connotes the basic principles of the RAM scheme. In the right-hand side of the equation, the "*Thmax-Thavg*" term denotes the flow's throughput dynamics and "*(Thmax-RsvRate)/(Thmax-Thavg)*" the flow's throughput gain. Finally "*RsvRate*" is the flow's reservation rate.

If a flow achieves an average throughput that is equal to its reservation rate (*RsvRate*) in the RAM scheme, the marking rate becomes the reservation rate. Also, if the average throughput is over the reservation rate, the marking rate will be lower than the reservation rate and proportional to the gap between the average throughput and the reservation rate. In contrast, if the average throughput is below the reservation rate, the marking rate will be higher than the reservation rate and also inversely proportional to the gap. However, it can happen that the result of equation 5 is negative or has an infinite value, when *Thmax* is below or equal to the *RsvRate*. This case denotes that the throughput does not reach its reservation rate as well as its target rate. The RAM scheme has an upper bound and a lower bound for the marking rate, in order to protect the marking rate and the throughput from experiencing heavy fluctuations. The upper bound is considered to be several times the reservation rate ("*a×RsvRate*") and the lower bound to be a half or less of the reservation rate ("*β×RsvRate*").

Consider the case of a non-responsive flow using UDP or a flow with low dynamics in throughput, the gap between *Thmax* and *Thavg* is very small. Thus, the value of equation 5 approaches zero, if the flow is not in an under-provisioned state. Also, the marking rate becomes the lower bound, because the value is under the lower bound and positive. If the flow is over-provisioned, however, the marking rate is increased over its reservation rate because *Thmax* is less than *RsvRate*. This operation can reduce the impact of non-responsive flows (UDP) on responsive flow (TCP).

3.1 RAM Schemes for Single Flow SLA

If a marker treats a single corresponding flow, then this marker can be located in a flow source or an edge router. The former case is referred to as source marking and the latter case as edge marking. An important feature of source marking is that the RAM marker can monitor the throughput accurately. This is because the ACK information can be provided to the marker. However, it is impossible for edge marking to estimate the throughput accurately, because of the absence of the ACK information. Thus, the edge marking RAM uses the sending rate of the flow source to measure the performance of the flow. Figure 2 and Figure 3 show the pseudo codes of the source marking algorithm and the edge marking algorithm, respectively.

```
When a packet loss happens
        PERIOD = now - ConsPoint
        Avg_Th = RecBytes/RERIOD
        Max_Th = CWnd*MSS/RTT
        PreMR = MR
        MR = (Max_Th - Avg_Th) /
             (Max_Th - RsvRate) * RsvRate
        If 0 <= MR < Low_Rate
        // Non-Responsive
             MR = RsvRate
        If MR > Max_Rate or MR < 0
                If PreMR >= RsvRate
                // Over-Subscribed
                        MR = RsvRate
                Else
                        MR = Max_Rate
        ConsPoint = now

RecBytes : Received Bytes for RERIOD
Avg_Th : Average Throughput
Max_Th : Maximum Temporary Throughput
MR : Marking Rate
Max_Rate : upper bound of marking rate
Rsv_Rate : Reservation Rate
```

Figure 2. the source marking algorithm of the RAM for single flow SLA

In the source marking case, the marking rate is updated whenever packet loss is detected. The maximum temporal throughput can be derived from the congestion window size at that time, because the congestion window is maximized just before congestion control backs it off. The average throughput also denotes the average acknowledged traffic between the previous congestion point and this congestion point (congestion period).

```
Whenever a short-term interval (RTT)
        If Max_Th > SndBytes1/Interval1
          Max_Th = SndBytes1/Interval1
        SndBytes1=0

Whenever a long-term interval (Average Congestion Inter-
val)
        Avg_Th = SndBytes2/Interval2
        PreMR = MR
        MR = (Max_Th - Avg_Th) /
             (Max_Th - RsvRate) * RsvRate
        If 0 <= MR < Low_Rate
        // Non-Responsive
             MR = RsvRate
        If MR > Max_Rate or MR < 0
```

```
        If PreMR >= RsvRate
        // Over-Subscribed
           MR = RsvRate
        Else
           MR = Max_Rate
     SndBytes2=0

SndBytes1 : Sent Bytes from source for Interval1
SndBytes2 : Sent Bytes from source for Interval2
Interval2 : RTT (average round trip time delay)
Interval2 : average congestion interval
```

Figure 3. the edge marking algorithm of the RAM for single flow SLA

The RAM marker that is located in the edge router can not detect the flow's packet losses. Thus, the edge marker has two time intervals to implement for the RAM scheme. Every average round trip time, the edge marker calculates the temporal throughput as the average sending rate for the round trip time and compares this with previous ones, in order to determine the maximum value. Also, every average congestion interval, the average sending rate for the interval is calculated. Other behaviors of this marking scheme are the same as that of the source marking RAM scheme.

3.2 RAM Scheme for Aggregated Flows SLA

It is more general that an SLA includes multiple flows and a single traffic profile of aggregated flows. In this case, a marker must be located at the edge router to support multiple flows from different sources. The general marking scheme does not maintain each flow state and marks packets as IN, in proportion to the sending rates of the flows. The RAM scheme should also be carried out at the edge router. The RAM marker generates a RAM instance corresponding to each flow. This RAM instance determines the marking rate depending on RAM strategy (equation 5) and marks packets according to this marking rate. The Reservation rate of each flow is considered as the *traffic profile rate for aggregated flows / the number of flows*. The marking algorithm for each flow is the same as the edge marking scheme of single flow (Figure 3). In particular, this scheme also uses the sending rate of each flow as information to determine the marking rate of the flow. This is because it is difficult and heavy load of processing complexity in edge router to evaluate throughput of TCP flow accurately.

4 Performance Study

In order to confirm the performance effect of the RAM scheme as against previous approaches, two experiments have been performed, using ns-2, on several simulation models. The first experiment is the evaluation of inter-SLA unfairness and the second experiment tests that the RAM scheme alleviates the impact of non-responsive flow(s). The third experiment tests the intra-SLA unfairness using the RAM scheme.

4.1 Evaluation of Inter-SLA Unfairness

We first examine the effectiveness of the RAM scheme on the inter-SLA unfairness when there are only TCP flows. Both RAM scheme types, source marking RAM and edge marking RAM, are considered in this experiment. Figure 4 depicts the testbed topology for this experiment. In this topology, there are 60 TCP flows and each edge router is connected with 10 TCP sources. Each TCP flow of s1-x, s2-x, s3-x, s4-x, s5-x and s6-x has its profile rate of respectively 0.01Mbps, 0.02Mbps, 0.04Mbps, 0.08Mbps, 0.16Mbps and 0.32Mbps, respectively. A RIO at C1 router is set at [50/80/0.02] and [10/40/0.5].

Figure 5 shows the average throughputs of 10 TCP flows with the same reservation rate when the source marking RAM (S_RAM) rule, the edge marking RAM (E_RAM), and the general marking rule (TSW2CM), respectively, are applied. In the TSW2CM, low profile (0.01, 0.02, 0.04, or 0.08 Mbps) flows achieve throughputs that are near or over their target rates, regardless of network load. Also, these throughputs are higher than those obtained when the S_RAM and E_RAM are used. When the TSW2CM is applied, however, the throughputs of high profile (0.08, 0.16, or 0.32 Mbps) flows are less than those obtained when the RAM rules are used, as well as being much less than the target rates. From this figure, we can see that the E_RAM can remedy the inter-SLA unfairness as well as the S_RAM does, because the result of the simulation using E_RAM is similar to that of the S_RAM, regardless of network load.

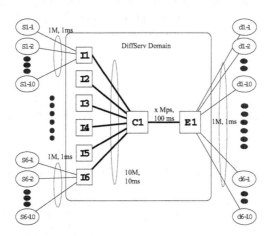

Figure 4. Experimental topology used to study inter-SLA unfairness

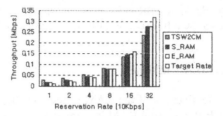

(a) Bottleneck link bandwidth is 6.3Mbps (Congestion)

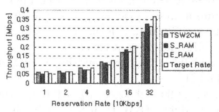

(b) Bottleneck link bandwidth is 9Mbps (Over-Provisioning)

Figure 5. Inter-SLA unfairness test results

4.2 Evaluation of the Impact of Non-responsive Flows

In this experiment, we test the robustness of the RAM scheme with regard to this non-responsive flow effect. The experimental testbed used for this experiment is similar to that shown in Figure 4, except that there are two ingress nodes (*I1, I2*). Also, it is considered that s1-x consists of long-lived FTP sources with a 0.16 Mbps profile rate, and s2-x becomes a CBR (Constant Bit Rate) source consisting of UDP protocol or long-lived FTP sources with 0.1 Mbps profile rates. The CBR sources have a 0.11Mbps generating rate and 0.1 Mbps profile rate.

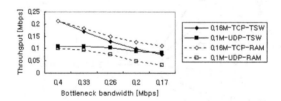

(a) RAM protocol and TSW2CM when there is UDP sources.

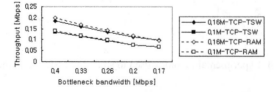

(b) RAM protocol and TSW2CM when there is no UDP source.

Figure 6. Evaluation of RAM scheme's effectiveness for non-responsive flows.

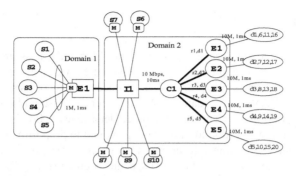

Figure 7. Experimental topology used to study the intra-SLA unfairness

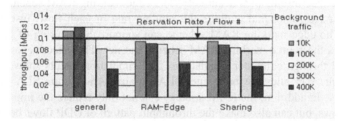

Figure 8. Average Throughputs of TCP flows included in a SLA, in the environment in which background traffic is TCP stream, all core-egress link bandwidth is 0.35 Mbps

Figure 6(a) shows that an unfairness problem occurs between TCP flows and UDP flows and that the RAM scheme causes the average throughput of the UDP flows to decrease, following a contour similar to that observed for TCP flows, as compared to Figure 6(b). In Figure 6, legends denote reservation rate-protocol-marking rule.

4.3 Evaluation of Intra-SLA Unfairness

In this experiment, we consider the case in which an SLA includes multiple flows. Figure 7 depicts the experimental topology used to test the intra-SLA unfairness. In this topology, domain 1 and domain 2 establish an SLA, in which 500Kbps is allocated as the traffic profile of the aggregated flows from sources (s1, s2, s3, s4 and s5). Thus, these flows should be remarked at the E1 according to the SLA. The other flows are used for background traffic and assumed to be coming into domain 2 from different domains, except for domain1. The background sources (s6, s7, s8, s9, and s10) have different marking rates (10, 100, 200, 300, and 400 Kbps respectively). In order that each flow included in the SLA experiences different network status, flows1, 2, 3, 4 and 5 are routed through different routes with different congestion levels within domain 2. The congestion levels are controlled by the background flows. In this experiment, we monitored the average throughputs of TCP flows included in the SLA, in the case when general TSW2CM, RAM-edge marking, or equal sharing marking (Sharing) is used. Figure 8 shows the simulation results. In this result, Jain's fairness index value of TSW2CM, RAM, and equal sharing scheme are 0.93, 0.98, and 0.97, respectively. From these results, we confirmed that the RAM scheme can reduce intra-SLA unfairness as well as inter-SLA unfairness.

5 Conclusion

In this paper, we described serious problems, inter-SLA unfairness and intra-SLA unfairness, witch arise when provisioning an Assured Service and also unfairness between TCP and UDP flows. We proposed an approach which consists of a modified marking rule, called RAM (Rate Adaptive Marking), for the Assured Service, to alleviate the unfairness problems. The RAM scheme determines a flow's marking rate periodically, in inverse proportion to throughput gain, and in proportion to throughput dynamics and to the reservation rate during a previous congestion period. We described two scenarios to implement the RAM scheme, source marking RAM and edge marking RAM. The source marking RAM scheme cooperates with TCP congestion control to calculate throughputs accurately. The edge marking RAM scheme operates at an ingress edge router and exploits the sending rate of each flow, instead of the throughput. The edge marking RAM scheme can support the marking of aggregated flows, whereas the source marking RAM scheme can not. The effectiveness of RAM was evaluated through computer simulations. The simulation results showed that the RAM technique can alleviate both inter-SLA unfairness and intra-SLA unfairness simultaneously. In addition, the RAM strategy can not only reduce the impact of non-responsive flows, but can also make the throughput pattern of UDP flows be similar to that of TCP flows.

References

1. Yeom, I., Reddy, A. L. N.: Realizing Throughput Guarantees in a Differentiated Service Network. In Proc. ICMCS'99, June (1999) 372-376
2. Ibanez, J. Nichols, K.: Preliminary Simulation Evaluation of an Assured Service. Internet Draft, Aug.(1998)
3. Yeom, I., Reddy, A. L. N.: Impact of marking strategy on aggregated flows in a differentiated services network. In Proc. IWQoS 1999 , May(1999)
4. Kim, H.: A Fair Marker. Internet Draft, April(1999)
5. Fang, W., Perterson, L.: TCP mechanisms for a diff-serv architecture, Tec. Rep
6. Cnodder, S. D., Elloumi, O., Pauwels, K.,: Rate adaptive shaping for the efficient transport of data traffic in diffserv networks. Computer Networks, Vol. 35(2001) 263-285
7. Li, N., Borrego, M., Li, S.: A rate regulating traffic conditioner for supporting TCP over Diffserv. Computer Commnications, Vol. 23(2000) 1349-1362
8. Park, W. H., Bahk, S., Kim, H.: A modified RIO algorithm that alleviates the bandwidth skew problem in Internet Differentiated Service. In Proc. ICC'00 (2000) 1599-1603
9. Feng, W., Kandlur, D., Saha, D., Shin, K.: Adaptive Packet Marking for Providing Differentiated Services in the Internet. In Proc. ICNP'98, Austin, TX, Oct. (1998) 108-117
10. Nam, D-H., Choi, Y-S. Kim, B-C., and Cho, Y-Z.: A traffic conditioning and buffer management scheme for fairness in differentiated services. in proc. ICATM 2001, April (2001)
11. Baines, M.: Using TCP Model To Understand Bandwidth Assurance in a Differentiated Services Network, In proc. IEEE Infocomm 2001 (2001)1800-1805
12. Padyhe, J., Firoiu, V., Townsley, D., Kurose, J.: Modeling TCP Throughput: A Simple Model and its Empirical Validation. CMPSCI Technical Report TR 98-008, University of Massachussetts, MA (1999)

Counters-Based Modified Traffic Conditioner

Maria-Dolores Cano, Fernando Cerdan, Joan Garcia-Haro,
Josemaria Malgosa-Sanahuja

Department of Information Technologies and Communications
Polytechnic University of Cartagena
Campus Muralla del Mar s/n (Ed. Hospital de Marina)
30202 Cartagena, Spain
Ph. +34 +968 32 5953, Fax +34 +968 32 5338
{mdolores.cano, fernando.cerdan, joang.haro, josem.malgosa}@upct.es

Abstract. Traffic conditioners play a key role in implementing the Assured
Service in the framework of the DiffServ approach. In this paper, we propose a
traffic conditioner for the Internet Assured Service called Counters-Based
Modified (CBM) that strictly guarantees target rates and performs a fair share
of the excess bandwidth among TCP sources. The fairness in the outbound
bandwidth distribution is met by probabilistically dropping OUT packets in the
traffic conditioner. To determine the dropping probability of an OUT packet,
some sort of signaling is needed. Although, it results more feasible than other
proposed intelligent traffic conditioners. The CBM traffic conditioner is
evaluated under different conditions by simulation using TCP Reno sources.

1   Introduction

The Assured Forwarding Per-Hop Behavior (AF-PHB) [1] is one of the IETF PHBs
for Differentiated Services with the status of proposed standards. The idea behind AF-
PHB is to ensure a minimum throughput (target rate or contracted rate) to a
connection, while enabling consuming more bandwidth if the network load is low. To
achieve this goal, packets of individual flows are marked belonging to one of the four
independently forwarded AF classes. Within each AF class an IP packet can be
assigned one of three different levels of drop precedence. In case of congestion,
DiffServ nodes try to protect packets with a lower drop precedence value from being
lost by preferably discarding packets with a higher drop precedence value. Note that
minimum throughput is also called in-profile bandwidth or inbound bandwidth, and
excess bandwidth can be also referred as outbound bandwidth along this study.

Despite of the abundant literature written about the AF-PHB (e.g. [2] to [9]), none
solution has been found to face up its two goals, assuring the inbound bandwidth and
offering a fair distribution of the excess bandwidth if available. Two different
concepts can be understood as fairness in the outbound bandwidth sharing. The first
considers fairness as the even distribution of excess bandwidth among all connections
that compose the aggregate. The second defines fairness as a proportional distribution
of the outbound bandwidth with respect to the contracted rate. In this paper we adopt
the first definition. It should be remarked that despite DiffServ mechanisms are not
implemented to provide and end-to-end service, it has sense to study the performance

B. Stiller et al. (Eds.): QofIS/ICQT 2002, LNCS 2511, pp. 57-67, 2002.
© Springer-Verlag Berlin Heidelberg 2002

of TCP connections in terms of throughput excluding retransmitted packets, which is usually called *goodput*.

The Counters-Based (CB) traffic conditioner developed in [10] has been demonstrated to perform comparatively better than other traffic conditioners. This mechanism based on counters guarantees the in-profile bandwidth allocation in scenarios with variable round trip times and different target rates. Its easy configuration and high accuracy make it suitable for general use. Only two counters are needed to implement this algorithm, C1 and C2, and no parameter configuration is required. It also includes a simple mechanism to avoid accumulation of "credits" when a source stops transmitting data, for instance when a time out expires. From the comparative simulation study carried out in [10], this traffic conditioner together with RIO [2] over performs the two classical Time Sliding Window [2] (TSW)-RIO and Leaky Bucket (LB)-RIO mechanisms in terms of guaranteeing inbound bandwidth, with fluctuations in the achieved rate that do not exceed 1% of the connection target rate. Nevertheless, it also presents problems regarding the excess bandwidth sharing among sources as previous proposals.

In this paper, we introduce an alternative approach for achieving fairness in the excess bandwidth distribution among TCP sources for the Internet Assured Service. Starting from a high accuracy in assuring inbound bandwidth provided by the CB algorithm [10], we meet the fairness in the outbound bandwidth distribution adding a probabilistically dropping of OUT packets in the traffic conditioner. We call this new version of the CB algorithm the Counters-Based Modified (CBM) traffic conditioner. With this modification, complexity remains at the assured service capable host before the RIO buffer management scheme. To determine the dropping probability of an OUT packet, it is assumed that the traffic conditioner knows the amount of excess bandwidth and the average Round Trip Time (RTT) of all connections. Although it implies using some sort of signaling, it is more feasible than other proposed traffic conditioners such as [3], [4], [7], [8] or [9]. Along this paper we describe the CBM characteristics and study its performance with the RIO buffer management scheme throughout simulations. In addition, CBM accomplishment is compared with its precursor, the CB mechanism, and the two deeply studied algorithms TSW and LB. As we show later in simulation results, it is possible to afford fairness in the excess bandwidth sharing by using the CBM traffic conditioner without losing accuracy in assuring contracted rates.

The rest of this paper is organized as follows. Section 2 describes the CBM traffic conditioner implementation. Section 3 presents the scenarios and assumptions for simulations. In Section 4, simulation results are shown and discussed. The paper concludes in Section 5 summarizing the most important facts.

2   The Counters-Based Modified Traffic Conditioner

Assuming that all packets have a similar size, if all sources introduce the same number of out-of-profile packets into the network, then each source can get the same portion of excess bandwidth. This ideal behavior is affected by the odd characteristics of each TCP connection, like different RTT or target rates among others, and the interaction with the RIO buffer management scheme in the router. To overcome these influences, we suggest that connections that are sending OUT packets beyond their ideal fair quota should be penalized. This penalty is based on probabilistically dropping OUT packets in the traffic conditioner. The arising question is how to select what OUT packets should be dropped, and what ones should be added to the aggregate.

To solve it, we have studied the behavior of the excess bandwidth distribution from a different perspective. In simulation results, we have observed the number of out-of-profile packets generated between consecutive in-of-profile packet arrivals. Hence, we can state that:

– (i) A source with small target rate generates more OUT packets between two consecutive IN packets than a source with higher target. TCP sources transmit at link rate, so the smaller the target the more OUT packets are injected into the network. For this reason, these sources can get more network resources.

– (ii) The faster time response of the TCP sources with small RTT makes them inject more traffic, i.e., more OUT packets. Therefore, a source with small RTT is able to generate more OUT packets between two consecutive IN packets than a source with higher RTT.

An example illustrating this fact is depicted in Figures 1 and 2. Simulations to obtain these figures have been done using the CB algorithm and RIO. Eight sources generate TCP traffic at link rate, where each source has a contracted rate of 1-1-2-2-3-3-4 and 4 Mbps respectively. The RTT for each connection ranges from 10 to 80 ms at increments of 10 ms. The x-axis represents the time in seconds, and the y-axis the number of OUT packets between two consecutive IN packets. Observing both figures, the source with lower RTT and smaller target is injecting more OUT packets into the network. Furthermore, total number of OUT packets generated by source number 1, with target rate of 1 Mbps and RTT of 20 ms, is 6,031 packets; whereas for source number 7, with target rate of 4 Mbps and RTT of 80 ms, it is nearly the third part 2,331 packets.

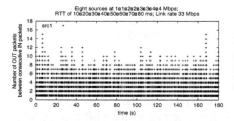 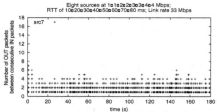

Fig. 1. OUT packets between IN packet tagging events for source 1 (6,031 OUT packets)

Fig. 2. OUT packets between IN packet tagging events for source 7 (2,331 OUT packets)

From these observations, the idea suggested in this paper is explained as follows (see Figure 3). The Counters-Based Modified (CBM) traffic conditioner, which is placed next to the TCP source (out of the reach of the final user), has a variable that counts the number of packets that have been marked as OUT between two consecutive IN packets. Every time a packet is marked as OUT, the CBM traffic conditioner checks this variable. If the variable does not exceed a minimum value *min*, then the OUT packet is injected into the network. If it exceeds a maximum value *max*, then the OUT packet is dropped. Finally, if the variable remains between *min* and *max*, the OUT packet is dropped with probability *p*.

To tune the *max* and *min* parameters we follow equations (1) and (2), where MSS stands for Maximum Segment Size. The excess bandwidth (BW_{excess}) could be seen as another TCP source whose maximum TCP window size is determined by the product BW_{excess} times $RTT_{average}$. Therefore we set the *max* limit to this value. A source that injects a number of OUT packets close to this limit would consume almost the entire excess bandwidth. In addition, if this limit is exceeded the source could even steal part of the guaranteed bandwidth, therefore the source cannot inject OUT packets beyond this *max* value. Another characteristic of this limit is that allows sources to increase the consumed excess bandwidth in case other sources finish their connections. In the extreme situation where only one source remains active it could use almost all the excess bandwidth, which is a reasonable behavior. It is well known that in TCP/IP, a simple additive increase and multiplicative decrease algorithm satisfies the sufficient conditions for convergence to an efficient state of the network, and it is used to implement congestion avoidance schemes. For this reason, a practical *min* value is half the *max* value.

```
Initially:                                                    (continue)
  Counter1=1                          if counter1>0
  Counter2=link_rate/target_rate        packet marked as IN
  Counter3=0                            counter3=0
  Calculate  the  values  for  the      counter1--
  probability p and the limits max    else
  and min                               packet marked as OUT
                                        counter3++
For each unit of time:                  if time>start_dropping_time
  Counter2--                              if counter3>max
  if counter2 <= 0                          OUT packet is dropped
    Counter1++                            else if counter3>min
    counter2=link_rate/target_rate          OUT   packet  is  dropped
  if there is a packet arrival            with probability p
                      (continue)          otherwise     OUT     packet
                                          accepted
```

Fig.3. Simplified pseudo-code of the CBM traffic conditioner algorithm

The estimation of RTT can be obtained by periodically signaling from the router device. The TCP protocol implements an algorithm that estimates the RTT of the current connection. This estimation is periodically sent to the router device, which calculates the average RTT. This value is then returned to the traffic conditioner, where packets are marked and/or dropped. Notice that per-flow state monitoring in the router is not required, in the sense that the router does not contain information on each individual active packet flow. It only has to periodically assess the RTT average with the information that receives from the TCP connections, and once performed, these values are not stored anywhere unlike traffic conditioner implementations from [3], [4], [8] and [9].

$$max = \left\lceil \frac{Bandwidth_{excess} \cdot RTT_{average}}{MSS} \right\rceil \tag{1}$$

$$min = \left\lceil \frac{max}{2} \right\rceil \tag{2}$$

The dropping probability p is shown in equation (3). Each source has a different value of p, between 0 and 1, based on its contracted rate. From statements (i) and (ii), it is intuitive to apply an equation in the form $p=1-x$, where x is *target_rate/link_rate*, thus connections with small target rates drop more OUT packets. However, once the *max* threshold is established, the traffic conditioner causes the lost of all OUT packets over the *max* limit. The fact of dropping a packet makes the source to slow down, so other sources can introduce more traffic into the network, that is more OUT packets. If an equation that favors sources with large target rates is employed, then we are penalizing sources with small targets in excess; thus, when they recover from the lost, buffer resources are being consumed by sources with high targets. This situation causes new losses and makes sources with small targets to slow down again, originating the opposite effect stated in (i) and (ii), which is not desirable either. Therefore, we should use an equation for the dropping probability that gives a little more preference to connections with small targets.

We first evaluated a lineal equation such as $p=x$, and simulations showed that the CBM performed a fairer distribution of the excess bandwidth than the CB, albeit still away from the ideal behavior. To observe how the shape of the equation could influence the CBM performance, we conducted simulations with $p=2*x/(1+x)$ and $p=x/(2-x)$, two curves that give preference to connections with small targets over connections with high targets but in a non-linear way. It is important to state that small differences in p value may cause big performance differences because of the TCP congestion algorithm. From these results, we experienced that expression (3) is the most adequate equation in the performance of the CBM traffic conditioner. Notice that equation (3) is only applied when the number of OUT packets is in the interval (*min*, *max*).

$$p = 2 \cdot \frac{target\_rate/link\_rate}{1 + target\_rate/link\_rate} \qquad (3)$$

Finally, if the dropping process starts at the same time for all connections, then connections with larger RTT are adversely affected because of the slower time response. As a result, each traffic conditioner starts the process when a random multiple of its RTT has elapsed.

3   Scenario for Simulations

The topology selected for our simulations is illustrated in Figure 4. TCP traffic is generated by eight TCP Reno sources transmitting at the link rate, which has been set to 33 Mbps. The simulation tool used in this work for the sliding window protocol of TCP Reno sources was developed in [11], and was applied to validate the analytical study carried out in [12]. As a first insight, we employ a large packet size of 9,188 bytes, which corresponds to classical IP over ATM (e.g. Differentiated Services over MPLS, where the use of the ATM technology seems inherent).

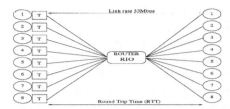

Fig. 4. Topology for simulations, the bottleneck is the router device (T≡Traffic Conditioner)

A router located inside the network, buffers and forwards the aggregated traffic. The queue management employs RIO with parameters [40/70/0.02] for IN packets and [10/40/0.2][1] for OUT packets. Weight_in and Weight_out RED parameters to calculate the average queue size were chosen equal to 0.002 as recommended in [13].

We consider five different scenarios in an undersubscribed situation (traffic load ≤ 60%), whose characteristics are included in Table 1. The oversubscribed scenario (traffic load > 60%) is less interesting in this study since the excess bandwidth represents a very small portion of the total available bandwidth. Simulation results have a confidence interval of 95% that has been calculated with a normal distribution function using 30 samples, with an approximate value of ±0.002 for all fairness calculations, and ±0.01 for the achieved target rates.

Table 1. Description of scenarios A through E employed in simulations (TR≡Target Rate)

	A	B	C	D	E
Link rate (Mbps)	33	33	33	33	33
TR src#0 to 7 (Mbps)	2.4	1-1-2-2-3-3-4-4	2.4	1-1-2-2-3-3-4-4	4-4-3-3-2-2-1-1
RTT src#0 to 7 (ms)	50	50	10-20-30-40-50-70-80	10-20-30-40-50-70-80	10-20-30-40-50-70-80
Σ target rates (Mbps)	19.2	20	19.2	20	20
BW$_{excess}$ (Mbps)	13.8	13	13.8	13	13
RTT$_{average}$ (ms)	50	50	45	45	45
max (#OUT packets)	9	8	8	7	7
min (#OUT packets)	5	4	4	4	4

In scenario A, all connections have same RTT and same contracted rates, which makes this situation both ideal and infrequent in real frameworks. It is expected to obtain the best simulation results in this scenario, which has been usually studied in most papers. In scenario B, all connections have same RTT but different contracted rates. With the introduction of different target rates we try to be closer to a real environment with Quality of Service. Scenario C is the opposite of scenario B, since all connections have different RTT but same contracted rates, hence we can analyze the effect of the RTT on the CBM traffic conditioner performance.

Scenario D is the worst and most complex case under study. All connections have different RTT and different contracted rates, where sources with small targets have small RTT. Therefore, it implies that these TCP connections are favored as reflected

---

[1] [minth, maxth, maxp]

in [3], [10] and [14]. Finally, in scenario E all connections have different RTT and different contracted rates (sources with small targets have large RTT). This is also a representative case, however assigning large round trip times to connections with small target rates avoids favoritism, as it occurs in scenario D.

4   Simulation Results

In this section, we present and discuss simulation results carried out in the scenarios described earlier. Firstly, it is shown how the new mechanism can control the number of OUT packets transmitted over the network leading to a fair share of the excess bandwidth. Moreover, we demonstrate that our scheme does not affect the CB performance presented in [10] regarding the in profile bandwidth assurance. We also present results of the interaction of Assured Service connections with Best-Effort connections competing for the outbound bandwidth.

4.1   OUT Packets Dropping

As indicated in Section 2, by setting the thresholds *max* (eq. 1) and *min* (eq. 2), and making use of the dropping probability *p* (eq. 3), the CBM traffic conditioner controls the number of out-of-profile packets injected into the network by each source. This effect can be observed in Figures 5 and 6. Simulations to obtain these figures have been done in scenario D (see Table 1), which is a usual environment in Internet because of the miscellaneous characteristics of each connection.

Figure 5 represents the number of OUT packets between consecutive IN packets arrivals using the CB traffic conditioner; that is, without applying the probabilistically dropping of out of profile packets. It corresponds to source number 1 with a target rate of 1 Mbps and a RTT of 20 ms. Figure 6 illustrates the improvement concerning Figure 5 in controlling the number of OUT packets introduced in the network when the proposed CBM algorithm is adopted.

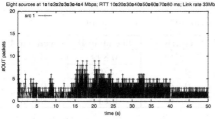 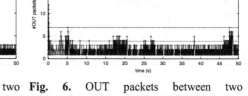

Fig. 5. OUT packets between two consecutive IN packets **without** OUT packet dropping in the traffic conditioner for source 1 (total OUTs=7,183 packets)

Fig. 6. OUT packets between two consecutive IN packets **with** OUT packet dropping in the traffic conditioner for source 1 (total OUTs=5,674 packets)

Comparing Figures 5 and 6, we observe that using the CBM algorithm we are able to obey sources with small targets and small RTT (e.g. source number 1) to generate less OUT packets. Likewise, with this mechanism we can increase the number of OUT packets injected into the network by connections with high target rates and large RTT. When the *max* number of OUT packets between two consecutive IN packets is exceeded, these packets are dropped. TCP connections reflect these drops slowing down, so more excess traffic (OUT packets) from other sources can be added to the aggregate. From these results, it can be presumed that the CBM traffic conditioner controls the number of out of profile packets that join the aggregate; hence, it can manage the sharing of excess bandwidth with the aim of providing a fair distribution as we illustrate in next section.

4.2   Fairness Index

To evaluate fairness we use the fairness index f shown in (4), where x_i is the excess throughput of source i, and n is the number of sources that compose the aggregate [15]. The closer to 1 in the f value, the more the fairness obtained.

$$f = \frac{\left(\sum_{i=1}^{n} x_i\right)^2}{n \cdot \sum_{i=1}^{n} x_i^2} ; f \le 1 \tag{4}$$

Table 2 depicts the different f values obtained from simulations, and compares them in the same scenarios to other traffic conditioners that do not implement probabilistic OUT packets dropping (CB, TSW and LB). Simulations for the TSW and LB traffic conditioners have been carried out taking into consideration the performance evaluation study from [10]. This research includes a TSW configuration guide, since one of the disadvantages of the TSW algorithm is the difficulty in adjusting all the parameters involved on it. Slight variations in the values of the TSW or LB parameters cause relevant differences in simulation results.

Fairness indexes included in Table 2 reveal that it is possible to assure fairness in the excess bandwidth sharing with the CBM traffic conditioner, achieving an f value close to 0.95. Although the LB and TSW algorithms attain a high f value in scenarios A and B respectively, it should be noted that using these mechanisms inbound bandwidths are not guaranteed. Therefore, the underlying idea of keeping all connections sending a similar number of OUT packets is presented as a comparatively improvement in the development of traffic conditioners for the Internet Assured Service.

Table 2. Fairness index in the five different scenarios (TC≡Traffic Conditioner)

TC – RIO	Scenario A	Scenario B	Scenario C	Scenario D	Scenario E
CBM	0.997	0.969	0.942	0.899	0.923
CB	0.854	0.855	0.781	0.708	0.836
TSW	0.582	0.807	0.631	0.489	0.562
LB	0.853	0.687	0.740	0.817	0.832

4.3   Interaction of Assured Service Sources with Best-Effort Sources Using CBM

In this subsection, best-effort (BE) sources compete with Assured Service (AS) sources for the available excess bandwidth. We use the topology shown in Figure 4, where the first four connections have an Assured Service and the last four connections belong to the best-effort class. The fact of being best-effort implies that all packets generated by these sources are considered as out of profile, and they do not have contracted target rates. We have conducted simulations for the five scenarios explained in Section 3 with slight modifications commented below. Link rate is kept at 33 Mbps.

In scenario A, the AS sources have a target rate of 5 Mbps, and all sources (included the BE ones) have a RTT of 50 ms. From equations (2) and (3), the limits *max* and *min* are 9 and 4 respectively. Ideally, each connection should get 1.625 Mbps of the excess bandwidth. Figure 7 depicts the achieved *goodput* of BE connections, where it is seen how these sources obtain nearly the same portion of the excess bandwidth after a transient interval. In this environment, we reach a fairness index of 0.906. Scenario B is like scenario A, but the four AS connections have contracted rates of 4-5-6 and 7 Mbps each. In this case, where thresholds *max* and *min* are 7 and 4 packets, the *f* value is nearly 0.87.

In scenario C, the AS sources have a target rate of 5 Mbps with RTT values that range from 10 ms to 40 ms at increments of 10 ms. Moreover, the BE connections have a RTT that varies from 50 ms to 80 ms at intervals of 10 ms. The limits *max* and *min* take a value of 7 and 4 packets. Figure 8 shows the *goodput* of BE sources in scenario C. In this situation, the excess bandwidth is 13 Mbps, thus the ideal *goodput* for BE connections is 1.625 Mbps. As depicted in this figure, BE sources achieve a *goodput* close to the ideal value with a difference of 0.5 Mbps between the maximum and minimum reached *goodputs*. The effect of having different values of RTT is hardly noticeable in the distribution of the outbound bandwidth, which is reflected in a fairness index of 0.847.

Finally, the most complex scenarios D and E also present an *f* value over 0.8. In scenario D, the four AS sources have contracted rates of 4-5-6 and 7 Mbps, and a RTT that goes from 10 ms to 40 ms at intervals of 10 ms. The RTT for the BE sources ranges from 50 ms to 80 ms in increments of 10 ms. Scenario E only differs from D in the target rates of the AS connections, being in this case 7-6-5 and 4 Mbps. The limits *max* and *min* take a value of 6 and 3 packets in both cases. Figure 9 shows the *goodput* of BE sources in scenario E. In this case, the difference between the maximum and minimum reached *goodput* is about 0.5 Mbps, but worse than case C (Figure 8).

Figure 10 displays the achieved rates for IN packets in scenario D to remark that the existence of best-effort sources does not influence the AS sources regarding the contracted rates with CBM. When you want to offer higher-quality connections for some customers, you need tools to limit the effect of malicious users within the best-effort class. This type of users only generates out of control OUT packets that difficult the provisioning of a consistent network service. We show that the robustness of the couple CBM-RIO makes the entire service structure resistant to malicious users who try to maximize the bandwidth they attain from the network, since all AS sources get their target rates and also benefit from the excess bandwidth quite closely to the ideal behavior.

5   Conclusions

In this paper, we introduce a modification to the Counters-Based traffic conditioner that fulfills a fair distribution of the outbound bandwidth and guarantees target rates, called Counters-Based Modified (CBM). The CBM ability of controlling the number

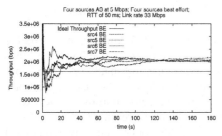

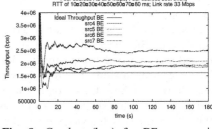

Fig. 7. *Goodput* (bps) for BE sources in scenario A

Fig. 8. *Goodput* (bps) for BE sources in scenario C

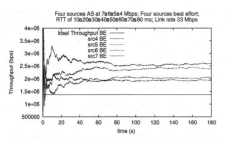

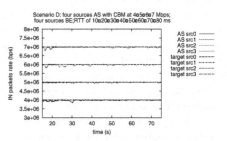

Fig. 9. *Goodput* (bps) for BE sources in scenario E

Fig. 10. Achieved rates for IN packets in scenario D, where AS and BE sources coexist

of out-of-profile packets that each source introduces in the aggregate helps to fair distribute outbound bandwidth, since excess bandwidth is occupied with this type of packets. The CBM traffic conditioner reaches this objective discarding out-of-profile packets before joining the aggregated, with a probability that depends on the target rate, the excess bandwidth and an estimation of the average RTT of all connections. We present simulation results in miscellaneous TCP environments (different target rates, different round trip times and share of resources with best-effort connections), showing that CBM can assure fairness in excess bandwidth sharing achieving a fairness index over 0.9. Results with CBM are also compared with other traffic conditioner implementations such as Time Sliding Window and Leaky Bucket, being illustrated that the CBM gets a comparatively better accomplishment. In addition, we have shown that in situations where Assured Service sources and best-effort sources coexist, the couple CBM-RIO is robust enough when possible best-effort users try to get more network resources than allowed. The high accuracy in guaranteeing the inbound bandwidth, the low complexity introduced, and the good value of the fairness index obtained in simulation results, lead us to believe that it is a feasible election in the Assured Service implementation with DiffServ.

Acknowledgements

This work was supported by the Spanish Research Council under grant FAR-IP TIC2000-1734-C03-03.

References

1. J. Heinanen, F. Baker, W. Weiss, J. Wroclawski, "Assured Forwarding PHB Group", RFC 2597, June 1999.
2. D. Clark and W. Fang, "Explicit Allocation of Best-Effort Packet Delivery Service", IEEE/ACM Transactions on Networking, Vol. 6 No. 4, pp. 362-373, August 1998.
3. W. Lin, R. Zheng, J. Hou, "How to make assured service more assured", Proceedings of the 7th International Conference on Network Protocols (ICNP'99), pp. 182-191, Toronto, Canada, October 1999.
4. B. Nandy, N. Seddigh, P. Pieda, J. Ethridge, "Intelligent Traffic Conditioners for Assured Forwarding Based Differentiated Services Networks", Proceedings of Networking 2000, Paris, France, pp.540-554, May 2000.
5. Elloumi O, De Cnodder S, Pauwels K, "Usefulness of the three drop precedences in Assured Forwarding Service", Internet draft, work in progress, July 1999
6. M. Goyal, A. Durresi, P. Misra, C. Liu, R. Jain, "Effect of number of drop precedences in assured forwarding", Proceedings of Globecom 1999, Rio de Janeiro, Brazil, Vol. 1(A), pp. 188-193, December 1999.
7. H. Kim, "A Fair Marker", Internet draft, work in progress, April 1999.
8. I. Alves, J. De Rezende, L. De Moraes, "Evaluating Fairness in Aggregated Traffic Marking", Proceedings of IEEE Globecom'2000, San Francisco, USA, pp. 445-449, November 2000.
9. I. Andrikopoulos, L. Wood, G. Pavlou, "A fair traffic conditioner for the assured service in a differentiated services internet", Proceedings of IEEE International Conference on Communications ICC2000, New Orleans, LA, Vol. 2, pp. 806-810, June 2000.
10. Maria-Dolores Cano, Fernando Cerdan, Joan Garcia-Haro, Josemaria Malgosa-Sanahuja, "Performance Evaluation of Traffic Conditioner Mechanisms for the Internet Assured Service", in Quality of Service over Next-Generation Data Networks, Proceedings of SPIE Vol. 4524, pp. 182-193, 2001.
11. F. Cerdan, O.Casals, "Performance of Different TCP Implementations over the GFR Service Category", ICON Journal, Special Issue on QoS Management in Wired &Wireless Multimedia Communications Network, Vol.2, pp.273-286, Baltzer Science.
12. V. Bonin, O.Casals, B. Van Houdt, C. Blondi, "Performance Modeling of Differentiated Fair Buffer Allocation", Proceedings of the 9th International Conference on Telecommunications Systems, Dallas, USA.
13. S. Floyd and V. Jacobson, "Random Early Detection Gateways for Congestion Avoidance", IEEE/ACM Transactions on Networking, Vol. 6 No 4, pp. 397-413, August 1993.
14. N. Seddigh, B. Nandy, P.Pieda, "Bandwidth Assurance Issues for TCP flows in a Differentiated Services Network", Proceedings of IEEE Globecom'99, Vol. 3, pp. 1792-1798,Rio de Janeiro, Brazil, December 1999.
15. R. Jain, "The Art of Computer Systems Performance Analysis", John Wiley and Sons Inc., 1991.

High Quality IP Video Streaming with Adaptive Packet Marking

Sebastian Zander, Georg Carle

Fraunhofer FOKUS
Kaiserin-Augusta-Allee 31
10589 Berlin, Germany
{zander, carle}@fokus.fhg.de
http://www.fokus.fhg.de/glone

Abstract. The transmission of high quality video streams over IP networks becomes more and more attractive for providing IP based TV or video on demand services. Still a reasonable high bandwidth is required for achieving high quality streaming. Instead of using hard reservations to guaranty the quality of the video transmissions we present an adaptive streaming algorithm based on adaptive packet marking which provides soft guarantees. A prototype streaming server and client have been developed supporting RTP-encapsulated MPEG-2 transmission, into which the adaptive streaming algorithm has been integrated. The software has been successfully used in a testbed, demonstrating that the algorithm is effective and has very promising properties.

1 Introduction

IP-based transmission of high quality video streams becomes more and more attractive especially for applications like video on demand, remote teaching, surveillance and Internet TV. Despite the wide distribution of MPEG-4 and its impressive compression capabilities still a reasonably high bandwidth is required for achieving a high video quality. With MPEG-4 a bandwidth of about 1-2 Mbit/s is needed for good quality. A higher quality can be achieved with MPEG-2 encoded video at a bandwidth of approximately 8 Mbit/s, as is used for DVD encoding. This is a relatively high data rate compared to typical access and wide-area data rates. In the case of network congestion, the video playback will either have dropouts if UDP is used as transport protocol, or the video will freeze when the playout buffer is empty if TCP is used. Such a quality degradation is not acceptable in many cases.

To guarantee the transmission quality, resource reservation such as IntServ [RFC2205] or DiffServ [RFC2475] can be used. These solutions are typically used in combination of a hard reservation of the maximum bandwidth of the flow to be transported. In the case of high bandwidth requirements such as MPEG-2 streaming with 8 Mbit/s, using resource reservation typically will be a costly alternative to a low price best effort service.

This paper describes an approach for providing high quality video streaming which uses expensive guaranteed resources only when necessary to maintain quality. Cheap

B. Stiller et al. (Eds.): QofIS/ICQT 2002, LNCS 2511, pp. 68-77, 2002.

bandwidth (i.e. a best effort service) is used as much as possible. The adaptive streaming algorithm presented in this paper is based on adaptive packet marking. It operates with two service classes. Best effort bandwidth is used for video streaming as much as possible. If there is not enough bandwidth available (which is detected by low fill level of the receiver playout buffer), the algorithm uses additional bandwidth from a guaranteed service based on the currently available best effort bandwidth and the receivers playout buffer fill level. The paper is focused on applications with a fairly large end-to-end delay budget such as video on demand, which allow for a playout buffer delay of a few seconds to compensate for the bandwidth fluctuations in the network.

We decided to implement and try our algorithm in a real DiffServ [RFC2475] enabled test network. We implemented a prototype streaming server and client software which is able to stream DVD quality MPEG-2 videos, as there is now free open source software available for providing high quality video streaming. The adaptive streaming algorithm has been integrated into this prototype.

The remaining parts of the paper are structured as follows. Section 2 discusses related work. Section 3 introduces the algorithm used for adaptive video streaming. Section 4 outlines the prototype implementation which adaptively streams MPEG-2 videos in a DiffServ enabled network. Section 5 presents evaluation results obtained by using the prototype and section 6 concludes the paper and outlines future work.

2   Related Work

The approach to use a mixture of best effort and guaranteed service for video streams was postulated in [RaTh98]. Besides the basic idea, this paper only describes a general network architecture supporting layered video but contains no solution for realizing adaptive streaming. [FeKS98] examines adaptive priority marking for providing soft bandwidth guarantees. Their proposed algorithm is similar to the one proposed in this paper. The main difference is that [FeKS98] proposes a general algorithm for use with TCP while our algorithm is specific for video transmission considering application feedback. [FeKS98] is focused on the optimal marking and overload behavior. In contrast, the question we want to answer is the achievable ratio of best effort and guaranteed packets. In [FeKS98], simulation is used to prove that the generic adaptive packet marking algorithm works. Our goal is to demonstrate that the algorithm works in a real network in combination with a real streaming application, and that the algorithm reacts fast enough on typical changes of network conditions.

A related but better understood problem is the smoothing of variable bit rate video streams. [SZKT96] and [ReTo99] describe possible solutions for smoothing the video stream to be transmitted by calculating a bandwidth plan before the video is sent over the network. [RSFT97] contains a proposal for a smoothing technique applicable for live streams.

[VFJF99] provides a good overview about the existing techniques for adaptive video streaming. It describes layered encoding, adaptive forward error correction and smoothing. [RNKA97] describes an adaptive video streaming service which adapts to the quality of the transmission. An algorithm chooses the video adaptation which achieves the best quality under the current conditions. No paper known to the authors describes an algorithm like the one proposed in this paper.

Our prototype software is based on protocols standardized by the IETF such as the Real Time Streaming Protocol (RTSP) [RFC2236] for controlling the playback and the Real Time Protocol (RTP) [RFC1889] for transporting the real-time data. For the transport of MPEG-2 over RTP our work is based on the MPEG-2 standard and the MPEG-2 RTP encoding as defined in [RFC1890], [RFC2250] and [RFC2343].

3   Adaptive Video Streaming

For guaranteeing the quality of a video transmission a simple solution is to reserve network resources according to the maximum video bandwidth for the full duration of the video transmission. Such a hard reservation will guarantee the quality of the video transmission but is much more expensive than using a best effort service. If the video stream is of variable bit rate some of the reserved resources will not be used but typically must be paid for. We assume that costs of using the guaranteed service class will depend on the maximum bandwidth of the guaranteed service class that can be used, and on the data volume sent using the guaranteed service class. By using a smoothing scheme, the maximum data rate which has to be reserved can be reduced, thereby reducing costs.

In existing streaming approaches that use a guaranteed service class, the complete video is transmitted over the guaranteed service (1. in Figure 1). When using a best effort service where the available bandwidth is lower than the required bandwidth for certain time intervals, buffer underuns occur at the receiver, impairing the visual quality (2. in Figure 1).

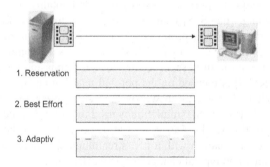

Figure 1: Adaptive Streaming

During the times when the best effort service class provides insufficient bandwidth, the required additional bandwidth is sent over the guaranteed service class (3. in Figure 1). If there is not enough guaranteed bandwidth available the video can not be transmitted without using some quality degradation scheme but this is considered out of scope of this paper. The main properties of our adaptive streaming are:

- As much data as possible is sent using cheap best effort traffic.
- When the available bandwidth of the best effort service is insufficient (i.e., low fill level of the playout buffer) the missing bandwidth is sent using a guaranteed channel.
- As long as sufficient guaranteed bandwidth is available (i.e. no blocking happens when using reserved bandwidth), no quality degradation becomes visible, and the video has full DVD-like quality.

The algorithm we developed for realizing the adaptive video streaming consists of two parts: rate control and adaptive marking. The sender controls both the sender rate and the adaptive marking based on the video which is streamed and the feedback information from the receiver.

Sender Rate Control

The sender estimates the data rate it has to send over time based on the timestamps contained in the video file. In the simplest case the send rate s is constant (see Figure 2) which is the case for DVD encoded MPEG-2. In case the send rate is not constant we use linear regression to estimate the gradient s. Videos with non constant bitrate could also be smoothed before transmission ([SZKT96]). Figure 2 illustrates the algorithm. At time T the current position in the video is L. Since L is below $s{\cdot}T$ we are below the average video rate. ρT is the time until the rate control function is called again and is estimated during runtime. The amount of data to send within ρT is ρL and can be calculated in a straightforward manner.

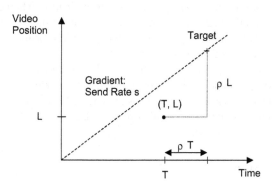

Figure 2: Sender Timing Algorithm

In a real network the current possible sending rate may be larger or smaller than s depending on the congestion. To compensate for the bandwidth fluctuation, a playout buffer is used at the receiver. At the beginning of a video transmission the playout buffer is filled to level P_0. This level can be adjusted by selecting a maximum startup delay. Afterwards the receiver starts the playback. The current buffer level is P and P_T is the desired target fill level during playback. The deviation from the target buffer fill level is $\rho P = P_T - P$. This value is send from the receiver to the sender in regular intervals ρT_F. The rate for the additional data to send is $s_A = \rho P / \rho T_F$. shows the sender algorithm in meta code. The amount of data to send S is divided into n packets of equal size. The size of the packets should be near but below the Maximum Transfer Unit (MTU) of the path between sender and receiver to avoid fragmentation.

```
SEND(T,L)
    S₀ = s·(T+ΔT)-L

    Sₐ= sₐ·ΔT

    S=S₀+Sₐ
    READ_AND_SEND(S)
```

Figure 3: Sender Rate Control Algorithm

Adaptive Marking

We assume the availability of two service classes: guaranteed and best effort. The best effort service class should be used as much as possible but in case the receiver runs out of data we transmit data over the guaranteed service. A packet will be marked with probability p as guaranteed class. Therefore p is the ratio of guaranteed packets while $1-p$ is the ratio of best effort packets. We set $p=min(0, S_A)/S_0, 0{\leq}p{\leq}1$ at each interval ρT_F. The rationale for this is that S_0 is the number of bytes in an ideal network where we can send with constant rate s and S_A gives us an indication whether our real rate is smaller or larger. Using the ratio of S_A and S_0 means that we increase the percentage of guaranteed packets with increasing non negative S_A which means the receiver buffer level is decreasing below the target fill rate.

Note that in case there is not enough bandwidth available in the guaranteed class a frame based marking scheme would lead to much better performance at the receiver. However in the opposite case the probabilistic scheme is much easier to implement, needs less CPU time and allows a smoother partitioning.

4   Implementation

Instead of simulation we decided to implement and try the algorithm in a real network. We have developed a prototype implementation of a MPEG-2 video server and client in which we integrated the adaptive streaming algorithm introduced in section 3. Our implementation is based on standard protocols such as RTSP [RFC2326] and RTP [RFC1889]. It supports the basic RTSP capabilities and MPEG-2 program and transport stream encoding over RTP. Both UDP and TCP can be selected as transport protocols. The adaptive streaming algorithm works only with TCP. For sending RTP packets over TCP connections we send the length of each RTP packet followed by the packet as proposed in [UDPT]. For feedback information we use the RTCP protocol [RFC1889]. Since RTCP as defined in the standard has non-deterministic and fairly long feedback intervals we have implemented a fast feedback mechanism which allows to have regular feedback in the order of 100 ms. The receiver sends the standard RTCP feedback messages, and additional feedback messages about the playout buffer state. The adaptive streaming algorithm which is built into the server uses this information. If TCP is used as transport protocol there are two separate connections: one for RTP and one for RTCP.

For adaptive streaming the sender has to mark packets as either best effort or guaranteed service. To be generic the software implemented by us has no direct interface to some QoS functionality. Instead we use the RTP marker bit (m-bit) which is included in every packet header to signal the service class. Obviously this only works with two classes. If more than two classes would be needed a custom RTP header extension could be defined. The m-bit is examined at the edge router by a NetFilter [NetFilter] classifier module running in the Linux kernel which has been implemented by us. We use the DiffServ [RFC2475] model for providing QoS. Based on the RTP m-bit the DiffServ Code Point (DSCP) will be set for each of the packets. The DSCP is later used for scheduling at the core routers. The classifier is optimized so that all retransmitted TCP packets are marked as guaranteed class. The software has been implemented in C/C++ and is running under the Linux operation system. Figure 4 shows a screenshot with the video and the statistics generated (see section 5).

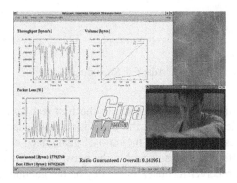

Figure 4: Adaptive Streaming Implementation

Since the Internet is a highly heterogeneous network it is questionable whether all routers will ever support QoS. Our application uses TCP as transport protocol and relies on DiffServ enabled router's for marking and scheduling. If no end-to-end QoS is supported there is no negative impact on the network because a normal best effort TCP connection is used. In our solution there is only one TCP connection for the video data which is multiplexed on two service classes. The advantage is that considering the number of connections it is not different to traditional applications and therefore it does not behave more aggressive as usual. However it needs to be investigated how this influences TCP congestion control. As shown in [FeKS98], pure marking results in a percentage of marked packets that is too high considering the fair share. Our algorithm prevents bursts of guaranteed packets by incrementally increasing and decreasing the percentage to be marked as suggested in [FeKS98].

5   Evaluation

The implementation has been tested in a DiffServ setup in our testbed. All machines in the testbed are Linux PCs equipped with Fast Ethernet network adapters. The server runs the MPEG-2 video server and streams MPEG-2 videos encoded with 8 Mbit/s bit rate. The client runs the prototype video client and has a MPEG-2 hardware decoder card. The routers run Linux 2.4 and are DiffServ enabled. One router is configured as edge router and marks the packets with a DSCP according to the RTP m-bit. The second router is a DiffServ core router performing the scheduling. Two classes are configured: Expedited Forwarding (EF) as guaranteed class, and a best effort class. In our setup we do not use background traffic. Instead we emulate congestion by randomly dropping packets in the best effort class. Our loss emulation can drop packets with uniform or exponential loss distribution. We only use exponential loss in the evaluation. To measure the distribution of the data onto the two classes we use a separate meter connected to the same hub as the server. Figure 5 shows the testbed setup.

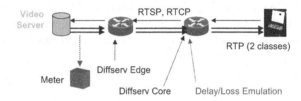

Figure 5: Adaptive Streaming Testbed

We show the results of three different tests. In the first test we prove that the implementation works as desired and show the behavior of the adaptive streaming algorithm. In the second test we show how the ratio of guaranteed class to overall volume behaves in case of different loss rates within the network. In the third test we investigate the impact of the receiver feedback frequency on the guaranteed/overall ratio.

Algorithm Behavior

In this trial an 8 Mbit/s video with a duration of approximately 4.5 minutes is transmitted over TCP from the sender to the receiver using the adaptive streaming algorithm. The receiver has a playout buffer of size 8 Mbyte and sends a feedback message every second. The mean loss was set to 3% and the maximum loss to 6% (exponential distribution). Figure 6 shows the throughput over time for both service classes (G = Guaranteed, BE = Best Effort) and Figure 7 shows the absolute volume transmitted. The ratio of guaranteed to overall volume is only 0.36. This means that only 36% of the traffic was sent using the guaranteed service class instead of 100% in case of a hard reservation. In this trial all video frames arrived at the receiver in time; no errors occurred during the playback.

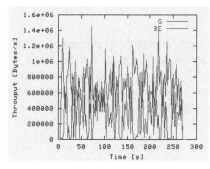

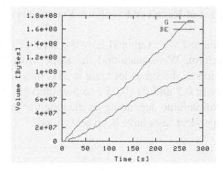

Figure 6: Throughput   **Figure 7**: Volume

Impact of Loss Rate

In this test the same video as before is transmitted while emulating different loss rates (congestion) in the network. The receiver buffer size is 8 Mbyte and the receiver feedback is sent every second. We investigated the algorithm with mean loss rates of 0%, 2%, 3%, 5%, 10% and 15%. The maximum loss rate is always twice the mean loss rate, and exponential distribution is used. Figure 8 shows the ratio of guaranteed to overall volume depending on the loss rate. For each loss rate we conducted three tests. In all tests all video frames arrived at the receiver in time; no playback errors occurred. However at 15% mean loss rate the receiver buffer size had to be increased to 12 Mbyte to avoid loosing data. To prevent data loss the buffer size must be increased with increasing mean loss rate but we have not performed investigations for loss rate higher than 15%.

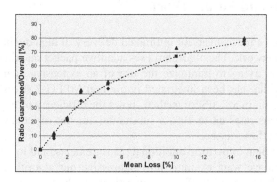

Figure 8: Impact of Mean Loss Rate

The figure shows that for small loss rates the gain of using adaptive streaming is quite high (only 20% needs to be transmitted over guaranteed service) while for large loss rates (>10%) the ratio is still over 70% but moving asymptotically to 100%.

Impact of Receiver Feedback Frequency

In this test we examined how the feedback frequency affects the performance of the algorithm. We transmitted the same video as before, the receiver buffer size is 8 Mbyte and the mean loss rate is 3% (maximum 6%). We vary the feedback frequency between 0.2 and 20 (0.05 s to 5 s interval). Figure 9 shows the ratio of guaranteed to overall volume depending on the receiver feedback frequency. For each frequency two independent tests have been performed.

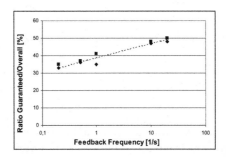

Figure 9: Impact of Feedback Frequency

The figure shows that the ratio is increasing with increasing feedback frequency (plotted with log scale). This means that the algorithm looses performance in case of frequent feedback because it is reacting too fast. We found that the smallest frequency which can be used is the playout time of half of the receiver buffer size.

6   Conclusions and Future Work

We have developed a mechanism for optimized delivery of high quality video streams over an IP based network. The algorithm uses adaptive packet marking for guaranteeing soft QoS while trying to minimize the use of an expensive guaranteed network service. The algorithm has been integrated into a real MPEG-2 streaming implementation. The evaluation in a DiffServ testbed shows the efficiency of the proposed algorithm. For mean loss rates up to 10% a substantial amount of bandwidth can be obtained from a best effort service.

In the future we plan to perform additional measurements, and to improve the algorithm. In particular we want examine how aggressive our algorithm behaves in the presence of TCP background traffic and investigate on how much guaranteed bandwidth is needed to satisfy a certain number of clients to be able to create admission control rules. Furthermore we want to make the algorithm to be aware of the MPEG-2 frame structure, in order to improve the mapping of data onto the different service classes and to allow an open loop solution without feedback messages from the receiver. Another goal is to better smooth the usage of guaranteed bandwidth by estimating the network conditions of future time intervals based on measurements of past time intervals.

7   Acknowledgements

This work was partially funded by the German Research Network (DFN Verein) project "Gigamedia" (see http://www.fokus.fhg.de/glone/gigamedia/).

The authors would like to thank Alexander Pikovski for his contribution to the implementation of the adaptive video streaming application described in this paper and the anonymous reviewers for their valuable comments.

References

[FeKS98]   W. Feng, Dilip D. Kandlur, D. Saha, K. Shin: "Adaptive Packet Marking for Providing Differentiated Services in the Internet", Proc. of Int. Conf. on Network Protocols, October 1998.

[NetFilter]   http://netfilter.samba.org/

[RaTh98]   R. Ramanujan, K. Thurber: "An Active Network Based Design for a QoS Adaptive Video Multicast Service", SCI'98, 1998.

[ReTo99]   J. Rexford, D. Towsley: "Smoothing Variable-Bit-Rate Video in an Internetwork", IEEE/ACM, 1999.

[RFC 2205]   Braden, B., Ed., et. al., "Resource Reservation Protocol (RSVP) - Version 1 Functional Specification", RFC 2205, September 1997.

[RFC1889]   H. Schulzrinne, S.Casner, R. Frederick, V. Jacobson: „RTP: A Transport Protocol for Real-Time Applications", RFC1889, Januar 1996.

[RFC1890]   H. Schulzrinne: „RTP Profile for Audio and Video Conferences with Minimal Control", RFC1890, Januar 1996.

[RFC2250]   D. Hoffman, G. Fernando, V. Goyal, M. Civanlar: „RTP Payload Format for MPEG1/MPEG2 Video", RFC2250, Januar 1998.

[RFC2326]   H. Schulzrinne, A. Rao, R. Lanphier: „Real Time Streaming Protocol (RTSP)", RFC2326, April 1998.

[RFC2343]   M. Civanlar, G. Cash, B. Haskell: „RTP Payload Format for Bundled MPEG", RFC2343, Mai 1998.

[RFC2475]   S. Blake, D. Black, M. Carlson, E. Davies, Z. Wang, W. Weiss: "An Architecture for Differentiated Service", RFC 2475, December 1998.

[RNKA97]   R. Ramanujan, J. Newhouse, M. Kaddoura, A. Ahamad, E. Chartier, K. Thurber: "Adaptive Streaming of MPEG Video over IP networks", IEEE LCN'97, November 1997.

[RSFT97]   J. Rexford, S. Den, J. Dey, W. Feng, J. Kurose, J. Stankovic, D. Towsley: "Online Smoothing of live variable-bit-rate video" in Proc. Network and OS Support for Digital Audio and Video, pp. 249-257, May 1997.

[SZKT96]   J. Salehi, Z. Zhang, J. Kurose, D. Towsley: "Supporting Stored Video: Reducing Rate Variability and End-to-End Resource Requirements through Optimal Smoothing", ACM SIGMETRICS, 1996

[UDPT]   http://www.cs.columbia.edu/~lennox/udptunnel/

[VFJF99]   B. Vandalore, W. Feng, R. Jain, S. Fahmy: "A Survey of Application Layer Techniques for Adaptive Streaming of Multimedia", Journal of Real Time Systems, April 1999.

SBQ: A Simple Scheduler for Fair Bandwidth Sharing Between Unicast and Multicast Flows

Fethi Filali and Walid Dabbous

INRIA, 2004 Route des Lucioles, BP-93
06902 Sophia-Antipolis Cedex, France
{filali, dabbous}@sophia.inria.fr

Abstract. In this paper, we propose a simple scheduler called SBQ (Service-Based Queuing) to share the bandwidth fairly between unicast and multicast flows according to a new definition of fairness referred as *the inter-service fairness*. We utilize our recently proposed active queue management mechanism MFQ (Multicast Fair Queuing) to fairly share the bandwidth among all competing flows in the multicast queue.

The simulation results obtained for very heterogeneous sources and links characteristics suggest that, on the one hand, our scheduler achieves the expected aggregated bandwidth sharing among unicast and multicast service, and on the other hand the multicast flows remain TCP-friendly.

1 Introduction

It is widely accepted that one of the several factors inhibiting the usage of the IP multicast is the lack of good, deployable, well-tested multicast congestion control mechanisms.

The precise requirements for multicast congestion control are perhaps open to discussion given the efficiency savings of multicast, but it is well known that a multicast flow is acceptable if it achieves no greater medium-term throughput to any receiver in the multicast group than would be achieved by a TCP flow between the multicast sender and that receiver. Such requirement can be satisfied either by a single multicast group if the sender transmits at a rate dictated by the lowest receiver in the group, or by a layered multicast scheme that allows different receivers to receive different numbers of layers at different rate.

The IETF orchestrated a very strong guideline for developing a TCP-friendly multicast congestion control scheme [14] regardless of the number of receivers in the multicast session. It is well-known that multiplicative decrease/linear increase congestion control mechanisms, and in particular TCP, lead to proportional fairness [9]. However, the authors of [2] have proven that if we treat the multicast flow as if it were a unicast flow, then the application of Kelley's model [9] shows that the larger the multicast group the smaller its share of the proportional bandwidth it would get.

In this paper, we develop a novel approach that helps multicast congestion control mechanisms to fairly exist with TCP protocol. Our approach is based

B. Stiller et al. (Eds.): QofIS/ICQT 2002, LNCS 2511, pp. 78–89, 2002.

on a new fairness notion, *the inter-service fairness*, which is used to share the bandwidth fairly between unicast and multicast services. In our fairness definition, the aggregated multicast traffic should remain *globally* TCP friendly in each communication link. In other words, the aggregated multicast average rate should not exceed the sum of their TCP-friendly rates. This approach allows ISPs to define their own intra-multicast bandwidth sharing strategy which may implement either a multicast pricing policy [8] or an intra-multicast bandwidth sharing strategy [10].

To implement our approach, we propose to use a two classes CBQ/WRR-like scheduler [6]: one for the unicast flows and the other one for multicast flows. We call our scheduler Service-Based Queuing (SBQ) because it distinguishes between two different transfer services: unicast and multicast services. SBQ integrates a method to dynamically vary the weights of the queues in order to match the expected bandwidth share between unicast and multicast flows. Upon a packet arriving, the router has to classify and redirect it to the appropriate queue.

We use simulation to evaluate the effectiveness and performance of our scheme for various sources including not only TCP, UDP, and multicast CBR sources, but also multicast sources that implement the recently proposed layered multicast congestion control FLID-DL [13]. Simulations are done for very heterogeneous network and link characteristics and with different starting time, finish time, packet size, rate, and number of receivers in the multicast sessions.

The body of the paper is organized as follows. Section 2 details the inter-service fairness notion. We explore and discuss our scheduler in Section 3. In Section 4, the simulation results will be presented. Section 5 concludes this paper by summarizing our findings and outlining future work.

2   Inter-service Fairness

In the informational IETF standard [14], the authors recommend that each end-to-end multicast congestion control should ensure that, for **each** source-receiver pair, the multicast flow must be TCP-friendly. We believe that this recommendation has been done because there is no network support to guarantee the tcp-friendliness and that it was an anticipated requirement which aims to encourage the fast deployment of multicast in the Internet.

We propose a new notion of the unicast and multicast fairness called: the *inter-service fairness* which is defined as follows.

Definition 1: *The Inter-service fairness: The multicast flows must remain* globally *TCP-friendly and not TCP-friendly for individual flows. In other words, we should ensure that the* sum of multicast flows rate *does not exceed the* sum of their TCP-friendly rate*.

The TCP throughput rate R_{TCP}, in units of **packets per second**, can be approximated by the formula in [7]:

$$R_{TCP} = \frac{1}{RTT\sqrt{q}(\sqrt{\frac{2}{3}} + 6\sqrt{\frac{3}{2}}q(1 + 32q^2))} \tag{1}$$

where R_{TCP} is a function of the packet loss rate q, the TCP round trip time RTT, and the round trip time out value RTO, where we have set $RTO = 4RTT$ according to [16]. Since the multicast analogue of RTT is not well defined, a target value of RTT can be fixed in advance to generate a target rate R_{TCP}.

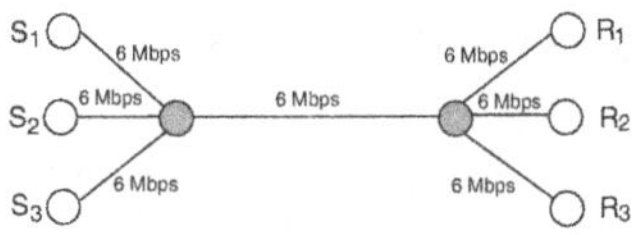

Fig. 1. A topology used as example for the inter-service fairness definition. All links have the same Round Trip Time (RTT).

We illustrate the inter-service fairness definition using the topology shown in Figure 1. We consider two multicast sessions, one from source S_1 sending to R_1 and the other one from source S_2 sending to R_2 and one unicast session from S_3 sending to R_3. Let r_1, r_2, and r_3, be the *inter-service fair share* of session S_1, S_2, and S_3, respectively. The tcp-friendly rate is equal to $\frac{6}{3} = 2$ Mbps given that all links have the same RTT. When applying the inter-service fairness definition to share the bandwidth between the three sessions, S_1 and S_2 will get together an aggregated fair share equal to 4 Mbps and session S_3 will get alone a fair share r_3 equal to 2 Mbps. Our definition of the inter-service fairness does not specify the way how the bandwidth should be shared among multicast competing flows. Therefore, each vector (r_1, r_2) where $r_1 + r_2 = 4$ Mbps is considered as a feasible intra-multicast fair share solution.

3   The SBQ Scheduler

3.1   Principals and Architecture

In order to implement the fluid model algorithm that integrates our inter-service fairness definition given in Section 2, we propose to use a CBQ-like [6] scheduler with two queues, one for each class as shown in Figure 2. We call our scheduler SBQ (Service-Based Queuing) because it differentiates between the packets according to their transfer service: unicast or multicast.

Before being queued, unicast and multicast packets are classified into two separated classes. We use the Weighted Round Robin (WRR) algorithm which is the most widely implemented scheduling algorithm as of date [6]. This is due to its low level of complexity and its ease of implementation which follows hence. In this scheduler, packets receive service according to the priority queue to which they belong. Each queue has an associated weight. In every round of service, the number of packets served from a queue is proportional to its associated weight and the mean packet size.

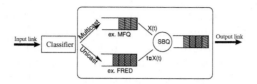

Fig. 2. SBQ scheduler architecture

As buffer management, we use our recently proposed MFQ [3] mechanism for the multicast queue and another queue management mechanism (RED, FRED, WRED, etc.) for the unicast queue.

Our scheduler operation is illustrated by the following example: consider a WRR server that uses two queues: U (for Unicast) and M (for Multicast) with weights 0.6 and 0.4, respectively. Let the mean packet size for U and M be 1000 Bytes and 500 Bytes, respectively. The weights are first normalized by the mean packet size to get 0.6 for class U and 0.8 for class M. The normalized weights are then multiplied by a constant to get the lowest possible integer. In this case we get 6 and 8 for U and M, respectively. This means that in every round of service 6 packets are served from queue U and 8 packets are served from queue M.

3.2   Scheduler Configuration

When configuring WRR to share the link bandwidth between the two classes. At time t we configure the multicast class (M queue) with a weight equal to $X(t)$ and the unicast class (U queue) with a weight equal to $(1 - X(t))$.

To implement our inter-service fairness notion defined in Section 2 and to be as close as possible to the fluid model algorithm developed in Section 3, we propose to update the weight $X(t)$ at time t as follows:

$$X(t) = \min\left(\frac{\sum_{i=1}^{i=m(t)} R_{TCP_i} * S_i}{C}, \frac{m(t)}{u(t) + m(t)}\right) \qquad (2)$$

where:

- R_{TCP_i} is the TCP-friendly throughput rate of the multicast flow i estimated using Eq. 1,
- S_i is the average packet size in bytes,
- C is the link capacity in **bytes per second**,
- $u(t)$ is the number of active unicast flows at time t,
- $m(t)$ is the number of active multicast flows at time t.

To be **globally fair** against TCP-connections, the sum of the rates of $m(t)$ active multicast flows must not exceed the sum of that of their $m(t)$ single TCP flows over large time scales. This corresponds to the first term of Eq. 2. The second term of this equation allocates instantaneous bandwidth fairly between unicast and multicast flows by sharing it proportionally to the number of flows in the two queues. Using this simplistic and efficient formula of $X(t)$, we can guarantee both short-term and long-term fairness. Indeed, the two terms used in $X(t)$ configuration allow us to ensure both short-term max-min fairness between active flows based on a local information about the number of flows and a long-term tcp-friendliness between active sessions based on a global information concerning the rates of multicast sources. The tcp-friendliness term (the first one) is useful when there is another bottleneck in the multicast delivery tree.

A possibility way to extend the configuration of $X(t)$ is to add a third term referring to the maximum portion of the link capacity that should not be exceeded by multicast flows. This value can be tuned by the ISP depending on a chosen policy used to handle multicast connections crossing its network.

Upon each change on the number of active flows in the unicast or multicast queue, the weights of both queues are updated to match the new fairness values. Both queues priority are set to 1.

To compute the value of $X(t)$, each SBQ router has to know the TCP-friendly rate of active multicast flows and maintain only the aggregated rate to be used in the first term of Eq. 2.

The TCP-friendly rate of a multicast session corresponds to the sending rate of the source. In single rate multicast transmission [18], every receiver periodically estimates its TCP-friendly reception rate using for example the formula 1 and reports this rate to the source. The source determines the lowest rate among all receivers rates and use it to send data downstream to all receivers. In the other hand, in multi-rate multicast transmission [12], the source sends data using several layers. For each layer, the source uses a specific sending rate depending on the data encoding scheme. The receivers join and leave the layers according to their reception TCP-friendly reception rates computed using Eq. 1.

For both cases, the source tcp-friendly throughput rate can be included in the IP packet header by the multicast source or the source's Designed Router. This technique is largely used by many other mechanisms such as CSFQ [15] for different purposes. Thus, a SBQ intermediate router gets the rates from the IP multicast packet headers and computes their aggregated value according to Eq. 2.

3.3   Weights Updating Time

In this sub-section, we try to answer the question: How often we update the weights ? In other words, what is the time-scale on which we should look at the bandwidth allocation. There is a tradeoff between complexity and efficiency when choosing the time-scale value. Indeed, larger time-scale are not suitable for short-lived TCP connections which are the most of TCP connections currently in the Internet[1]. In the other hand, what happiness on shorter time-scales if we consider fairness on longer time-scale. We do not claim that there is an optimal

---

[1] Internet traffic archive: http://www.cs.columbia.edu/~hgs/internet/traffic.html

value of the time-scale which can be applied for each type of traffic and which leads to both less complexity and good efficiency.

The updating time is designed to allow the update of weights to take effect before the value is updated again, it should be longer than the average round trip time (RTT) among the connections passing through the router. In the current Internet, it can be set in the range from 10 ms up to 1 sec, so this property can guarantee that the unfairness can't increase very quickly and make the queue parameters stable. But this parameter can't be set too big so that the scheduler weight can't be adaptive enough to the network dynamics.

3.4 Counting Unicast Connections

To update the value of the weight used in our scheduler computed using Eq. 2, we need to know the number of multicast and unicast flows. While the former is provided by MFQ, the latter could be obtained through the use of a flow-based unicast active queue management mechanism such as FRED [11]. However, unicast flow-based AQM are not available everywhere and most of current routers use a FIFO or RED [5] schemes which don't provide the number of unicast connections. That's why, we propose hereafter a simple method which we use to estimate the number of unicast connections in the unicast queue.

SBQ counts active unicast connections as follows. It maintains a bit vector called v of fixed length. When a packet arrives, SBQ hashes the packets connection identifiers (IP addresses and port numbers) and sets the corresponding bit in v. SBQ clears randomly chosen bits from v at a rate calculated to clear all of every few seconds. The count of set bits in v approximates the number of connections active in the last few seconds. The SBQ simulations in Section 4 use a 5000-bit v and a clearing interval (t_{clear}) of 10 seconds. The code in Algorithm given in [4] may under-estimate the number of connections due to hash collisions. This error will be small if v has significantly more bits than there are connections. The size the error can also be predicted and corrected; see [4]. This method of counting connections has two good qualities. First, it requires no explicit cooperation from unicast flows. Second, requires very little state: on the order of one bit per connection.

3.5 Deployment Issues

The deployment of CBQ/WRR-like algorithms in the Internet may raise some open questions for large deployment. The scalability issue is the main barrier of their large deployment in the Internet. We mean by the scalability, the ability of the mechanism to process a very large number of flows with different characteristics at the same time. We believe that our scheduler can be deployed in large networks thanks to two mainly key points: (1) it uses only two queues, so we need to classify only two types of service: unicast and multicast flows. This task has already been done in part by the routing lookup module before the packet being queued, and (2) all unicast flows are queued in the same queue.

It is important to note that we usually associate the flow-based mechanisms support with complexity and scalability problems since they require connection specific information. These concerns are justifiable only in point-to-point connections, for which routing tables do not maintain connection-specific state. In multicasting, routing tables keep connection specific state in routers anyway; namely, the multicast group address refers to a connection. Thus, adding multicast flow specific information is straightforward and increases the routing state only by a fraction.

Comparing to CBQ/WRR, our mechanism is less complex to be deployed in the Internet. It should be noted that even CBQ is now supported by large number of routers and many research works such as [17] demonstrate its deployment feasibility.

One major advantage of our approach is that it minimizes the complexity of designing multicast congestion control. Indeed, the network guarantees that the multicast flows will share fairly the bandwidth with competing unicast flows. Moreover, our scheme provides to the ISPs a flexible way to define and implement their own intra-multicast fairness strategy. The simulation results presented in the next section will confirm our claims.

4   Simulation Methodology and Results

We have examined the behavior of our scheme under a variety of conditions. We use an assortment of traffic sources and topologies. All simulations were performed in network simulator (ns-2).

4.1   Single Bottleneck Link

We validate our scheme for the topology of Figure 3 which consists of a single congested link connecting two routers n_1 and n_2 and having a capacity C equal to 1 Mbps and a propagation delay D equal to 1 ms.

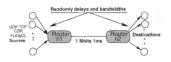

Fig. 3. A single congested link simulation topology

We configure our scheduler in the bottleneck link from router n_1 to router n_2. The maximum buffer size $qlim$ of both queues is set to 64 packets. Other links are tail-drop and they have sufficient bandwidth to avoid packets loss. We assume 32 multicast sources and 32 unicast sources that compete to share the link capacity. We index the flows from 1 to 64. The 32 multicast sources are divided as follows:

- Flows from 1 to 16: CBR sources. These sources implement no type of congestion control mechanism (neither application-level nor transport-level).
- Flows from 17 to 32: FLID-DL (Fair Layered Increase-Decrease with Dynamic Layering) [13] sources. The FLID-DL simulation parameters are the same as those recommended in [13][2].

Each source uses 17 layers encoding, each layer is modeled by a UDP traffic source.

As outlined earlier, we use MFQ [3] as the active queue management in the multicast queue. Without loss of generality we utilize a receiver-dependent logarithm policy (the LogRD policy) proposed in [10] to share the bandwidth between multicast flows. This policy consists in giving to the multicast flow number i a bandwidth fraction equal to $\frac{1+\log n_i}{\sum_j (1+\log n_j)}$, where n_j is the number of receivers of flow j.

The 32 unicast sources are composed as follows:

- Flows from 33 to 48: UDP sources.
- Flows from 49 to 64: TCP sources. Our TCP connections use the standard TCP Reno implementation provided with ns.

Unless otherwise specified, each simulation lasts 100 seconds and the weight updating period is set to 2 sec. Other parameters are chosen as follows:

- packet size: the packet size of each flow is randomly generated between 500 and 1000 bytes.
- starting time: the starting time of each flow is randomly generated between 0 and 20 seconds before the end of the simulation.
- finish time: the finish time of each flow is randomly generated between 0 and five seconds before the end of the simulation.
- rate: the rate of both unicast and multicast UDP flows is randomly generated between 10 Kbps and 100 Kbps.
- number of receivers: the number of downstream receivers of the 32 multicast sessions is randomly generated between 1 and 64.

The four first unicast UDP and multicast UDP flows are kept along the simulation time (start time = 0 sec, and finish time = 100 sec) to be sure that the link will be always congested. Each one of these flows is sending at a rate equal to $\frac{1\,Mbps}{8} = 125\,Kbps$. Initially (at $t = 0$ sec), the weight X is set to 0.5.

In Figure 4(a), we plot the variation of unicast and multicast queue weights in function of the simulation time. As we can easily see the value of weights change during the simulation because they depend on the number of active unicast and multicast flows in the queues. The multicast weight increases when the unicast weight decreases and vice versa.

---

[2] These parameters are used in the ns implementation of FLID available at http://dfountain.com/technology/library/flid/.

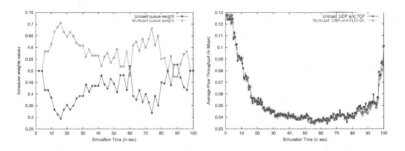

(a) Scheduler weights variation in function of the simulation time

(b) The variation of the unicast and multicast average rate in function of the simulation time over 500 msec

Fig. 4. Weights and average rates variation

We now interest in the inter-service fairness defined in Section 2 which is the main performance metric of our scheme. To this end, we compare the average unicast an multicast rates over 500 *msec* of simulations. We show in Figure 4(b), the variation of the unicast and the multicast average rate in function of the simulation time.

As we can observe the results match exactly what we expect. Two main observations can be done from the plots of Figure 4(b). Firstly, there is a fluctuation on the average aggregated rate for unicast and multicast flows. Secondly, the multicast average rate is very close to the unicast average rate. This demonstrates the ability of SBQ to share the bandwidth fairly between unicast and multicast flows according to our inter-service fairness notion defined in Section 2.

In order to evaluate the impact of the time-scale value on the performance of SBQ, we measure the normalized aggregated rate (average unicast rate/average multicast rate) obtained for various values of the time-scale. In Figure 5(a), we show the variation of this metric in function of the time-scale value. We can conclude that using a 1 sec time-scale, we can reach 93 % of the performance of SBQ. In other words, the use of an updating period equal to 1 sec leads to a good tradeoff between complexity and efficiency.

We claimed earlier that our scheduler is flexible in the sense that its performance is independent of the intra-multicast fairness function. To argument this affirmation, we compare in Figure 5(b) the normalized rate over 500 *msec* of simulation time for linear (LIN), logarithm (LOG), and receiver-independent

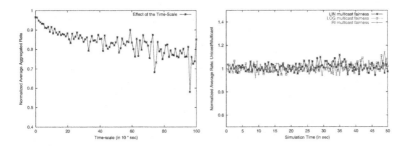

(a) Sensitivity of SBQ performance on the time-scale value

(b) The normalized rate when
modifying the intra-multicast
fairness function over 500 msec

Fig. 5. Sensitivity of SBQ performance on time-scale value and fairness function

(RI) bandwidth sharing policies which are defined in [10][3]. As we can see the
normalized average rate is always varying around 1 (between 0.82 and 1.18) for
the three cases.

We interest in the bandwidth shared between multicast flows obtained by
MFQ buffer management mechanism. The multicast flow number i is assumed
to have exactly i downstream receivers (members). We use a logarithm multicast
bandwidth allocation scheme. We vary the number of UDP unicast flows from
1 to 16 flows. In Figure 6(a), we show the bandwidth fair rate obtained for the
32 multicast flows. It is very clear that the shape of flows rate curves follow a
logarithm function of the number of downstream receivers.

4.2   Multiple Bottleneck Links

In this sub-section, we analyze how the throughput of multicast and unicast
flows is affected when the flow traverses L congested links. We performed two
experiments based on the topology of Figure 7. We index the links from 1 to L
and the capacity C_1 of the link number 1 is set to 1 Mbps.

We use the same source as the case of single bottleneck and we measure
the aggregated bandwidth received by each service type: unicast and multicast
type in function of the number of congested links. A link j, $2 \leq j \leq 10$, is kept
congested by setting its capacity C_j to $C_{j-1} - 50$ Kbps.

---

[3] Assume that we have n active multicast flows and denote by n_i the number of
downstream receivers of flow i. The RI, LIN, LOG bandwidth sharing policies consist
to give to flow i a bandwidth share equal to $\frac{1}{n}$, $\frac{n_i}{\sum_j n_j}$, and $\frac{1+\log n_i}{\sum_j (1+\log n_j)}$, respectively.

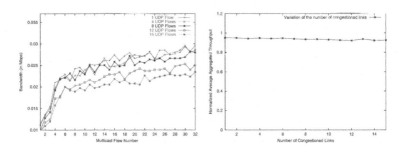

(a) Rates of multicast flows when using a logarithm multicast bandwidth sharing function

(b) The normalized rate as a function of the number of congestioned links

Fig. 6. SBQ Performance for multiple bottleneck links

We plot in Figure 6(b), the variation of the normalized average rate as a function of the number of congested links. As we can see, the normalized average rate remains close to 1 even when the number of congested links increases.

5   Conclusion and Future Work

In this paper, we have presented a CBQ-like scheduler for bandwidth sharing between unicast and multicast flows. The scheduler uses two queues: one for unicast and the other one for multicast. We used a simplistic and efficient dynamic configuration method of the WRR scheduler to achieve the expected sharing based on a new fairness notion called *the inter-service fairness*.

The buffer management mechanism used in the multicast queue was MFQ, a new scheme that we have proposed in [3] and which provides the expected multicast bandwidth sharing between multicast flows using a single FIFO queue.

To validate our scheme, we simulated a very heterogeneous environment with different types of sources, starting times, sending rates, delays, and packets sizes. We demonstrated that SBQ achieves the expected results in the sense that the bandwidth is shared fairly between unicast and multicast flows according to our new definition of fairness.

Future work could evaluate the performance for other types of multicast traffic that include others application-based or transport-based congestion control mechanisms. Another area of future study is to develop a new multicast congestion control mechanism that uses a small but efficient help from our scheduler.

Fig. 7. Topology for analyzing the effects of multiple congested links

References

1. J. Byers, M. Luby, M. Mitzenmacher, and A. Rege, *A digital fountain approach to reliable distribution of bulk data transfer,* In Proc. of ACM SIGCOMM'98, September 1998.
2. D. M. Chiu., *Some Observations on Fairness of Bandwidth Sharing,* In Proc. of ISCC'00, July 2000.
3. F. Filali and W. Dabbous, *A Simple and Scalable Fair Bandwidth Sharing Mechanism for Multicast Flows,* In Proc. of IEEE ICNP'02, November 2002.
4. F. Filali and W. Dabbous, *SBQ: Service-Based Queuing,* INRIA Research Report, March 2002.
5. S. Floyd, V. Jacobson, and V. Random, *Early Detection gateways for Congestion Avoidance,* IEEE/ACM TON, V.1 No.4, pp. 397-413, August 1993.
6. S. Floyd and V. Jacobson, *Link-sharing and Resource Management Models for Packet Networks,* IEEE/ACM TON, V. 3, No.4, 1995.
7. S. Floyd, M. Handley, J. Padye, and J. Widmer, *Equation-based congestion control for unicast applications,* In Proc. of ACM SIGCOMM'00, August 2000.
8. T. N.H. Henderson and S. N. Bhatti, *Protocol-independent multicast pricing,* In Proc of NOSSDAV'00, June 2000.
9. F. Kelly, A. Maulloo and D. Tan, *Rate control in communication networks: shadow prices, proportional fairness and stability,* Journal of the Operational Research Society 49, pp. 237-252, 1998.
10. A. Legout, J. Nonnenmacher, and E. W. Biersack, *Bandwidth Allocation Policies for Unicast and Multicast Flows,* IEEE/ACM TON, V.9 No.4, August 2001.
11. D. Lin and R. Morris, *Dynamics of Random Early Detection,* In Proc. of ACM SIGCOMM'97, September 1997.
12. M. Luby, V. Goyal, and S. Skaria, *Wave and Equation Based Rate Control: A massively scalable receiver driven congestion control protocol,* draft-ietf-rmt-bb-webrc-00.txt, October 2001.
13. M. Luby, L. Vicisano, and A. Haken, *Reliable Multicast Transport Building Block: Layered Congestion Control,* Internet draft: draft-ietf-rmt-bb-lcc-00.txt, November 2000.
14. A Mankin, et al., *IETF Criteria for evaluating Reliable Multicast Transport and Applications Protocols,* IETF RFC 2357, June 1998.
15. I. Stoica, S. Shenker, and H. Zhang, *Core-Stateless Fair Queueing: A Scalable Architecture to Approximate Fair Bandwidth Allocations in High Speed Networks,* In Proc. of SIGCOMM'98.
16. J. Padhye, V. Firoiu, D. Towsley, and J. Kurose, *Modeling TCP throughput: a simple model and its empirical validation,* In Proc. of ACM SIGCOMM'98.
17. D. C. Stephens, J.C.R. Bennet, and H. Zhang, *Implementing scheduling algorithms in high speed networks,* IEEE JSAC, V. 17, No. 6, pp. 1145-1159, June 1999.
18. J. Widmer and M. Handley, *TCP-Friendly Multicast Congestion Control (TFMCC): Protocol Specification,* draft-ietf-rmt-bb-tfmcc-00.txt, November 2001.

A Control-Theoretical Approach for Fair Share Computation in Core-Stateless Networks

Hoon-Tong Ngin and Chen-Khong Tham

National University of Singapore,
Department of Electrical and Computer Engineering,
4 Engineering Drive 3,
Singapore 117576, Singapore,
engp8555@nus.edu.sg and eletck@nus.edu.sg

Abstract. In this paper, we propose the use of a control theoretical approach for a fast and robust fair share computation for a class of fair queueing algorithms based on the Scalable Core (SCORE)/Dynamic Packet State (DPS) architecture. Our proposed approach is generic and can be applied to other SCORE/DPS fair queueing algorithms with little modifications. For purpose of illustration, we applied the Linear Quadratic (LQ) control method from the optimal control theory to our modification of Core-Stateless Fair Queueing (CSFQ). Simulation results show that this approach gives improved flow isolation between TCP and UDP flows when compared with CSFQ.

1 Introduction

The importance of fair bandwidth sharing in flow isolation and its ability to to greatly improve the performance of end-to-end congestion control algorithms have lead many researchers to study this area. Until recently, fair bandwidth allocations were best achieved using per-flow queueing mechanisms like Weighted Fair Queueing [1] and its many other variants [2]. These proposed mechanisms are usually based on a stateful network architecture that is a network in which every router maintains per flow state. In addition, most of these mechanisms require per-flow queueing. As there can be a large number of flows in the Internet, the complexity required to implement these mechanisms severely limits their deployment over high speed backbone core routers.

In order to reduce this complexity of doing per-flow queueing and maintaining per-flow state information new algorithms have been introduced. Stoica et al [3] was the first to propose a fair queueing scheduling algorithm that do not require per-flow state information, called Core-Stateless Fair Queueing (CSFQ). CSFQ is based on the Scalable Core (SCORE) architecture [3], which is similar to the Diffserv architecture [4]. The main idea of the SCORE architecture is to keep per-flow state at the edge routers and carry that information using Dynamic Packet State (DPS) in the packets at the core. Specifically in CSFQ, packets are labelled with their flow arrival rate at the edge, and they are dropped probabilistically when their arrival rate exceeds the fair share estimated by the

B. Stiller et al. (Eds.): QofIS/ICQT 2002, LNCS 2511, pp. 90–99, 2002.

core routers. Through extensive simulations, CSFQ was found to approach the fairness of Deficit Round Robin (DRR) [2] and offer significant improvements over FIFO and Random Early Drop (RED) [5]. Other similar algorithms based on the same SCORE/DPS architecture, like Rainbow Fair Queueing (RFQ) [6] and Tag-based Unified Fairness (TUF) [7], were subsequently proposed.

Unfortunately, CSFQ is known to function poorly under bursty traffic [6], [7], which can be attributed to the ad-hoc method used in its fair share estimation. Our main contribution in this paper is the use of a control-theoretical approach that can provide a fast and robust computation of the fair share value. This approach is generic and can be applied to other SCORE/DPS fair queueing algorithms with little modifications. For purpose of illustration, we applied the Linear Quadratic (LQ) control method from the optimal control theory to our modification of CSFQ. Simulations are then used to show the superior performance of our approach.

This paper is structured as follows: In Sect. 2, we describe how fair bandwidth sharing is achieved in the SCORE/DPS architecture and discuss in greater detail how CSFQ achieves this. In Sect. 3, we describe the control model of our proposed system and discuss key implementation issues involved. In Sect. 4, we evaluate our system using some simulations before we conclude in Sect. 5.

2   SCORE/DPS Fair Queueing

2.1   Objective

The SCORE network deals with traffic aggregates at the core routers and do not need to perform per-flow state management. It makes use of DPS to achieve a functional approximation of a stateful network. The primary objective of SCORE/DPS fair queueing algorithms is therefore, to provide flow isolation without the need to provide or maintain per flow state information. In other words, to achieve max-min fairness [8] with minimum implementation complexity. To achieve this objective, these algorithms make use of the following idea: Consider a link with capacity C serving N number of flows with rates given as $r_1, r_2, ..., r_N$. Assume weights w_i are assigned to different flows, such that a flow assigned a weight of 2 will get twice as much bandwidth compared with a flow with weight 1. Max-min fairness is then achieved when the fair share r_{fair} is the unique solution to

$$C = \sum_{i=1}^{N} w_i \, min(\frac{r_i}{w_i}, r_{fair}) \tag{1}$$

Note that when congestion occurs, weighted flow rates r_i/w_i above r_{fair} will be constrained to r_{fair}, while weighted flow rates r_i/w_i equal or below r_{fair} remain unchanged. On the other hand, when $C \geq \sum_{i=1}^{N} r_i$, then all flows can pass through unconstrained and r_{fair} becomes equal to the highest weighted flow rate r_i/w_i.

2.2   Core-Stateless Fair Queueing Framework

To facilitate our discussion, we consider how CSFQ achieves the above objective. CSFQ does it through the following four steps:

(1) When a flow arrives at the edge of the network, its rate r is estimated by exponential averaging calculated as:

$$r_{new} = (1 - e^{-\Delta t/\tau})/\Delta t + e^{-\Delta t/\tau} r_{old} \qquad (2)$$

where Δt is the packet inter-arrival time, and τ is a constant[1].

(2) The edge router then labels each packet of the flow with a state value that is proportional to this estimated rate. Due to the need to represent a large range of flow rates with a limited number of state values, a non-linear representation is used to limit the error of representation to a fixed range. For CSFQ, a simple floating point representation consisting of mantissa and exponential component is used [3].

(3) Inside the network, packets from different flows are interleaved together. Core routers use FIFO queueing and do not keep per-flow state. At the network core, each router estimates a fair share rate α, by approximating the aggregate traffic acceptance rate $F(.)$ of the router, by a linear function that intersects the origin with a slope of $\hat{F}/\alpha_{old}$, yielding

$$\alpha_{new} = \alpha_{old} \frac{C}{\hat{F}} \qquad (3)$$

where $\hat{F}$ is the estimated aggregate traffic acceptance rate computed using exponential averaging.

(4) When congestion occurs, packets of every flow i in the system are dropped with probability

$$Prob = max(0, 1 - \frac{\alpha}{r_i/w_i}) \qquad (4)$$

where r_i and w_i denote respectively the rate and weight of flow i found in the header of the packet. Finally, packets are relabelled using the minimum of the incoming flow rate and the router's estimated fair share rate α.

Steps (3) and (4) illustrates how max-min fairness is achieved using the CSFQ algorithm. When F is larger than C, α will be reduced due to the $C/\hat{F}$ ratio (see Eq. (3)), leading to a higher packet dropping probability for flows with r/w larger than α (see Eq. (4)). On the other hand, when F is smaller than C, α will be increased, thereby reducing the packet dropping probability. Therefore, the process of adjusting α so that F converges to C leads to max-min fairness because $F = \sum_{i=1}^{N} w_i \, min(\frac{r_i}{w_i}, \alpha)$ (see Eq. (1)). Fig. 1 illustrates how α will eventually converge to $\alpha_{final} = r_{fair}$ for the case when F is larger than C.

The method used in CSFQ for estimating fair share α is simple but does not function well when the aggregate incoming traffic is bursty [6], [7]. In fact, in order to minimize the negative effects of buffer overflow, Stoica included a

---

[1] A good discussion on what value to set for τ can be found in [3].

simple heuristic whereby each time the buffer overflows, α is decreased by a small fixed percentage, taken to be 1% in his simulations [3]. In addition, to avoid over-correction, it is ensured that during consecutive updates α does not decrease by more than 25%. The CSFQ algorithm is therefore, unable to quickly and robustly compute α for very bursty traffic, which is crucial for achieving max-min fairness.

3   Control Theoretical Approach for Fair Share Estimation

3.1   Closed-Loop Dynamics

We assume a system whereby the traffic and buffer occupancy are sampled and updated at periodic intervals T. Let $F(n)$ denote the aggregate traffic accepted during the n timeslot. In addition, let $Q(n)$ denote the buffer occupancy of the FIFO queue at the beginning of the n timeslot. The closed loop dynamics of our proposed system is therefore, given by

$$F(n+1) = Sat_0^{C_{in}} \left\{ F(n) - \sum_{j=0}^{J} \lambda_j (Q(n-j) - Q^0) - \sum_{k=0}^{K} \mu_k F(n-k) \right\} \quad (5)$$

$$Q(n+1) = Sat_0^B \{ Q(n) + F(n) - \psi \} \quad (6)$$

where

$$Sat_b^a(x) = \begin{cases} b \text{ if } x < b \\ a \text{ if } x > a \\ x \text{ otherwise} \end{cases} \quad (7)$$

The other parameters involved in the computations above are:
(1) Update/sampling period T: this corresponds to the duration of a single timeslot, where computations of F and Q are done periodically.
(2) Buffer threshold Q^0: this is the buffer occupancy value that the controller tries to achieve and corresponds to the desired steady state buffer occupancy value. In general, Q^0 is chosen to satisfy two objectives, it must be small enough to minimize queueing delay but large enough to ensure full utilization of the link capacity.
(3) C_{in}: this is the maximum aggregate traffic that can be accepted in a single timeslot.
(4) Feedback control gains λ_j and μ_k: these values will be determined later using the LQ algorithm. It will be shown from steady state analysis that λ_j and μ_k must satisfy the following requirements:

$$\sum_{j=0}^{J} \lambda_j > 0, \qquad \sum_{k=0}^{K} \mu_k = 0$$

(5) J and K are non-negative integers, whose values are determined by the stability of the system and the aggregate traffic characteristics.

(6) Buffer size B: this is used to impose bounds on the buffer size of the switch.

(7) Traffic service rate ψ: this denotes the aggregate traffic serviced during a single timeslot. We have assumed that ψ is a constant equal to the output link capacity multiplied by T.

Next, we establish the steady state conditions of the system. Let F_s and Q_s be the steady state values corresponding to Eq. (5) and (6) under the assumption that the input traffic is constant. Therefore, we have

$$F_s = Sat_0^{C_{in}} \left\{ F_s - \sum_{j=0}^{J} \lambda_j (Q_s - Q^0) - \sum_{k=0}^{K} \mu_k F_s \right\} \tag{8}$$

$$Q_s = Sat_0^B \{ Q_s + F_s - \psi \} \tag{9}$$

which gives us

$$F_s = \psi \qquad \text{for} \quad 0 \leq Q_s \leq B \tag{10}$$

$$Q_s = Q^0 - \frac{\sum_{k=0}^{K} \mu_k}{\sum_{j=0}^{J} \lambda_j} F_s \quad \text{for} \quad 0 \leq Q_s \leq B \tag{11}$$

To ensure that $Q_s = Q^0$ for non-zero F_s, we have the constraint $\sum_{k=0}^{K} \mu_k = 0$, which in other words mean $\mu_K = -\sum_{k=0}^{K-1} \mu_k$. In addition, $\sum_{j=0}^{J} \lambda_j > 0$ is required to ensure the stability of the system.

By removing the saturation non-linearity of the dynamic Eq. (5) and (6), the linear part of Eq. (5) and (6) can be expressed in terms of state vector $Z(n)$ as

$$Z(n+1) = \begin{pmatrix} 1 & 0 & \cdots & 0 & 1 & 0 & \cdots & 0 \\ 1 & 0 & \cdots & 0 & 0 & 0 & \cdots & 0 \\ & \ddots & \ddots & \vdots & \vdots & \vdots & \ddots & \vdots \\ & & 1 & 0 & 0 & 0 & \cdots & 0 \\ 0 \cdots & & & 0 & 1 & 0 & \cdots & 0 \\ & & & & 1 & 0 & \cdots & 0 \\ & & & & & \ddots & \ddots & \vdots \\ & & & & & & 1 & 0 \end{pmatrix} Z(n) + \begin{pmatrix} 0 \\ 0 \\ \vdots \\ 0 \\ 1 \\ 0 \\ \vdots \\ 0 \end{pmatrix} u(n) \tag{12}$$

where

$$Z(n) = [(Q(n) - Q^0), \cdots, (Q(n-J) - Q^0), F(n), \cdots, F(n-K)]'$$

$$u(n) = -\sum_{j=0}^{J} \lambda_j [Q(n-j) - Q^0] - \sum_{k=0}^{K} \mu_k F(n-k)$$

$$= -GZ(n)$$

$$G = (\lambda_0, \cdots, \lambda_J, \mu_0, \cdots, \mu_K)$$

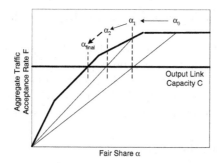

Fig. 1. Basic framework on how CSFQ estimates the fair share, α

with the prime symbol (') denoting the transpose operator. Note that the first matrix on the right-hand side of Eq. (12) is a $(J + K + 2) \times (J + K + 2)$ matrix, while the second matrix is a single column matrix consisting of a single '1' at the $J + 2$ position. A simplification of the above matrices is possible, but due to space constraint it will not be discussed here.

3.2   Gain Selection

The method for choosing the feedback controller gain G in order to achieve good dynamic behavior of the closed loop system is based on the Linear Quadratic (LQ) control method [9]. Given a system with state space model:

$$Z(n + 1) = AZ(n) + BU(n) \qquad (13)$$

the LQ design problem involves solving the control function $U(n)$ that minimizes the cost function:

$$L = \sum_{n=0}^{\infty} \left[Z^{'}(n)W_1 Z(n) + U^{'}(n)W_2 U(n) \right] \qquad (14)$$

where W_1 and W_2 are the design parameters of the LQ control problem that allow different emphasis to be placed on the states and inputs. The first term on the right is the cost associated with state deviation, while the second term is the cost associated with inputs to the system. Note that both W_1 and W_2 must be symmetric and positive definite.

The optimal solution to the minimization problem, obtained using dynamic programming [9] is

$$U(n) = -GZ(n)$$

where

$$G = W_2 + (B^{'}SB)^{-1}B^{'}SA$$

and S is the solution of the following matrix equation known as the *Riccati equation*:

$$A^{'}SA - S - A^{'}SB(W_2 + B^{'}SB)^{-1}B^{'}SA + W_1 = 0$$

The vector of closed-loop poles using the LQ algorithm are the eigenvalues of the matrix $A - BG$.

There are several tools available for solving this equation, but for high-speed implementations, a lookup table will be required to store the pre-computed values of feedback gain G as a function of pre-determined weighting matrices W_1 and W_2. We can limit the size of the lookup table by choosing W_1 and W_2 values that limit the output response characteristics of the system to a fixed range.

3.3   Implementation Issues

A good discussion on the stability of the Discrete Linear Quadratic Regulator (DLQR) system can be found in [9]. In general, the DLQR system is stable if there is a solution for the feedback controller gain matrix G and $\sum_{j=0}^{J} \lambda_j > 0$ (see Eq. (11)). In addition, this proposed system do not have long and uncertain feedback delays that make stability difficult to achieve.

Another issue is the non-linear region when the buffer remains at its lower boundary zero for an extended period of time due to lack of incoming traffic. In order to address this non-linear region, we replace the actual queue length Q with the concept of virtual queue (VQ) in our implementation. Specifically, non-linearity is removed when VQ is updated as follows: (a) When the buffer occupancy is greater than one packet, the virtual queue is equal to the actual queue length. (b) When the buffer occupancy is less than or equal to one packet, the virtual queue is computed by subtracting any excess bandwidth from the previous virtual queue value. Hence by using VQ, when the real queue underflows, the virtual queue will continue to decrease below 0, and F will increase at a much faster rate due to a larger error term.

Note that our proposed system is generic in nature and can be easily used to replace the original packet discarding algorithms for CSFQ, RFQ and TUF. We are unable to discuss this due to space constraint. Comparing with the other approaches, this proposed system is simple and more straightforward because we do not have to apply additional codes for traffic saturation conditions [3], [6].

4   Simulations

In the following section, we illustrate our approach with examples and show some simulation results associated with them to evaluate the performance of our proposed approach. Due to the space limitation, we are unable to show all our simulation evaluations. Specifically, we will compare the performance of DRR, RED and CSFQ with our modified CSFQ and RFQ, which we will call the Control-theoretical Approach to CSFQ (CA-CSFQ) and the Control-theoretical Approach to RFQ (CA-RFQ) respectively. Deficit Round Robin (DRR) [2] is an efficient implementation of WFQ that requires per-flow queueing and is used as the benchmark for fair bandwidth sharing.

In the simulations, the output link capacity C is set at 10 Mbps and the link buffer size is set at 64 Kbytes. The packet size is fixed at 1 Kbytes. For CA-RFQ, we used the non-linear encoding algorithm proposed by Cao in [6] with

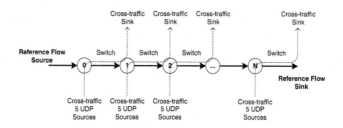

Fig. 2. Multiple congested link network configuration.

parameters $a = 3$, $b = 32$ and $P = 65$ Mbps. For CA-CSFQ and CA-RFQ, we set Q^0 at 20% of link buffer size, $\mu(n)$ at C and T at 1 msec. Detailed descriptions of other simulation parameters of CSFQ can be found in [3].

The control system in CA-CSFQ and CA-RFQ uses only $F(n)$, $F(n-1)$, $Q(n)$ and $Q(n-1)$. The cost function considered is

$$L = \sum_{n=0}^{\infty} \left[(Q(n) - Q^0)^2 + u^2(n) \right]$$

Solving the cost function above yields a feedback vector G given by

$$G = [1.4731, -1.0891, 1.0891, -1.0891].$$

4.1 Multiple Links

We first evaluate the performance of various algorithms over a multiple congested link network configuration (see Fig. 2). The number of nodes varies from 1 to 5. The output link capacity is 10 Mbps with propagation delay 1 msec. At each of the nodes, 5 cross-traffic flows carrying UDP traffic at an average of 4 Mbps each are connected. In the first experiment, the reference flow is a UDP flow transmitting at 4 Mbps. Fig. 3(a) shows the normalized throughput. DRR has the best performance, achieving almost perfect bandwidth sharing, while CSFQ, CA-CSFQ and CA-RFQ have comparable performance slightly inferior to DRR. RED has the worst performance.

In the second experiment, we change the reference flow to be a TCP flow. Fig. 3(b) shows the normalized bandwidth share it receives. The results show that the performance of CA-CSFQ and CA-RFQ are comparable, but slightly better than CSFQ. On the other hand, RED fails to protect the TCP flow.

4.2 Bursty Cross Traffic

Next, we look at the effects of bursty cross-traffic sources, which also indirectly shows the ability of each algorithm to quickly respond to transient load variations. Only the results for SCORE/DPS algorithms CSFQ, CA-CSFQ and CA-RFQ will be compared.

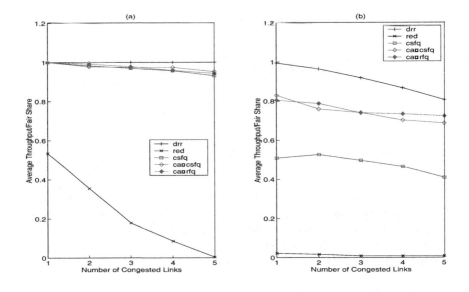

Fig. 3. (a) Normalized throughput of a UDP flow as a function of the number of congested links. Cross traffic are UDP sources sending at twice the fair share. (b) The same plot as (a) but with the UDP flow being replaced by a TCP flow.

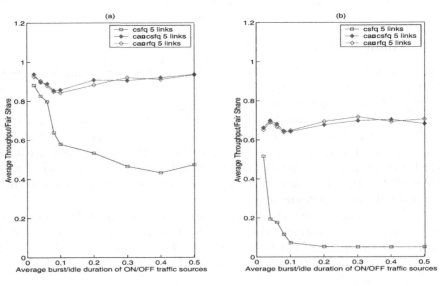

Fig. 4. (a) Normalized throughput of a UDP flow going through 5 congested links. Cross traffic are ON/OFF sources whose average rate is twice the fair share. The burst and idle times vary between 20 msec to 0.5 sec. (b) The same plot as (a) but with the UDP flow being replaced by a TCP flow.

The simulations uses the same topology as Fig. 2, but the UDP sources that form the cross traffic are now replaced with ON/OFF sources. The burst (ON) and idle (OFF) time periods are both exponentially distributed with the same average chosen between 20 msec and 0.5 sec. The cross traffic's average intensity remains unchanged from the previous set of simulations, that is the ON/OFF sources send at 4 Mbps during the ON period. The results in Fig. 4 showed that as the ON/OFF time periods reach the critical 100 msec, the performance of CSFQ becomes seriously affected. We believe the reason is because the fair share estimation algorithm in CSFQ is unable to quickly converge to the correct fair share value. In comparison, both CA-CSFQ and CA-RFQ are still able to achieve a reasonable normalized throughput.

5 Conclusion

This paper presents a formal method for designing the packet discarding component of SCORE/DPS fair queueing algorithms, that can be applied to CSFQ, RFQ and TUF. Using a control theoretical approach, a generic control system is developed for the CSFQ core router dynamics. In particular, we demonstrate how an optimal control approach can be used to design a stable system that allows for arbitrary control of the core router's performance. Compared with the original CSFQ algorithms, the resulting controller is simple and more straightforward because we do not have to apply additional codes for traffic saturation conditions. A number of simulation results have been presented to show that the resulting controller yields better results and is more control responsive than the original CSFQ algorithms.

References

1. Demers, A., Keshav, S., Shenker, S.: Analysis and simulation of a fair queueing algorithm. Proc. ACM SIGCOMM (1989) 1–12
2. Shreedhar, M., Varghese, G.: Efficient Fair Queueing Using Deficit Round Robin. Proc. ACM SIGCOMM (1995) 231–243
3. Stoica, I.: Stateless Core: A Scalable Approach for Quality of Service in the Internet. PhD. Dissertation, CMU-CS-00-176 (2000)
4. Blake, S., Black, D., Carlson, M., Davies, E., Wang, Z., Weiss, W.: An Architecture for Differentiated Services. IETF RFC 2475 (1998)
5. Floyd, S., Jacobson, V.: Random early detection gateways for congestion avoidance. IEEE/ACM Trans. on Networking, Vol. 1 (1993) 397–413
6. Cao, Z.R., Wang, Z., Zegura, E.: Rainbow Fair Queueing: Fair Bandwidth Sharing Without Per-Flow State. Proc. IEEE INFOCOM (2000) 922–931
7. Clerget, A., Dabbous, W.,: TUF: Tag-based Unified Fairness. Proc. IEEE INFOCOM (2001) 498–507
8. Bertsekas, D., Gallager, R.: Data Networks. Prentice-Hall (1987)
9. Anderson, B.D.O., Moore, J.B.: Optimal Control: Linear Quadratic Methods. Prentice-Hall (1990)

SOS: Sender Oriented Signaling for a Simplified Guaranteed Service

Evgueni Ossipov and Gunnar Karlsson

Department of Microelectronics and Information Technology
KTH, Royal Institute of Technology, Sweden
{eosipov,gk}@imit.kth.se

Abstract. A resource reservation scheme is an important mechanism of providing guaranteed QoS to applications. Today the only protocol that is standardized by IETF is the resource reservation protocol RSVP. Development of the next generation of signaling protocols is still open for research and development. We now propose SOS – a simple signaling protocol for guaranteed service connections. It overcomes poor scalability of RSVP and is simpler than existing proposals: Our protocol does not require per-flow soft states in core routers; it is robust and can handle losses of all types of signaling messages. Simple operations in the routers allow processing of 700 thousand messages per second.

1 Introduction

The Internet has to support real-time communication and multimedia services, which might be the core of future network services. These services require that the delivery of information to a client must satisfy certain quality requirements. They can be expressed in terms of loss rate, maximum end-to-end delay, and maximum jitter. We are developing a service architecture proposed by Karlsson and Orava [1] consisting of three quality classes, as illustrated in Fig. 1. The aim of this architecture is using the strength of existing QoS architectures (*intserv* [2] and *diffserv* [3]) to provide similar services in a simpler and more scalable way. The definition of the service classes is similar to the one of *intserv*: The *guaranteed service* (GS) class provides deterministic guarantees, namely constant throughput, absence of packet loss due to queue overflows in routers, and tightly bounded delay. Flows of the *controlled load service* (CLS) class have bounded packet loss and limited delay. The third class is the customary *best effort* (BE). The flows of this class have no guarantees; they obtain leftover capacity in the routers, and may be lost due to congestion. Both GS and CLS are allocated restricted shares of link capacity so that BE traffic always has a possibility to get through. Best effort traffic is also using all unused portions of capacity reserved for CLS and GS traffic. The work in our group that is related to CLS is reported in [4 – 6].

A resource reservation (or QoS signaling) scheme is an important mechanism for providing guaranteed service to applications. We have therefore developed a simple signaling protocol by which applications may explicitly reserve capacity in all nodes

B. Stiller et al. (Eds.): QofIS/ICQT 2002, LNCS 2511, pp. 100-114, 2002.
© Springer-Verlag Berlin Heidelberg 2002

along a path. The description of the guaranteed service class as well as the basic operations of the signaling protocol is described in [7].

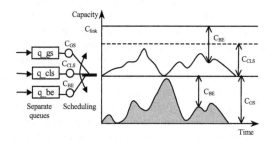

Fig. 1. The network architecture.

The work presented in this paper further develops the signaling protocol for the guaranteed service. Our contribution is as follows. We provide a complete description of the protocol operations in the end systems and the routers. We introduce two garbage collection schemes, which make our signaling protocol able to handle losses of all types of messages. We report the results of experiments, which we are conducted to evaluate the influence of garbage collection on call blocking.

The remainder of the paper is organized as follows. Section 2 presents an overview of the related work. Section 3 describes the network architecture, gives definitions of the major actors and states our assumptions. In Section 4 the description of the SOS protocol is presented. Section 5 gives detailed description of two garbage collection schemes. Section 6 describes performance parameters and simulation results. We state open issues and summarize our work in Section 7.

2 Related Work

Today the only signaling protocol that is standardized by IETF is the resource reservation protocol RSVP [8]. The major concern of the research community about RSVP is its scalability for large IP networks. The thesis that RSVP does not scale well because it installs per-connection states in routers led to the appearance of new signaling schemes [10 – 14]; however none of them found practical application in the Internet. Common opinion of the research community is that there is a need for a new signaling protocol, which can overcome the inflexibility of RSVP. The recently created IETF working group Next Steps in Signaling (NSIS) [9] works on specification of the next generation signaling. In this section we give an overview of the signaling protocols that appeared recently.

It is worth to mention the work in [15, 16]. The authors advocate the ability of RSVP to scale in the large network by performing flow aggregation on the boundaries of administrative domains and optimization of its implementation in all network devices. The work in [17, 18] suggests an extension to RSVP, which improves its performance.

YESSIR (Yet another Sender Session Internet Reservations) [10] differs from RSVP in the following ways. It is an in-band protocol, meaning that it uses signaling messages of the already existing protocol RTP. This implies easier implementation of the protocol in the end systems. It is sender oriented, so that the resource reservation request is initiated from the sender. The developers argue that the sender-oriented nature of the protocol significantly reduces its implementation complexity. Like RSVP, YESSIR relies on per-connection soft states in the routers.

ST-II is a set of protocols for real-time communications [13, 26]. The setup protocol of ST-II is another sender oriented scheme for bandwidth reservation. The protocol was designed for point-to-multipoint real time communications. The main idea of the protocol is to install multicast forwarding state and reserve resources along each next hop subnet. The protocol uses six types of messages: The CONNECT message issued by a sender is used to establish a connection, the ACCEPT message issued by a receiver to signal success of the reservation, DISCONNECT/REFUSE messages to explicitly teardown a connection; two other types are used to change the connection parameters and to establish the reservation for new multicast members. The robustness of the protocol is assured by reliable message retransmission, and a special protocol for ST-routers to recover from path changes.

Boomerang is another attempt to eliminate the complexity of RSVP [14]. It uses one message to set up a bi-directional soft-state reservation in the routers. The authors modify the Ping protocol, so that the reservation is included in the Ping message. Usage of the ICMP protocol as a transport for Boomerang reservations messages implies simplicity of its implementation in the end systems and in the routers.

The Flow Initiation and Reservation Protocol (FIRST) is a reservation protocol for multicast traffic [12]. It uses hard state in the routers and assumes that routes for guaranteed flows are fixed. In fact the idea of this protocol is very similar to RSVP: FIRST is receiver oriented, it uses some kind of PATH messages, which establish the states and RESV message to actually reserve the resources in the routers. The connection states in FIRST are hard, meaning that there are no update messages for the established connection. A state of the particular connection should be explicitly deleted from the routers by a termination message. This protocol remained on the level of a proposal without further analysis.

The Ticket Signaling Protocol (TSP) is a lightweight resource reservation protocol for unicast IP traffic [11]. The resources can be reserved on a per-connection basis without introducing per-flow states in the network. The main idea of this protocol is usage of a so-called ticket. When the reservation request propagates through the network it can be either accepted or rejected in each router. When the reservation reaches the receiver it issues an acknowledgment that contains the traffic specification for the connection. The sender transforms this message into a ticket. This is a refresh message transmitted periodically through the network to confirm the usage of the resources.

Our protocol differs from these approaches in the following ways. It uses a simpler traffic descriptor: We describe a connection solely by its peak rate. The protocol does not use per-flow soft state in the routers. Instead we introduce soft state for the aggregate reservation on a network link. Two garbage collection schemes, which we have developed, make our protocol more robust than earlier proposals. The recently proposed scheme RMD (resource management in diffserv) [19] is similar to our approach in the way it uses aggregated soft state in the core routers. However, the way RMD handles fault situations is more complex than our protocol.

The comprehensive work in [20] as well as efforts of ATM forum [21] deal with signaling for QoS routing, while our protocol uses existing routing algorithms. Other approaches for the resource provisioning are centralized schemes (bandwidth brokers), like in [22], which are out of scope for our work.

3 Description of the Network Architecture

For the purpose of our presentation we consider a network as shown in Fig. 2. We define the following actors:

End systems (ES). These are client systems, which use certain services in the network, and the servers, which offer the services to the clients. Sufficient capacity for a particular GS flow is provided by explicit reservation in all nodes along a path. The ES's use the SOS protocol to make reservations, as described below.

Access gateways (or ingress nodes) (AG). These are routers located on the border between the backbone and the access networks of an ISP. An AG is the first IP-level node of a GS flow. In our architecture AG's keep per-connection soft state, perform policing and billing.

Core nodes (or routers). These are all routers apart from AG's. They provide IP-level packet forwarding, and do not keep per-connection soft state.

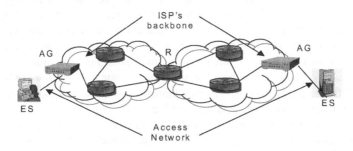

Fig. 2. The network architecture.

3.1 Assumptions Used in This Paper

We base this work on the following assumptions.
- Each GS connection is described solely by a peak rate as traffic descriptor. The signaling protocol assumes the reservation state to be purely additive: The reserved rate of an outgoing link of a router equals to the sum of the incoming rates for the link. This is enforced by non-work-conserving scheduling [7, 23].
- We also assume that re-routing does not occur in the network. Although we are aware of the problem that routes in the real Internet may change, we leave this problem for our future work.
- We consider only unicast communication.

- We do not consider wireless links.
- Security, multicast, policing and billing are out of scope for this paper and will also be considered in our future work.

4 The Signaling Protocol

We have the following design goals for our signaling protocol: It should not require per-connection soft-states in the routers; the time to process a signaling message should be small. The protocol should be robust and tolerate loss of all types of messages.

The idea behind the signaling is simple and uses the additivity of the reservation state. We keep track of available capacity for each network link in a variable C_{GS}, and have three public state variables for *requested, confirmed* and *refreshed* capacity, R_r, R_c and R_{ref}. All users are able to modify these variables by means of the signaling protocol. A router therefore has four state variables for each outgoing link. To make a reservation, the router simply adds the requested rate in the signaling message to the requested capacity variable, R_r. This permits the requestor to send data with this rate. The router does the reverse operation to release capacity when the end system terminates the connection by an explicit tear down message. Let us define the requested rate r as a multiple of Δ, where Δ bits per second is a minimum value (quantum) of reservation. The following messages are needed for the signaling:

- Reservation message $mR(r)$ for the amount of capacity r. Messages of this kind may be discarded by the routers as a result of admission control.
- Confirmation message $mC(r)$ from the initiating party for the reserved capacity. Messages of this type might be lost due to link failure or bit errors.
- Refresh message $mRef(r)$ for the amount of capacity r. Messages of this type might be lost due to link failure or bit errors.
- Tear down message $mT(r)$, which indicates that the initiating party finishes the connection and releases the reserved capacity. Messages of this type might be lost due to link failure or bit errors.

The service specifies that all signaling messages are carried in the GS part of the link capacity, and they are therefore not subjected to congestion-induced loss.

The continued work on the protocol specification includes finding the best possible way of encoding the signaling messages in the fields of the IP header. We believe that the length of a signaling message could be as small as an IP header, 20 bytes, considering that we only need to represent four distinct messages and one rate value. The message types could, for instance, be encoded as DSCPs and the reservation values could be encoded in the "Fragment offset" field or be represented by the packet's length. We will report the details regarding messages encoding scheme in our future publications.

4.1   Time Cycles and Synchronization of Network Devices

For the purposes of long-term garbage collection (see Section 5.2) we specify that all network devices operate in cycles of fixed duration T_c. The beginning of each cycle is synchronized in all devices by using SNTP [25]. All events of sending and receiving a signaling message take place in certain strictly defined time intervals within a cycle. Fig. 3 shows how a cycle is divided into intervals in the end systems and the routers. Times T_c, T_{Rref} and T_{margin} are the constants specified for the routers, other time intervals are computed in relation to them. Precise definition of each time interval is given in the corresponding sections (for the end systems Section 4.2 and for the routers Section 5.2).

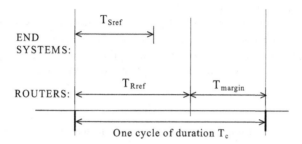

Fig. 3. Time intervals in the end systems and routers.

4.2   End System Behavior

In order to establish a guaranteed service connection, the initiator sends *mR(r)* towards the destination. After issuing this message a waiting timer is started with a value *T*. This is the allowed waiting time for an acknowledgment at the sender and it is a constant specified for the service. By defining *T* as maximum *RTT* we ensure correct handling of delayed acknowledgments: If no acknowledgment arrives during *T* seconds, the initiator sends a new reservation request. The establishment of a connection is repeated until an acknowledgment arrives or a sender decides to stop. If an acknowledgement was delayed by the receiving end system and the sender times out before receiving this message, it will issue a new reservation message. The acknowledgment for the old reservation will be ignored. The core routers will handle this case as a lost reservation, and will clean up the reserved capacity during the garbage collection (see Section 5.1). Reservations may be sent during whole duration of the cycle T_c (see Fig.3).

Each signaling message has a unique identifier. An acknowledgment is issued by the receiver when it receives any signaling message except a tear down, and it is encoded as a zero-rate reservation (i.e. *mR(0)*) with the same identifier as in the signaling message. For a sender an acknowledgment indicates a successful transmission of a message along the path. If expected acknowledgment did not arrive to the sender, this situation is interpreted as an error and will be handled as shown in Table 1.

When an acknowledgment for a reservation is received the source then sends a confirmation message towards the destination, and may immediately start to send data at a rate below r b/s.

During the lifetime of the established connection the sender periodically emits refresh messages $mRef(r)$. The time of the next refresh message is computed from the moment of sending a $mR(r)$ or previous $mRef(r)$ by the following formula:

$$T_{mRef} = \delta + Uniform[T_i, T_i + T_{Sref}] + \varepsilon \cdot \tag{1}$$

where δ is the time from the moment of sending previous message until the end of the current cycle, T_i is the beginning of the next cycle. T_{Sref} is the time until which the sender may transmit refresh messages. It is computed as $T_{Sref} = T_{Rref} - T/2$; ε is an error of synchronization introduced by SNTP. In simple words Eq. 1 says that a sender refreshes a GS connection once per cycle. If the sender does not receive an acknowledgment for issued $mRef()$ then, according to Table 1, the sender must interrupt the connection without issuing a tear down message, and setup a new reservation.

Table 1. Handling of errors at the sender.

Event	Action
Acknowledgment for $mR()$ did not arrive during timeout T.	Retransmit reservation after T.
Acknowledgment for $mC()$ did not arrive during timeout T.	Interrupt ongoing session without issuing a tear down.
Acknowledgment for $mRef()$ did not arrive during timeout T.	Interrupt ongoing session without issuing a tear down.

When the initiating party wishes to end the session, it sends a teardown message, freeing the reserved capacity on the path. The diagram in Fig. 4 represents the behaviors of the sender and the receiver in the case where no message is lost in the network.

4.3 Router Behavior

First, when a router receives $mR(r)$, it performs an admission control based on the requested rate. The router will drop a reservation if it exceeds the remaining capacity ($r > C_{GS} - R_r$). Otherwise it increases R_r and R_{ref} by the bit rate requested in the reservation message.

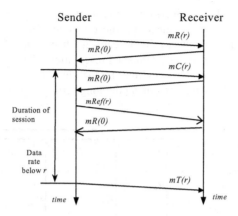

Fig. 4. Lossless behaviors of the sender and the receiver.

When getting a confirmation message, the router increases the value of R_c by the rate r. In reaction to the arrival of a refresh message the router increments R_{ref} variable by the bit rate specified in the refresh message. When the router gets a teardown message, it decreases both R_r and R_c by the rate in the message. The summary of a router's behavior is presented by the pseudo-code in Fig. 5.

```
void server() {
  //types of messages: 0-mR, 1-mC, 2-mRef 3-mT
  r=0; //requested capacity
  Rr=0;//reserved capacity
  Rc=0; //confirmed capacity
  Rref=0; //refreshed capacity
  while (1) {
    int type=receive();
    switch type {
      0: if(r>Cgs-Rr) //Admission control
           drop(message);
         else Rr +=r; Rref +=r;
      1: Rc +=r;
      2: Rref +=r;
      3: Rr-=r; Rc-=r;
    } //switch type
    if (Rr == Cgs) {
      wait(W);
      /* function wait process only messages of types 1 and 2 for
the duration of W seconds */
      if (Rr!=Rc) garb_collect(SHORT_); //See Section 5.1 for
detailed description
    } //if (Rr == Cgs)
    if (END_OF_CYCLE)
      if (Rr!=Rref) garb_collect(LONG_); //See Section 5.2 for
detailed description
  } //while (1)
} //void server()
```

Fig. 5. Algorithm for processing signaling messages in a router.

As one could notice from the algorithm, if a router gets an acknowledgment message in the form of $mR(0)$, it will not change the state variables. However, acknowledgments are carried in the unallocated part of the GS capacity, and might therefore be lost due to buffer overflow if that part is nil. Fig. 6 shows the capacity allocation for the signaling messages and the dynamic behavior of R_r and R_c. The message $mR(r)$ is carried in the available capacity from which it reserves a share; $mC(r)$, $mRef(r)$, and $mT(r)$ use the established reservation. Thus, the messages are not subjected to congestion; however they may be lost due to a link failure or bit errors. The router will handle this situation during the long-term garbage collection described in Section 5.2.

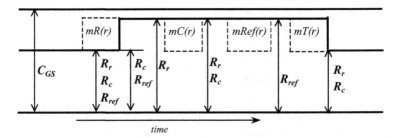

Fig. 6. Dynamic of R_r, R_{ref} and R_c.

5   Garbage Collection

There are several cases when the routers need garbage collection to re-establish the reservation levels, which may be inconsistent due to losses of signaling messages. A reservation message is discarded when a router blocks a reservation request. This means that there will be pending reservations upstream along the path of the message. Although confirmations and teardowns are carried in the reserved path of capacity, messages of these types might be lost either due to a link failure or bit errors. In case the loss is of a confirmation message, a sender cannot issue a teardown for this connection, since it is impossible to deallocate the capacity on the unconfirmed part of the path. If a teardown message is lost, all the nodes after the point of failure will never receive the message and therefore the amount of capacity, which was supposed to be deallocated, will remain in use. These situations can potentially lead to a state in which all the capacity in a router is unavailable for new reservations. Some important observation can be made on frequency of losses due to admission control ($mR()$) and link failures or bit errors ($mC()$, $mT()$) and the relationship between them:

- The reasons for losses of $mR()$ on one hand and $mC()$ and $mT()$ on the other are unrelated.
- Losses due to link or bit errors are more rare events than those due to admission control.

Therefore, we decided to develop two independent garbage collection processes. The first one is more reactive to handle losses of reservations: It will be activated as soon as some losses of $mR()$ happen due to admission control. We call this scheme the short-term garbage collection. The second scheme, which we call the long-term garbage collection, will be activated less frequently and will handle losses of other types of messages.

5.1 The Short-Term Garbage Collection

In Section 4 we state that confirmation messages may not be lost due to congestion. Since $mC()$ will not be lost, all successfully acknowledged reservations are expected to be confirmed. The garbage collection starts when all GS capacity is reserved $Rr=C_{GS}$. The router will therefore expect a confirmation for the reserved capacity within a waiting time $W>T$. If some of the requested capacity is not confirmed after W, the output port will free the capacity $Rr-Rc$. Detailed example of the short-term garbage collection is given in [7].

   Example. Consider the following case. The capacity variable of the outgoing link is $C_{GS}=5\ Mb/s$. There was no reservation request before T0, so $Rr=Rc=0$.

   Table 2 shows the dynamic behavior of R_r and R_c. Steps T1 to T3 show a successful cycle of a reservation for the amount of 1 Mb/s. The router gets a confirmation message at T2 that makes R_r and R_c equal. This indicates success of the reservation. A tear down message arrives at time T3. At T4 we have a 4 Mb/s requested reservation, and so far have only 2 Mb/s been confirmed. At time T5 the router accepts a 1 Mb/s reservation. Since the amount of requested capacity becomes equal to the available capacity C_{GS}, the garbage collection is started. During the time from T5 to T6, which equals W seconds, the router accepts only confirmation and tear down messages; all reservation messages will be discarded. There are several possibilities of what messages may arrive to the router during the garbage collection. One of the cases will be if the router gets confirmations for all unconfirmed capacity. This indicates that all the capacity on the link is in use and the garbage collection is ended without action. Another case is when only some or no confirmations will arrive. In this case the garbage collection procedure at time T6 will realize that the amount of capacity R_r-R_c= 5-2 =3 Mb/s is not confirmed and will free it. The router will then start accepting new reservations again.

Table 2. Bookkeeping of parameters for reservations.

Time Point	Event	C_{GS}	R_r	R_c
T0	Initial settings	5	0	0
T1	MR(1) arrives	5	1	0
T2	MC(1) arrives	5	1	1
T3	MT(1) arrives	5	0	0
....				
T4	Settings after some time	5	4	2
T5	MR(1) arrives	5	5	2
T5'	Garbage Collection starts [Accept only mC() and mT()]			
T6	Settings after Garbage Collection [Accept all messages]	5	2	2

5.2 The Long-Term Garbage Collection

The absence of per-flow states in the network nodes is the major difficulty when designing the long-term garbage collection scheme. The lack of knowledge about ongoing connections in the network makes it impossible to identify the flow for which a confirmation or a teardown was lost inside the network, and where exactly the message was lost. We have however developed a garbage collection scheme that uses soft-state for the aggregated reserved capacity.

The idea of the long-term garbage collection is based on precise definition of time intervals during which the senders can issue signaling messages. Duration of the cycles T_c and the interval within a cycle during which refresh messages may arrive to a router T_{Rref} are constants defined for the signaling protocol. Since losses of packets in the Internet due to link failures or bit errors are relatively rare events we expect T_c to be in order of several minutes. The value of T_{Rref} must be calculated so that the reminder of the cycle (T_{margin}) will be long enough to process signal messages that arrived before T_{Rref} and are still waiting in the buffer.

The definition of T_{Sref} given in Section 4.1 ensures that refresh messages from senders will not arrive to a router after T_{Rref}. This means that by the end of any chosen cycle the reservations for all ongoing connections will be refreshed. Then at the end of a current cycle the router's state variable R_r will contain an amount of actually reserved capacity, and the variable R_{ref} will contain an amount of refreshed capacity. The decision about the garbage collection is done by comparison of these variables. If R_r is larger then R_{ref} this indicates losses of $mC()$ or $mT()$. Therefore, the garbage should be cleaned: $R_r = R_{ref}$; $R_c = R_r$. The refreshed capacity variable should be reset at the end of the garbage collection ($R_{ref}=0$).

In fact the long-term garbage collection can handle losses of all types of signaling messages including losses of reservations. However, the length of the cycles is larger than the time of activation for the short-term scheme. We have evaluated the influence of the long-term garbage collection on the call blocking probability for the cases where it would be implemented alone and in combination with short-term scheme. The results of these experiments are reported in Section 6.

6   System Requirements and Performance Parameters

In this section we present simulations to evaluate the influence of the two garbage collection schemes (see Section 5) on call blocking probability. We show that although the long-term scheme implemented alone introduces much higher blocking probability than implemented alone short-term scheme, their combination eliminates this effect.

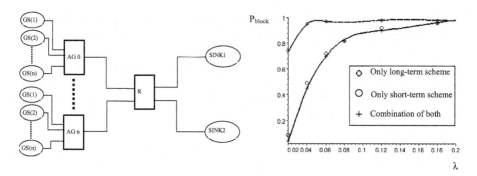

Fig. 7. Network topology for the simulations.   **Fig. 8.** The call blocking probability with the short-term and the long-term schemes, and their combination, as a function of the arrival rate of reservations.

We implemented both schemes and their combination in the network simulator ns-2 [24] (we are working on an implementation of the proposed architecture in a Linux-based router, and will report the results in our future publications).

We used a two-stage network as shown in Fig. 7. At the first stage, n groups of n sources gain access to n access gateways AG1...AGn. The capacity of all links is $(m \cdot \Delta)$ b/s, where Δ is an amount of capacity needed for one call, and m is the maximum number of calls on the link. At the second stage, calls from n access gateways arrive to a router R. The router has two output links. Sources GS1, at the first stage, try to place a call to SINK1, and the rest of the sources establish connections with SINK2. The calls from every source arrive to the access gateways as a Poisson stream with intensity λ calls per second. The duration of each call is exponentially distributed with a mean μ^{-1} seconds. If a call is blocked the source will try it again after the timeout T. We evaluate the blocking probability for sources GS1 on the path towards SINK1.

The following settings were used for our experiments: The capacity C_{GS} of all links is 9 Mb/s, m=3 and $\Delta = 3$ Mb/s, the number of sources and access gateways n=5, the timeout $T = 0.1$ seconds, the mean duration of a call μ^{-1} =5 seconds, the mean arrival rate λ is varied in the interval [0.02,0.2]. The duration of a cycle T_c=2 seconds; therefore, all sources will issue in average two refresh messages per connection.

We intentionally choose the parameters so that the call blocking probability would be high. Our task is to capture general behaviors of the two schemes for garbage collection and their influence on each other. We do not show the performance of these schemes in the real environment, this is a subject for our future work.

Fig. 8 shows the call blocking probability with the short-term, the long-term garbage collection schemes, and their combination. Simulations show that implemented alone, the long-term scheme introduces much higher blocking probability than the short-term scheme. This is due to the difference in the time of activation of the garbage collection. The time of activation of the long-term scheme (the end of each cycle) is much larger than for the short-term scheme. By separating time scales of the two garbage collection schemes we obtain the resulting blocking probability of their combination being the smallest of individual probabilities. This was shown in the second experiment. The resulting call blocking probability in the case of combination of the two schemes is the same as in the case of the short-term scheme alone.

The important conclusion of our experiments is that combining the short-term garbage collection scheme with the long-term one we drastically increase the robustness properties of our protocol and that in this case the long-term scheme does not increase the call blocking probability.

6.1 Buffer Dimensioning

Since the long-term garbage collection can handle losses of all types of messages the size of buffers for signaling messages can be chosen arbitrarily. However, we decided to dimension buffers so that losses of refreshes, confirmations, and teardowns would happen only due to link failures or bit errors, and not due to buffer overflow. We have assumed that there is a minimum reservation quantum. The worst case is when all capacity available for the guaranteed service will be reserved by the smallest quantum. In this case the total number of supported connections will be equal to C_{GS} $/\Delta$. Since refresh, tear down and confirmation messages can be issued only for successful reservations, the total number of loss-intolerant messages will be three times the number of connections. If we assume that the call processing is equally fast in all routers then it is easy to show that a router with n ports requires $Bmax$ packets of buffering per output port:

$$B_{\max} = \frac{3C_{GS}(n-2)}{\Delta(n-1)}. \tag{2}$$

6.2 Message Processing Rate

We have implemented the signaling protocol in the network simulator ns-2. Our goal was to observe the performance of our protocol in a real router; therefore, we measured the actual time the processor spends on processing every signaling message and not the local ns-2 time, since it is independent of the computer power. In a series of simulations on a processor Intel Pentium III 500 MHz we obtain the following results. The time of processing one reservation message is 2 microseconds, a tear down message takes 1.18 microseconds, and a confirmation message 1.18 microseconds, and a refresh message 1.35 microseconds. Although the operations of reservation and tear down are similar, recall that the router performs admission control for reservations. This explains the slightly different processing times for these types of messages. The average rate of message processing is therefore *700000* messages per second. Assuming for example a link capacity equal to 100 Mb/s and a reservation quantum of 10 kb/s, then the maximum number of supported connections will be 10000 and each connection could issue 70 messages per second without overloading a processor of a power we have used. A buffer for 30000 messages is needed to hold signaling messages. This corresponds to 600 kB of memory if each message is 20 bytes.

7 Summary and Continuation of Work

Summarizing the description of our signaling protocol, we have developed a new simple signaling protocol for the guaranteed service. This protocol has the following properties.

- It is sender oriented.
- It uses four types of messages $mR()$, $mC()$, $mRef()$ and $mT()$ that can be encoded in the IP packet header.
- The protocol does not use per-flow state in core routers and the simple operations allow fast message processing.

The ongoing work on the protocol continues in the following ways. We are conducting more experiments for evaluation of the blocking probability. The extended results will allow finer adjustment of the time parameters defined for our protocol in order to minimize the blocking probability. In our future publications we will report the details regarding messages encoding scheme. We develop a mechanism in the signaling protocol that will handle re-routing. Obviously forwarding rerouted GS traffic without reservation is unacceptable. The design in this paper was concerned only with unicast communication. Another direction of our work is to consider multicast GS flows. We are currently working on an implementation of the described architecture in a Linux-based platform.

References

1. G. Karlsson, F. Orava, "The DIY approach to QoS", Seventh International Workshop on Quality of Service (IWQoS), 1999, pp. 6 – 8.
2. R. Braden, D. Clark, S. Shenker, "Integrated services in the Internet architecture: an overview", RFC1633, IETF, June 1994.
3. S. Blake, D.Black, M.Carlson, E.Davies, Z. Wang, W. Weiss, "An architecture for differentiated services", RFC 2475, IETF, December 1998 .
4. V.Elek, G.Karlsson, R. Rönngren, "Admission control based on end-to-end measurements", in Proc. of INFOCOM 2000, pp. 623 – 630.
5. I. Mas Ivars, G. Karlsson, "PBAC: probe based admission control", Second International Workshop on Quality of Future Internet Services (QoFIS), pp. 97 – 109, 2001.
6. I. Mas Ivars, G. Karlsson, "Probe-based admission control for multicast", in Proc. of Tenth International Workshop on Quality of Service (IWQoS), 2002.
7. E. Ossipov, G. Karlsson, "A simplified guaranteed service for the Internet", In Proc. of Seventh International Workshop on Protocols for High-Speed Networks (PfHSN), 2002.
8. R. Braden, L. Zhang, S. Berson, S. Herzog, S. Jamine, "Resource reservation protocol (RSVP)", RFC 2205, IETF, September 1997
9. IETF Working group Next Steps in Signaling (NSIS), http://www.ietf.org/html.harters/nsis-charter.html.
10. P. Pan, H. Schulzrinne, "YESSIR: a simple reservation mechanism for the Internet", Computer Communication Review, vol. 29, No. 2, April 1999
11. A. Eriksson, C.Gehrman, "Robust and secure light-weight resource reservation for unicast IP traffic", Sixth International Workshop on Quality of Service (IWQoS), 1998, pp. 168 – 170.

12. T.W.K. Chung, H.C.B. Chang, V.C.M. Leung, "Flow initiation and reservation tree (FIRST)", In Proc. ofIEEE Conference on Communications, Computers, and Signal Processing, pp. 361 – 364, 1999.
13. L. Delgrossi, L. Berger, "Internet stream protocol version 2 (ST2) protocol specification – version ST2+", RFC 1819, IETF, August 1995.
14. G. Feher, K. Nemeth, M. Maliosz, I. Cselenyi, J. Bergkvist, D. Ahlard, T. Engborg, "Boomerang - a simple protocol for resource reservation in IP networks", in Proc. of IEEE Workshop on QoS Support for Real-Time Internet Application, June 1999.
15. Y. Bernet, P. Ford, R. Yavatkar, F. Baker, L. Zhang, M. Speer, R. Braden, B. Davie, J. Wroclawski, E. Felstaine, "A framework for integrated services operation over diffserv networks", RFC 2998, IETF, November 2000
16. M. Menth, R. Martin, "Performance evaluation of the extensions for control message retransmissions in RSVP", In Proc. of Seventh International Workshop on Protocols for High-Speed Networks (PfHSN), 2002.
17. M. Karsten, "Experimental extensions to RSVP – remote client and one-pass signaling", Ninth International Workshop on Quality of Service (IWQoS), 2001.
18. M. Karsten, J. Schmitt, N. Berier, R. Steinmetz, "On the feasibility of RSVP as general signaling interface", First International Workshop on Quality of Future Internet Services (QoFIS), 2000.
19. L.Westberg, A. Csaszar, G. Karagiannis, A. Marquetant, D.Partain, O.Pop, V. Rexhepi, R. Szabo, A.Takacs, "Resource management in diffserv (RMD): a functionality and performance behavior overview", In Proc. of Seventh International Workshop on Protocols for High-Speed Networks (PfHSN), 2002.
20. R.A. Guerin and A. Orda, "Networks with advance reservations: the routing perspective", in Proc. of INFOCOM 2000, pp. 118 – 127.
21. ATM Forum, http://www.atmforum.com
22. K. Nichols, V.Jacobson, L.Zhang, "A two-bit differentiated services architecture for the Internet", RFC 2638, IETF, July, 1999.
23. M. Mowbray, G. Karlsson, T. Köhler, "Capacity reservation for multimedia traffics", Distributed Systems Engineering, vol. 5, 1998, pp. 12 – 18.
24. Network Simulator ns-2, http://www.isi.edu/nsnam/ns/.
25. D.Mills, "Simple Network Time Protocol (SNTP) Version 4 for IPv4, IPv6, and OSI", RFC 2030, IETF, October, 1996.
26. CIP Working Group, "Experimental Internet stream protocol, version 2 (ST-II)", RFC 1190, IETF, October 1990.

The Performance of Measurement-Based Overlay Networks

Daniel Bauer, Sean Rooney, Paolo Scotton, Sonja Buchegger, and Ilias Iliadis

IBM Research, Zurich Laboratory
Säumerstrasse 4
8803 Rüschlikon, Switzerland
<dnb,sro,psc,sob,ili>@zurich.ibm.com

Abstract. The literature contains propositions for the use of overlay networks to supplement the normal IP routing functions with higher-level information in order to improve aspects of network behavior. We consider the use of such an overlay to optimize the end-to-end behavior of some special traffic flows. Measurements are used both to construct the virtual links of the overlay and to establish the link costs for use in a link-state routing protocol. The overlay attempts to forward certain packets over the least congested rather than the shortest path. We present simulation results showing that contrary to common belief overlay networks are not always beneficial and can be detrimental.

Keywords: Overlay network, QoS Routing, Link measurement

1 Introduction

Quality of Service (QoS) in large networks is achievable through the presence of control logic for allocating resources at network nodes coupled with inter-router coordination protocols. The various approaches — ATM, DiffServ, IntServ — differ in the trade-off between the precision with which the behavior of flows can be specified and the cost of the additional control logic. However, none of the approaches are widely used in the public Internet. Increased network capacity has meant that the benefits of resource guarantees are reduced and consequently outweighed by the management overhead. Moreover, for HTTP-type request/response traffic this is unlikely to change as the majority of the delay incurred is in the servers [1] rather than in the network, so network guarantees for such flows are of marginal importance.

Applications in which the timeliness of the arrival of data is important, such as continuous media streams, distributed games and sensor applications, would benefit from resource guarantees. Whereas the fraction of Internet traffic that such applications constitute may increase, it is unlikely that this increase will be sufficient to force internet service providers (ISPs) to instrument flow or aggregated flow guarantees. Moreover, it would involve the difficult coordination of policy between the border gateways of autonomous systems of different ISPs.

B. Stiller et al. (Eds.): QofIS/ICQT 2002, LNCS 2511, pp. 115–124, 2002.
© Springer-Verlag Berlin Heidelberg 2002

In an overlay network higher-layer control and forwarding functions are built on top of those of the underlying network in order to achieve a special behavior for certain traffic classes. Nodes of such a network may be entities other than IP routers, and typically these networks have a topology that is distinct from that of the underlying physical network. Nodes in the overlay network use the IP network to carry both their data and control information but have their own forwarding tables as a basis for routing decisions. Examples of overlays are the Gnutella file-sharing network and the Mbone multicast network.

Our approach is to treat traffic requiring guarantees as the special case rather than the common one. This special traffic is forwarded between network servers with hardware-based packet forwarding across a dedicated overlay. We call these servers *booster boxes* [2]. The routing logic between the booster boxes uses dynamic measurements and prediction to determine the least congested path over the overlay. Traffic that is carried over the booster overlay network is called overlay traffic.

While it is trivial to describe simple idealized scenarios in which overlays bring a gain, the more pertinent question is whether and under what circumstances measurement-based overlay networks are beneficial in realistic networks. The focus of this paper is on the applicability and performance of a measurement-based overlay network.

The remainder of the paper is organized as follows. After a review of related work in Section 2, we outline the general architecture of such an overlay network of booster boxes in Section 3. In Section 4 we describe our detailed simulation of its behavior in diverse scenarios. The simulation results and discussion can be found in Section 5, followed by the conclusions in Section 6.

2   Related Work

The resilient overlay network (RON) architecture [3] addresses the problem that network outages and congestion result in poor performance of IP routing and long recovery time due to slow convergence of Internet routing protocols such as BGP. RON uses active probing and passive monitoring in a fully meshed topology in order to detect network problems within seconds. Experimental results from a small real-world deployment of a RON have been obtained demonstrating fast recovery from failure and improved latency and loss rates. Note that the authors do not claim that their results are representative of anything other than their deployment and no general results for different topologies, increased RON traffic, etc., have been published.

The Detour [4] framework pointed out several routing inefficiencies in the Internet and mainly attributed them to poor routing metrics, restrictive routing policies, manual load balancing, and single-path routing. By comparing actual routes with measurement traces between hosts, Savage et al. found that in almost every case there would have been a better alternative path. They envision an overlay network based on IP tunnels to prototype new routing algorithms on top of the existing Internet; however, they concede that measurement-based adaptive routing can lead to instability and, to the best of our knowledge, no evaluation of the overlay performance has been published.

A different application of overlay networks is content-based navigation in peer-to-peer networks. The goal of content-addressable networks such as Chord, CAN [5], Tapestry, and Pastry [6] is efficient, fault-tolerant routing, object location and load-balancing within a self-organizing overlay network. These approaches provide a scalable fault-tolerant distributed hash table enabling item location within a small number of hops. These overlay networks exploit network proximity in the underlying Internet. Most use a separate address space to reflect network proximity.

Although these overlay networks have been shown to work in some specific cases, no extensive simulations or practical measurements on a wide range of topologies have been carried out.

3   Architectural Overview

In this section we briefly outline the overlay architecture which we evaluate by simulation. The overlay network consists of a set of booster boxes interconnected by IP tunnels. Packets are forwarded across virtual links, i.e. the IP tunnels, using normal IP routing. The IP routers are not modified and are entirely unaware of the existence of the overlay network.

Booster boxes that are directly connected across a virtual link are called peers. Note that a single virtual link may correspond to multiple IP paths. Booster boxes peer with other booster boxes with which they are likely to have good connectivity. This is determined using pathchar [7] and/or packet tailgating [8]. Pathchar provides more information than packet tailgating about the entire path but has more restrictive assumptions. Both techniques are known to fail beyond a certain threshold number of hops, because of error amplification. We therefore restrict the hop count of the virtual links.

Although the establishment of a virtual link between two booster boxes is asymmetrical, both sides must agree to the peering. We use this, together with the fact that boosters box have good knowledge of the links to which they are directly attached, to determine the accuracy of the link measurement.

The techniques for determining link characteristics require the transmission of a large number of packets and accordingly take a significant amount of time to determine a result. They are adequate for the construction of the overlay network but not for the transient state links measurements used to make forwarding decisions. Booster boxes, on the other hand, measure the current latency and loss probability of the virtual links by periodically exchanging network probes with their peers. This is similar to the Network Weather Service described in [9].

Booster boxes maintain the overlay network forwarding tables using a link-state routing protocol. If the link state on a booster box changes significantly, the forwarding tables are recomputed. Packets are forwarded between booster boxes using classical encapsulation techniques.

4   Overlay-Network Simulation

The simulation process contains the following steps. First, we generate a physical network topology using the Brite [10] topology generator. The result is a graph consisting of nodes that represent autonomous systems (ASs) and of edges that represent network links with certain capacities and delays. In a second step, we populate the network with applications with four sources sending a constant stream of packets to a single sink using UDP as the transport protocol; this corresponds to a sensor-type application. To generate "background" traffic and thus congestion we add several TCP sources. The result of this step is a TCL script that is fed into the NS-2 network simulator [11]. Then, we create an overlay network by adding booster boxes to the network topology. Finally, a small fraction of the applications are reconfigured such that they send traffic over the overlay network. The result of this last step is another TCL script, executable by the NS-2 network simulator.

The traffic generated by the applications is analyzed. In particular, we are interested in the average packet-drop ratio, i.e. the ratio of dropped packets to the total number of packets sent. For a given topology, we obtain three results. The first is used as reference and is obtained when no overlay network is present. With an overlay network present, we obtain a second drop ratio of those applications that do not use the overlay, and a third result which is the drop ratio of those applications that use the overlay network. Figure 1 shows the four steps involved and the resulting two experiments.

Physical Network Topology. We generate random topologies using the Waxman model with parameters set to $\alpha = 0.9$ and $\beta = 0.2$. A high value for α was chosen to prioritize local connections. The ratio of nodes to links is 1:2. The link capacity varies randomly from 1 to 4 Mb/s and link propagation delay in the range of 1 to 10 ms. We consider network sizes of 100, 200, and 400 ASs. These are rather small compared to the Internet but we are constrained by the performance of the NS-2 tool. It would perhaps be more realistic to use the power law [12] rule of Internet ASs, but our networks are too small for this to be feasible.

We assume the physical topology to be invariant during a run of the simulation, i.e. nodes and links do not fail and therefore no dynamic routing protocol is needed. Given the fact that routing convergence in the Internet using BGP

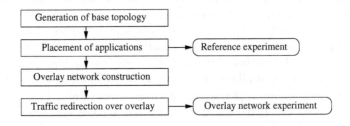

Fig. 1. Steps involved in a single simulation run

is rather slow [13] compared with the convergence time of the link-state routing protocol used in the overlay network, it would be expected that the overlay would react to failure more quickly than the physical network does.

Overlay Network Topology. The overlay network is constructed by adding booster boxes to the ASs with the highest degree of connectivity. This is to ensure that booster boxes are mainly placed in transit ASs. Booster boxes construct virtual links to the four closest neighboring booster boxes. Closeness is measured in terms of hop count, and the path capacity is used as a tie-breaker in the case of equal hop counts. In our simulation model, an AS is equipped with a booster box by creating an additional node of type "booster box" and connecting it to a single AS node using a high-capacity link of 100 Mb/s and a latency of 100 μs. Inter-AS communication has much longer delay and is more susceptible to packet loss than AS-booster box communication. We carry out our simulation across a small range of booster/AS ratios: 1:10, 1:5, and 1:2.

Traffic Characterization. Each application consists of four sources sending a constant bit-rate stream of UDP packets to a single sink. The bit rates of the applications follow a normal distribution with mean of 250 kb/s and standard deviation of 50 kb/s. The packet size is 576 bytes for all applications, as this is the predominant data packet size in the Internet. The background TCP traffic has exponentially distributed burst length. The idle time is also exponentially distributed with the same mean as the burst length. To reflect the diurnal nature of Internet traffic we introduce periods where the number of background TCP traffic sources is double. In different experiments, we vary the burst length of the background traffic such that their mean is either 1 ms, 10 ms, 100 ms, or 1000 ms. Figure 2 shows the effect of the diurnal traffic model over a link that carries 10 TCP streams with burst length 1 ms. The application and background traffic sources and sinks are allocated at the edge of the network. This is done by randomly distributing the sources and sinks among the 60% of the ASs that have the lowest connectivity. The number of application sources is 0.4 times the number of nodes, for background traffic this factor is 5. We do not attempt a realistic characterization of traffic produced by an AS, but simply try to ensure that congestion occurs at arbitrary times and for different periods.

Ratio of Overlay Traffic. A small fraction of traffic produced by the applications is sent over the overlay network. In our experiments, this ratio ranged from 5% to 12%. As the applications themselves only produce 5% of all the traffic, the ratio of overlay traffic to the total traffic is in the range of 0.3% to 0.6%.

Measurement of the Dynamic Metrics. In the experiments the cost of an overlay link is a linear function of the TCP smoothed RTT (SRTT) as measured between each booster box and its peers. Using the SRTT prevents the link cost from oscillating wildly. As the RTT is not updated using retransmitted packets [14] when TCP times out, and when therefore potentially a packet has been lost, we set the cost of the link to a much higher value than any observed RTT; in this way more weight is attached to loss than delay. We send a single 50-byte

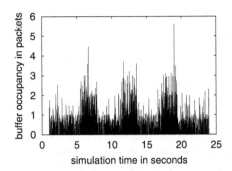

Fig. 2. Buffer occupancy per time for the diurnal traffic model

probe every 500 ms between peered booster boxes. For reasons of simplicity, the simulation does not use a predictive model for attempting to identify future values of the SRTT such as those described in [9].

Frequency of Virtual Routing Exchanges. The booster-box routing agents check every second their link-state values. They exchange link state updates with their peers if the costs associated with their links have changed.

Simulation Scenarios. We divided the experiments into three sets:

- *Light congestion*: 500 background traffic sources using a normally distributed bit rate with a mean of 250 kb/s and a standard deviation of 50 kb/s and 40 application sources are used. Only the data generated by four of the application sources is routed through the overlay network. In this scenario we have a maximum of 4% packet loss in the reference experiments.
- *Heavy congestion*: the mean rate and the standard deviation of the background traffic is doubled to 500 kb/s and 100 kb/s respectively. Here the packets losses in the reference experiment reach 36%. While such high losses are unusual, they have been observed on backbone routers [15].
- *Low overlay usage*: the same as *light congestion* except that for the applications using the overlay network, only the data generated by two of its sources is routed through the overlay network.

For each set we run ten experiments varying the number of booster boxes in the network and the burst length of the background traffic. Each experiment simulates 24 seconds of network operation. A different topology is generated for each experiment.

5   Results

We do not observe significant variation in the results owing to the different topologies sizes — perhaps due to the fact that they are only scale orders of difference — and therefore we only present those for 100 nodes. Tables 5 shows the experimental results for the three sets. The tables contain the following information for each booster box ratio and burst length couple:

- The column labeled "Ovl vs Ref" shows the percentage of the experiments in which the loss ratio of the traffic using the overlay network is smaller than that of the traffic in the reference experiments.
- The column labeled "Norm vs Ref" shows the percentage of the experiments in which the loss ratio of the traffic not using the overlay network is smaller than that of the traffic in the reference experiments.
- The three last columns show the average loss ratio and, in parenthesis, the standard deviation of all the experiments for traffic using the overlay network; traffic not using the overlay network, and traffic in the reference experiments.

We do not think the mean of the packet drop ratio is representative due to the high variance of the results. More interesting is the number of experiments which show a benefit. For example, the first row of the table reports an average drop rate of 8.5% for the overlay traffic, but only a drop rate of 3.6% for the reference traffic. On the other hand, in 40% of the experiments, the overlay traffic had a lower drop rate.

An overlay network is *beneficial* when the overlay traffic behaves better than the reference traffic and the non-overlay traffic behaves no worse than the reference traffic. An overlay network is *partially beneficial* when the overlay traffic behaves better than the reference traffic but the non-overlay traffic behaves worse than the reference traffic. An overlay network is *detrimental* when the overlay traffic behaves worse than the reference traffic.

All three cases are observed in the results. Our belief is that the detrimental behavior is due to an aggregation effect, in which flows that would have taken different paths over the physical network are forced to take the same one owing to lack of an alternate path in the overlay network. This causes unnecessary congestion and is strongly dependent on both physical and overlay network topologies.

We suppose that the additional traffic needed for overlay routing and measurement is the cause of the partially beneficial results. The traffic not using the overlay network is effected by this overhead without deriving any benefits from it. This effect is worsened by the fact that exchanges of link-state advertisements occur most often in the case of congestion.

The beneficial case is when the overlay succeeds in recognizing and routing around congestion without disrupting other traffic. In some cases both the traffic using the overlay network and that not using it benefit from the overlay, simply because they are routed through different paths.

In general, the highest benefits from the overlay network are observed in the sets of experiments with heavy congestion and low overlay usage. In these two cases the benefit seems to increase the more booster boxes there are. This is not true for the light congestion case. Another remark is that the benefit does not seem to be radically effected by the burstiness of the background traffic. We expected to see a significant difference as the burst length increases allowing more time for the overlay to detect the congestion and react. That this was not so could be due to an artifact of the relative durations of the congestion burst and measurement times. However, it is very difficult to characterize the circumstances in which the overlay is beneficial, as we observe a high variance in the experimental results. This fact is depicted in Fig. 3 for the three simulation scenarios. It compares the drop ratios of the overlay network traffic with those of the reference traffic. A "+1" indicates a better result for the overlay case, a "0" the case where both are equal within ± 5%, and a "-1" the case where the reference traffic had a lower drop ratio. A clear tendency can only be observed in the heavy congestion and to a lesser extend the low overlay usage scenarios

Table 1. Simulation results

#BBoxes	Burst	Ovl vs Ref	Norm vs Ref	Drops Ovl		Drops Norm		Drops Ref	
light congestion									
10	1	40	40	8.5	(6.7)	3.1	(0.7)	3.6	(1.8)
10	10	30	60	8.1	(6.9)	2.5	(0.7)	3.1	(1.9)
10	100	50	70	7.8	(6.9)	2.2	(0.8)	2.8	(1.9)
10	1000	50	60	6.4	(6.2)	1.9	(0.7)	2.5	(1.9)
20	1	80	10	3.8	(6.4)	4.3	(1.7)	4.0	(1.7)
20	10	80	20	3.0	(6.1)	3.8	(1.8)	3.4	(1.8)
20	100	90	20	2.4	(5.8)	3.3	(2.0)	3.0	(1.9)
20	1000	90	30	2.5	(6.0)	3.0	(1.8)	2.8	(1.7)
50	1	70	10	3.7	(2.6)	4.0	(1.7)	3.4	(1.6)
50	10	40	0	3.4	(2.4)	3.4	(1.8)	2.9	(1.6)
50	100	50	20	3.0	(2.7)	2.9	(1.9)	2.5	(1.6)
50	1000	50	20	2.6	(2.3)	2.6	(1.8)	2.2	(1.5)
heavy congestion									
10	1	80	10	20.8	(13.0)	35.9	(5.0)	33.5	(4.8)
10	10	70	0	14.3	(8.4)	20.9	(4.6)	19.2	(4.2)
10	100	50	10	14.0	(8.1)	17.1	(4.5)	15.6	(3.8)
10	1000	80	0	12.0	(7.4)	17.1	(4.9)	15.7	(4.0)
20	1	80	30	26.9	(16.5)	37.4	(7.0)	36.0	(7.3)
20	10	60	20	18.7	(16.0)	22.1	(6.7)	21.2	(7.0)
20	100	60	20	17.2	(14.8)	18.5	(6.4)	17.5	(6.4)
20	1000	60	40	16.3	(14.1)	17.8	(6.1)	17.4	(6.0)
50	1	100	60	13.7	(6.6)	36.1	(6.8)	35.3	(5.6)
50	10	100	40	5.7	(3.4)	21.2	(6.2)	20.8	(5.7)
50	100	100	40	5.5	(3.0)	17.8	(5.9)	17.2	(5.4)
50	1000	100	40	5.2	(2.3)	18.0	(5.8)	17.5	(5.3)
low overlay usage									
10	1	50	20	5.3	(4.0)	3.9	(2.1)	3.9	(2.3)
10	10	40	10	5.6	(4.1)	4.0	(2.4)	3.7	(2.3)
10	100	50	10	4.9	(4.2)	3.5	(2.3)	3.3	(2.2)
10	1000	60	20	3.7	(3.7)	2.9	(2.0)	2.8	(2.0)
20	1	40	20	4.3	(3.8)	3.4	(1.8)	3.2	(1.7)
20	10	40	10	3.9	(3.2)	3.0	(1.9)	2.7	(1.8)
20	100	40	10	3.7	(4.0)	2.5	(1.9)	2.2	(1.7)
20	1000	40	20	3.1	(3.5)	2.2	(1.8)	1.9	(1.7)
50	1	90	0	0.9	(1.0)	4.4	(1.6)	3.4	(1.3)
50	10	90	0	0.7	(0.6)	3.8	(1.7)	2.9	(1.4)
50	100	90	0	0.7	(0.9)	3.0	(1.6)	2.3	(1.4)
50	1000	80	0	0.5	(0.7)	3.0	(1.7)	2.1	(1.4)

when a large number of booster boxes is used. In the other cases, the behavior exhibits no clear trend.

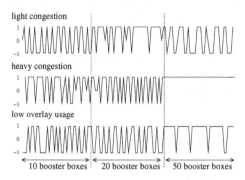

Fig. 3. High Variance in Simulation Results

A distinction can be made between two types of parameters that influence the performance of the overlay: those under the control of the booster-box operators, such as frequency of measurements, routing, or peering strategy, and those not under the control of the booster-box operators, such as network topology or pattern of background traffic. For the first set it is feasible that extensive simulation for precise scenarios would allow useful heuristics to be derived, e.g. never send more than 10% of the total traffic over the overlay. The second are, in general, unknown to the operator. This leads us to conclude that overlays of the type described here need to be reactive, i.e. they need to test the network state and only be activated when they can bring benefit and deactivated otherwise.

While it would be unwise to attach too much importance to simulations that might produce very different results by the simple modification of one parameter, the results show that in the situations tested overlaying can cause significant deterioration of the network. The scenarios may or may not be realistic; however, more simulations and modeling are necessary to better understand the behavior of overlays before they can be deployed.

6   Conclusion

We have outlined an architecture for measurement-based overlay networks that allows certain traffic flows to be privileged over others. We present results from simulation showing how this architecture might behave in the public Internet. We found that while the overlay can be beneficial, it often is detrimental. The circumstances under which the undesired behavior occurs are difficult to characterize and seem very sensitive to small changes in the parameters. As some of these parameters are not under the control of the overlay supervisors, we conclude that such overlays need to be reactive. As a final remark we suggest that proponents of overlay networks need to investigate the effect of their deployment, not only in simple, idealized scenarios, but on the network as a whole.

References

1. Cleary, J., Graham, I., McGregor, T., Pearson, M., Ziedins, I., Curtis, J., Donnelly, S., Martens, J., Martin, S.: High Precision Traffic Measurement. IEEE Communications Magazine **40** (2002) 167–183
2. Bauer, D., Rooney, S., Scotton, P.: Network Infrastructure for Massively Distributed Games. In: NetGames 2002 – First Workshop on Network and System Support for Games, Braunschweig, Germany (2002)
3. Andersen, D.G., Balakrishnan, H., Kaashoek, M.F., Morris, R.: Resilient Overlay Networks. In: Proc. 18th ACM Symposium on Operating Systems Principles, Banff, Canada (2001)
4. Savage, S., Anderson, T., Aggarwal, A., Becker, D., Cardwell, N., Collins, A., Hoffman, E., Snell, J., Vahdat, A., Voelker, G., Zahorjan, J.: Detour: a Case for Informed Internet Routing and Transport. Technical Report TR-98-10-05, University of Washington (1998)
5. Ratnasamy, S., Francis, P., Handley, M., Karp, R., Shenker, S.: A Scalable Content Addressable Network. In: ACM SIGCOMM. (2001) 161–172
6. Rowstron, A., Druschel, P.: Pastry: Scalable, distributed object location and routing for large-scale peer-to-peer systems. In: IFIP/ACM International Conference on Distributed Systems Platforms (Middleware). (2001) 329–350
7. Jacobson, V.: How to Infer the Characteristics of Internet Paths. Presentation to Mathematical Sciences Research Institute (1997) ftp://ftp.ee.lbl.gov/pathchar/msri-talk.pdf.
8. Lai, K., Baker, M.: Measuring link bandwidths using a deterministic model of packet delay. In: ACM SIGCOMM 2000, Stockholm, Sweden. (2000) 283–294
9. Wolski, R., Spring, N., Hayes, J.: The Network Weather Service: A Distributed Resource Performance Forecasting Service for Metacomputing. In: Journal of Future Generation Computing Systems. (1998)
10. Medina, A., Lakhina, A., Matta, I., Byers, J.: BRITE: Universal Topology Generation from a User's Perspective. Technical Report BUCS-TR2001 -003, Boston University (2001)
11. Fall, K., Varadhran, K.: The ns manual (formerly ns Notes and Documentation. http://www.isi.edu/nsnam/ns/ns-documentation.html (2002)
12. Faloutsos, M., Faloutsos, P., Faloutsos, C.: On Power-Law Relationships of the Internet Topology. In: ACM SIGCOMM, Harvard University, Cambridge, US (1999) 251–262
13. Labovitz, C., Ahuja, A., Bose, A., Jahanian, F.: Delayed Internet Routing Convergence. In: ACM SIGCOMM 2000, Stockholm, Sweden. (2000) 175–187
14. Karn, P., Partridge, C.: Improving Round-Trip Time Estimates in Reliable Transport Protocols. Computer Communications Review **17** (1987) 2–7
15. Floyd, S., Paxson, V.: Difficulties in Simulating the Internet. IEEE/ACM Transactions on Networking **9** (2001) 392–403

Using Redistribution Communities
for Interdomain Traffic Engineering*

Bruno Quoitin, Steve Uhlig, and Olivier Bonaventure

Infonet group
University of Namur, Belgium
(bqu,suh,obo)@infonet.fundp.ac.be
http://www.infonet.fundp.ac.be

Abstract. In this paper, we present a traffic engineering technique that can be used by regional ISPs and customer networks. On the basis of various characterizations of ASes in today's Internet we show the requirements of the small ASes. Then we detail the methods that such ASes currently use to engineer interdomain traffic. We present an analysis of real routing tables showing that a lot of ISPs rely on the BGP Community attribute to build scalable traffic engineering configurations. We also show that this solution suffers from several drawbacks that limit its widespread utilization. To avoid the problems of such a technique, we propose the redistribution communities, a special type of non transitive extended community attribute and show that the cost of implementing this solution is small.

1 Introduction

Initially developed as a research network, the Internet has been optimized to provide a service where the network does its best to deliver packets to their destination. In the research Internet, connectivity was the most important issue. During the last years, we have seen a rapid growth and an increasing utilization of the Internet to carry business critical services such as e-commerce, Virtual Private Networks and Voice over IP. To efficiently support those services, several Internet Service Providers (ISP) rely on traffic engineering techniques to better control the flow of IP packets.

During the last years, several types of traffic engineering techniques have been developed [ACE+01]. Most of these techniques have been designed for large IP networks that need to optimize the flow of IP packets inside their internal network. These techniques are of very limited use for small IP networks that constitute most of the Internet today. For these networks, the costly resource that needs to be optimized is usually their interdomain connectivity. In this paper, we try to fill this gap by proposing a simple technique that can be used to provide useful traffic engineering capabilities targeted at, but not limited to, those small ISPs.

This document is organized as follows. We first discuss in section 2 the requirements for implementable interdomain traffic engineering techniques targeted at small ISPs. Then, we describe in section 3 the existing interdomain traffic engineering techniques. In section 4, we describe the redistribution communities that can be used to solve most of the traffic engineering needs of small ISPs.

---

* This work was supported by the European Commission within the IST ATRIUM project.

B. Stiller et al. (Eds.): QofIS/ICQT 2002, LNCS 2511, pp. 125–134, 2002.
© Springer-Verlag Berlin Heidelberg 2002

2 Interdomain Traffic Engineering for Small ISPs

The Internet is currently composed of about 13.000 Autonomous Systems (AS) [Hus02] and its organization is more complex than the research Internet of the early nineties. Those 13.000 AS do not play an equal role in the global Internet. ASes can be distinguished on the basis of various characteristics like the connectivity one AS has with its peers, the services provided by one AS to its peers and the behaviour of the users inside the networks of one AS.

First, ASes can be distinguished on the basis of their connectivity. [SARK02] has shown that there are two major types of interconnections between distinct ASes: the *customer-provider* and the *peer-to-peer* relationships. The *customer-provider* relationship is used when a small AS purchases connectivity from a larger AS. In this case, the large AS agrees to forward the packets received from the small AS to any destination and it also agrees to receive traffic destined to the small AS. On the other hand, the *peer-to-peer* relationship is used between ASes of similar size. In this case, the two ASes exchange traffic on a shared cost basis. According to [SARK02], the *customer-provider* relationship is used for about 95 % of the AS interconnections in today's Internet.

Relying on this connectivity, [SARK02] makes a first characterization of ASes. There are basically two types of ASes: transit ASes that consitute the core of the Internet and regional ISPs or customer networks. The core corresponds to about 10 % of the ASes in the Internet and can be divided in three different subtypes (*dense, transit* and *outer core* depending on the connectivity of each AS). Regional ISPs and customer networks correspond to 90 % of the Internet and they maintain only a few *customer-provider* relationships with ASes in the core and some *peer-to-peer* relationships with other small ASes.

In this paper, we do not address the traffic engineering needs of ASes in the core but rather requirements of small ASes. The interested reader is referred to [FBR02] for a discussion of the needs of ASes in the core.

A second important element used to characterize an AS is the type of customer it serves. If the AS is mainly a content provider, it will want to optimize its outgoing traffic since it generates more traffic than it receives. On the other hand, if the AS serves a population of SMEs (Small and Medium Enterprises), dialup, xDSL or cable modems users, it will receive more traffic than it sends. Such ASes will typically only need to control their incoming traffic.

Another point to consider is the "topological distribution" of the interdomain traffic to be engineered [UB02a]. Although the Internet is composed of about 13.000 ASes, a given AS will not receive (resp. transmit) the same amount of traffic from (resp. towards) each external AS. The characteristics of the interdomain traffic seen from a customer AS have been analyzed in details in [UB02b]. In this paper, we have analysed the characteristics of all the interdomain traffic received by two small ISPs based on traces collected during one week. The first trace was collected at the interdomain routers of BELNET, an ISP providing access to universities and research labs in Belgium in December 1999. The second trace was collected during one week in April 2001 at the interdomain routers of YUCOM, a Belgian ISP providing a dialup access to the Internet. This study revealed two important findings that are summarized in figure 1. First, the left part of the figure shows the percentage of the number of IP addresses that are reachable from the BGP routers of the studied ASes at a distance of x AS hops. This figure shows that for both studied ASes, most reachable IP addresses are ony a few AS hops away. Second, the right part of figure 1 shows the cumulative distribution of the traffic sent by each external AS during the studied week. The figure shows that for both ASes, a

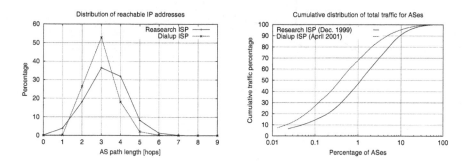

Fig. 1. BGP routing tables (left) and cumulative distribution of total traffic (right)

small percentage of external ASes contribute to a large fraction of the incoming traffic. Hence, by influencing this limited percentage of ASes a large fraction of the traffic can be engineered. Similar findings were reported in [FBR02] for an AS of the dense core.

3   Interdomain Traffic Engineering Today

In this section, we review the traffic engineering techniques that are in use today in the global Internet. Since these techniques rely on a careful tuning of the BGP routing protocol, we first briefly review its operation.

3.1   Interdomain Routing

The Border Gateway Protocol (BGP) [Ste99, RL02] is the current de facto standard interdomain routing protocol. BGP is a *path-vector protocol* that works by sending *route advertisements*. A route advertisement indicates the reachability of one IP network through the router that advertises it either because this network belongs to the same AS as this router or because this router has received from another AS a route advertisement for this network. Besides the reachable network, each route advertisement also contains attributes such as the AS-Path which is the list of all the transit ASes that must be used to reach the announced network.

A key feature of BGP is that it supports routing policies. That is, BGP allows a router to be selective in the route advertisements that it sends to neighbor BGP routers in remote AS. This is done by specifying on each BGP router a set of input and output filters for each peer.

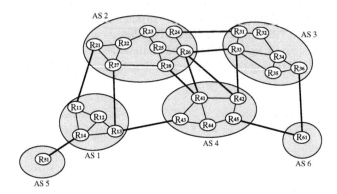

Fig. 2. A simple Internet

3.2   BGP-based Traffic Engineering

The BGP-based traffic engineering solutions in utilization today rely on a careful tuning of the BGP decision process [1] that is used to select the best-route towards each destination. This process is based on a set of criteria that act as filters among all the BGP routes known by the router.

Control of the Outgoing Traffic   The control of the outgoing traffic is often a requirement for content providers that wish to optimize the distribution of their content. For this, they can rely on the weight and the `local-pref` attribute to control the routes that will be chosen for the packets that leave each BGP router of the content provider. The actual distribution of the outgoing traffic will depend on the quality of the setting of the weight and the `local-pref` on the BGP routers of the AS. The setting of these two parameters can be done manually based on the knowledge of the interdomain links or automatically with tools that rely on traffic measurements.

Control of the Incoming Traffic   A customer AS serving a large number of individual users or small corporate networks will typically have a very assymetric interdomain traffic pattern with several times more incoming than outgoing traffic. These ASes typically need to optimize their incoming traffic only. For this, a first method that they can use is to announce different route advertisements on different links. For example in figure 2, if AS1 wanted to balance the traffic coming from AS2 over the links $R_{11} - R_{21}$ and $R_{13} - R_{27}$, then it could announce only its internal routes on the $R_{11} - R_{21}$ link and only the routes learned from AS5 on the $R_{13} - R_{27}$ link. Since AS2 would only learn about AS5 through router R_{27}, it would be forced to send the packets whose destination belongs to AS5 via router R_{27}.

---

[1] Due to space limitations, we cannot detail the BGP decision process in this paper. A description of the BGP decision process may be found in [FBR02, Hal97, QUPB02].

A variant of the selective advertisements is the advertisement of more specific prefixes. This advertisement relies on the fact that an IP router will always select in its forwarding table the most specific route for each packet (i.e. the matching route with the longest prefix). This fact can also be used to control the incoming traffic. In the following example, we assume that prefix 16.0.0.0/8 belongs to AS3 and that several important servers are part of the 16.1.2.0/24 subnet. If AS3 prefers to receive the packets towards its servers on the R_{24}-R_{31} link, then it would advertise both 16.0.0.0/8 and 16.1.2.0/24 on this link and only 16.0.0.0/8 on its other external links. An advantage of this solution is that if link R_{24}-R_{31} fails, then subnet 16.1.2.0/24 would still be reachable through the other links. However, an important drawback of advertising more specific prefixes is that it increases the number of BGP advertisements and thus the size of the BGP routing tables ([BNC02]).

Another method would be to allow an AS to indicate a ranking among the various route advertisements that it sends. Based on the BGP decision process, one possible way to introduce a ranking between routes to influence the selection of routes by a distant AS is to artificially increase the length of the AS-Path attribute. Coming back to our example, AS1 would announce the routes learned from AS5 on links $R_{11} - R_{21}$ and $R_{13} - R_{27}$, but would attach a longer AS-Path attribute (e.g. AS1 AS1 AS1 AS5 instead of AS1 AS5) on the $R_{13}-R_{27}$ link. The required amount of prepending is often manually selected on a trial and error basis. The manipulation of the AS-Path attribute is often used in practice ([BNC02]). However, it should be noted that this technique is only useful if the ASes that we wish to influence do not rely on local-pref and weight.

Community-Based Traffic Engineering In addition to these techniques, several ASes have been using the BGP Community attribute to encode various traffic engineering actions [QB02]. This attribute is often used to add markers to announced routes and to simplify the implementation of scalable routing policies on BGP routers. The community attribute is a transitive attribute that contains a set of community values, each value being encoded as a 32 bits field. Some community values are standardized (e.g. NO_EXPORT), but the Internet Assigned Numbers Authority (IANA) has assigned to each AS a block of 65536 community values. The community values are usually represented as ASx:V where ASx is the AS number to which the community belongs and V a value assigned by ASx. The community attribute is often used to encode the following traffic engineering actions [QB02]:

1. Do not announce the route to specified peer(s);
2. Prepend n times the AS-Path (where we have found values for n generally ranging from 1 to 3) when announcing the route to specified peer(s);
3. Set the local-pref value in the AS receiving the route [CB96];

In the first case, the community is attached to a route to indicate that this route should not be announced to a specified peer or at a specified interconnection point. For example, in figure 2, AS4 could configure its routers to not announce to AS1 routes that contain the 4:1001 community. If AS4 documents the utilization of this community to its peers, AS6 could attach this value to the routes advertised on the R_{45}-R_{61} ink to ensure that it does not receive packets from AS1 on this link.

The second type of community is used to request the upstream AS to perform AS-Path prepending for the associated route. To understand the usefulness of such community values, let us consider again figure 2, and assume that AS6 receives a lot of traffic from AS1 and AS2 and that it would like to receive the packets from AS1 (resp. AS2) on the R_{45}-R_{61} (resp. R_{36}-R_{61}) link. AS6 cannot achieve this type of traffic distribution by performing prepending itself. However, this would be possible if AS4 could perform the prepending when announcing the AS6 routes to external peers. AS6 could thus advertise to AS4 its routes with the community 4:5202 (documented by AS4) that indicates that this route should be prepended two times when announced to AS2.

Finally, the third common type of community used for traffic engineering purposes is to set the local-pref in the upstream AS.

Our analysis of the RIPE whois database [QB02] provides more details on the utilization of the community attribute to request a peer to perform path prepending, to set the local-pref attribute and to not redistribute the route. The survey indicates that the specified peer is usually specified as an AS number, an interconnection point or a geographical region.

4 Redistribution Communities

The community based traffic engineering solution described in the previous section has been deployed by at least twenty different ISPs, but it suffers from several important drawbacks that limit its widespread utilization. First, each AS can only define 65536 distinct community values. While in practice no AS today utilizes more than 65536 community values, this limited space forces each AS to define its own community values in an unstructured manner. [2] Second, each defined value must be manually encoded in the configurations of the BGP routers of the AS. Third, the AS must advertise the semantics of its own community values to external peers. Unfortunately, there is no standard method to advertise these community values. Some ASes define their communities as comments in their routing policies that are stored in the Internet Routing Registries. The RPSL language [AVG+99] used for these specifications does not currently allow to define the semantic of the community attribute values. Other ASes publish the required information on their web server or distribute it directly to their clients. This implies that an AS willing to utilize the traffic engineering communities defined by its upstream ASes needs to manually insert directives in the configurations of its BGP routers. Once inserted, these directives will need to be maintained and checked if the upstream AS decides for any reason to modify the semantics of some of its community values. This increases the complexity of the configuration of the BGP routers and is clearly not a desirable solution. A recent study has shown that human errors are already responsible

---

[2] We note however that facing the need for structured community values, some ASes like AS9057 have started to utilize community values outside their allocated space [QB02] and that other ASes are using community values reserved for standardization. This kind of selfish behavior is questionable in today's Internet, but it shows the operationnal need for more structured community values.

for many routing problems on the global Internet [MWA02]. An increasing utilization of community-based traffic engineering would probably cause even more errors.

A second drawback of the BGP community attribute is its transitivity. It implies that once a community value has been attached to a route, this community is distributed throughout the global Internet. To evaluate the impact of the communities on the growth of the BGP tables [Hus02], we have analyzed the BGP routing tables collected by RIPE RIS [RIS02] and the Route Views projects [Mey02][3] from January 2001 until now. A first observation of those BGP table dumps reveals that although most of the community values have a local semantics [QB02], a large number of community values appear in the BGP routing tables.

The evolution of the utilization of the communities reveals a sustained growth since the availability of the first dumps with community information in January 2001 (see figure 3). For instance, in recent dumps of routing tables provided by Route-Views

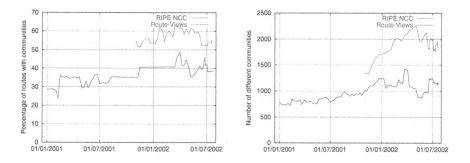

Fig. 3. Evolution of the utilization of the community attribute.

([Mey02]) at the beginning of the year 2002, the number of communities has increased to more than 2200 distinct values while more than 60% of the routes had at least one community attached and some routes can have up to 40 communities attached ! We could see the same evolution at other sites.

4.1   The Redistribution Communities

To avoid the problems caused by the utilization of the community attribute, we propose a new type of extended community attribute. The extended community attribute defined in [STR02] provides a more structured and larger space than the community attribute since each extended community value is encoded in an 8 octets field. The redistribution communities are non-transitive extended communities that can be used to encode a set of redistribution actions that are applicable to a set of BGP speakers. The current definition of the redistribution communities [BCH+02] supports the following actions:

- the attached route should not be announced to the specified BGP speakers.
- the attached route should only be announced to the specified BGP speakers.

---

[3] The Route-Views project started in November 2001

- the attached route should be announced with the NO_EXPORT attribute to the specified BGP speakers.
- the attached route should be prepended n times when announced to the specified BGP speakers.

Each redistribution community is encoded as an 8 octets field divided in three parts. The first octet is used to specify the type of non-transitive extended community [STR02]. The second octet is used to encode one of the four actions above and the last 6 octets encode a BGP_Speakers_Filter that determines the BGP speakers to which the action applies.

The BGP_Speakers_Filter field is used to specify the eBGP speakers that are affected by the specified action. There are two methods to specify the affected eBGP speakers. The first method is to explicitly list all those BGP speakers inside the BGP_Speakers_-Filters field of redistribution communities. In this case, the high order bit of the BGP_-Speakers_Filter field is set to 1. The second method is to explicitly list only the eBGP speakers that will not be affected by the specified action. In this case, the high order bit of the BGP_Speakers_Filter type field shall be set to 0. In the current specification [BCH+02], the BGP_Speakers_Filter can contain an AS number, two AS numbers or a CIDR prefix/length pair.

4.2   Implementation of the Redistribution Communities

In order to evaluate the cost of supporting the redistribution communities in a BGP router, we have modified the Zebra BGP implementation [Ish01]. The implementation of the redistribution communities requires two distinct functionnalities. The first one is to allow a network operator to specify the redistribution communities that must be attached to given routes and the second one is to influence the redistribution of the routes that have such communities attached.

First, in order to allow a network operator to attach redistribution communities to routes, we have extended the route-map statement available in the command-line interface (CLI) of Zebra. The route-map statement is an extremely powerful and versatile tool for route filtering and attribute manipulation that is composed of a filter and a list of actions. Our extension consists in the addition of a new action that can be used to attach a list of redistribution communities to routes that match the route-map filter. An example of a route-map using our new action is given below. The example presents the configuration in routers of AS6. This configuration attaches a redistribution community to every route announced to AS4. This community requests that AS4 prepend 2 times the AS-PATH of routes announced by AS6 when redistributing to AS2 (see example in section 1).

```
neighbor <as4-neighbor-ip> route-map prepend2_to_as2
route-map prepend2_to_as2 permit 10
  match ip address any
  set extcommunity red prepend(2):as(2)
```

Then, we have modified zebra so that redistribution communities are automatically taken into account. The implementation extracts the redistribution communities attached to the route and on the basis of their content, decides to attach the NO_EXPORT community, to prepend n times or to ignore the route when redistributing to specified peers. These modifications in the source code of Zebra were quite limited compared to the amount of work required to configure by hand redistribution policies similar to what redistribution communities provide. For instance, to establish the same configuration as shown above with a manual setup of communities, a lot more work is required.

5 Perspectives

Compared to the utilization of classical community values, the main advantages of the redistribution communities is that they are non-transitive and have a standardized semantics. The non-transitivity suppresses the risk of community-based pollution of routing tables while the standardized encoding allows simplification of the configurations of BGP routers and thus reduces the risk of errors [MWA02]. Furthermore, this will allow operators to provide services that go beyond the simple *customer-provider* and *peer-to-peer* policies currently found in today's Internet.

For example, BGP-based Virtual Private Networks [RR99] rely on communities to indicate where the VPN routes should be redistributed. The redistribution communities could be used to significantly reduce the configuration complexity of interdomain VPNs.

The redistribution communities could also be used to reduce the impact of denial of service attacks. For example, assume that in figure 2, AS6 suffers from an attack coming from sources located inside AS2. In order to reduce the impact of the attack, AS6 would like to stop announcing its routes towards AS2. With the standard BGP techniques, this is not possible while maintaining the connectivity towards the other ASes. With the redistribution communities, AS6 simply needs to tag its routes with a community indicating that they should not be redistributed towards AS2. If the attack originated from AS5, the redistribution communities would not allow AS6 to stop advertising its routes without also blocking traffic from sources like AS7. However, in this case, AS4 and AS1 might also detect the denial of service attack and could react with the redistribution communities.

In this paper, we have proposed a solution that allows an AS to influence the redistribution of its routes. It is nevertheless difficult to use a similar technique to influence the route redistribution farther than two AS hops away due to the variety and the complexity of the routing policies that might be found in the global Internet. However this is a first step towards a global interdomain level traffic engineering solution, which is our ultimate goal.

Acknowledgements

We would like to thank Russ White, Stefaan De Cnodder and Jeffrey Haas for their help in the development of the redistribution communities. The traffic traces were provided by Benoit Piret and Marc Roger. We thank, RIPE for their whois database and their RIS project and Route Views for their routing tables.

References

[ACE+01]   D. Awduche, A. Chiu, A. Elwalid, I. Widjaja, and X. Xiao. Overview and principles of internet traffic engineering. Internet draft, draft-ietf-tewg-principles-02.txt, work in progress, December 2001.

[AVG+99]   C. Alaettinoglu, C. Villamizar, E. Gerich, D. Kessens, D. Meyer, T. Bates, D. Karrenberg, and M. Terpstra. Routing Policy Specification Language (RPSL). Internet Engineering Task Force, RFC2622, June 1999.

[BCH+02]   O. Bonaventure, S. De Cnodder, J. Haas, B. Quoitin, and R. White. Controlling the redistribution of bgp routes. Internet draft, draft-ietf-ptomaine-bgp-redistribution-00.txt, work in progress, April 2002.

[BNC02]   A. Broido, E. Nemeth, and K. Claffy. Internet expansion, refinement and churn. *European Transactions on Telecommunications*, January 2002.

[CB96]   E. Chen and T. Bates. An Application of the BGP Community Attribute in Multihome Routing. Internet Engineering Task Force, RFC1998, August 1996.

[FBR02]   N. Feamster, J. Borkenhagen, and J. Rexford. Controlling the impact of BGP policy changes on IP traffic. Toronto, June 2002. Presented at NANOG25.

[Hal97]   B. Halabi. *Internet Routing Architectures*. Cisco Press, 1997.

[Hus02]   G. Huston. AS1221 BGP table statistics. available from http://www.telstra.net/ops/bgp/, 2002.

[Ish01]   K. Ishiguro. Gnu zebra – routing software. available from http://www.zebra.org, 2001.

[Mey02]   D. Meyer. Route Views Archive project. University of Oregon, http://archive.routeviews.org, January 2002.

[MWA02]   R. Mahajan, D. Wetherall, and T. Anderson. Understanding BGP misconfigurations. In *ACM SIGCOMM 2002*, August 2002.

[QB02]   B. Quoitin and O. Bonaventure. A survey of the utilization of the bgp community attribute. Internet draft, draft-quoitin-bgp-comm-survey-00.txt, work in progress, March 2002.

[QUPB02]   B. Quoitin, S. Uhlig, C. Pelsser, and O. Bonaventure. Internet traffic engineering techniques. Technical report, 2002. http://www.infonet.fundp.ac.be/doc/tr.

[RIS02]   Routing Information Service project. Rseaux IP Europens, http://www.ripe.net/ripencc/pub-services/np/ris-index.html, January 2002.

[RL02]   Y. Rekhter and T. Li. A border gateway protocol 4 (bgp-4). Internet draft, draft-ietf-idr-bgp4-17.txt, work in progress, January 2002.

[RR99]   E. Rosen and Y. Rekhter. BGP/MPLS VPNs. Request for Comments 2547, Internet Engineering Task Force, March 1999.

[SARK02]   L. Subramanian, S. Agarwal, J. Rexford, and R. Katz. Characterizing the internet hierarchy from multiple vantage points. In *INFOCOM 2002*, June 2002.

[Ste99]   J. Stewart. *BGP4 : interdomain routing in the Internet*. Addison Wesley, 1999.

[STR02]   S. Sangli, D. Tappan, and Y. Rekhter. Bgp extended communities attribute. Internet draft, draft-ietf-idr-bgp-ext-communities-05.txt, work in progress, May 2002.

[UB02a]   S. Uhlig and O. Bonaventure. Implications of interdomain traffic characteristics on traffic engineering. *European Transactions on Telecommunications*, January 2002.

[UB02b]   S. Uhlig and O. Bonaventure. A study of the macroscopic behavior of internet traffic. under submission, available from http://www.infonet.fundp.ac.be/doc/tr, January 2002.

A Multi-path Routing Algorithm for IP Networks Based on Flow Optimisation*

Henrik Abrahamsson, Bengt Ahlgren, Juan Alonso, Anders Andersson, and
Per Kreuger

SICS – Swedish Institute of Computer Science
first.lastname@sics.se

Abstract. Intra-domain routing in the Internet normally uses a single
shortest path to forward packets towards a specific destination with no
knowledge of traffic demand. We present an intra-domain routing algorithm based on multi-commodity flow optimisation which enable load
sensitive forwarding over multiple paths. It is neither constrained by
weight-tuning of legacy routing protocols, such as OSPF, nor requires
a totally new forwarding mechanism, such as MPLS. These characteristics are accomplished by aggregating the traffic flows destined for the
same egress into one commodity in the optimisation and using a hash
based forwarding mechanism. The aggregation also results in a reduction of computational complexity which makes the algorithm feasible for
on-line load balancing. Another contribution is the optimisation objective function which allows precise tuning of the tradeoff between load
balancing and total network efficiency.

1 Introduction

As IP networks are becoming larger and more complex, the operators of these
networks gain more and more interest in *traffic engineering* [3]. Traffic engineering encompasses performance evaluation and performance optimisation of
operational IP networks. An important goal with traffic engineering is to use the
available network resources more efficiently for different types of load patterns
in order to provide a better and more reliable service to customers.

Current routing protocols in the Internet calculate the shortest path to a
destination in some metric without knowing anything about the traffic demand
or link load. Manual configuration by the network operator is therefore necessary
to balance load between available alternate paths to avoid congestion. One way
of simplifying the task of the operator and improve use of the available network
resources is to make the routing protocol sensitive to traffic demand. Routing
then becomes a flow optimisation problem.

One approach taken by others [8, 9, 12] is to let the flow optimisation result in a set of link weights that can be used by legacy routing protocols, e.g.,
open shortest path first (OSPF), possibly with equal cost multi-path (ECMP)

---

* Supported in part by Telia Research AB.

B. Stiller et al. (Eds.): QofIS/ICQT 2002, LNCS 2511, pp. 135–144, 2002.

forwarding. The advantage is that no changes are needed in the basic routing protocol or the forwarding mechanism. The disadvantage is that the optimisation is constrained by what can be achieved with tuning the weights. Another approach is to use MPLS [4], multi-protocol label switching, for forwarding traffic for large and long-lived flows. The advantage is that the optimisation is not constrained, but at the cost of more complexity in the routing and forwarding mechanisms.

Our goal is to design an optimising intra-domain routing protocol which is *not* constrained by weight-tuning, and which *can* be implemented with minor modifications of the legacy forwarding mechanism based on destination address prefix.

In this paper we present a routing algorithm for such a protocol based on multi-commodity flow optimisation which is both computationally tractable for on-line optimisation and also can be implemented with a near-legacy forwarding mechanism. The forwarding mechanism needs a modification similar to what is needed to handle the ECMP extension to OSPF.

The key to achieve this goal, and the main contribution of this paper, is in the modelling of the optimisation problem. We aggregate all traffic destined for a certain egress into one commodity in a multi-commodity flow optimisation. This reduces the number of commodities to at most N, the number of nodes, instead of being N^2 when the problem is modelled with one commodity for each pair of ingress and egress nodes. As an example, the computation time for a 200 node network was in one experiment 35 seconds. It is this definition of a commodity that *both* makes the computation tractable, *and* the forwarding simple.

Another important contribution is the definition of an optimisation objective function which allows the network operator to choose a maximum desired link utilisation level. The optimisation will then find the most efficient solution, if it exists, satisfying the link level constraint. Our objective function thus enables the operator to control the trade-off between minimising the network utilisation and balancing load over multiple paths.

The rest of the paper is organised as follows. In the next section we describe the overall architecture where our optimising routing algorithm fits in. Section 3 presents the mathematical modelling of the optimisation problem. We continue with a short description of the forwarding mechanism in Sect. 4. After related work in Sect. 5 we conclude the paper.

2   Architecture

In this work we take the radical approach to completely replace the traditional intra-domain routing protocol with a protocol that is based on flow optimisation. This approach is perhaps not realistic when it comes to deployment in real networks in the near future, but it does have two advantages. First, it allows us to take full advantage of flow optimisation without being limited by current practise. Second, it results in a simpler overall solution compared to, e.g., the metric tuning approaches [8, 9, 12]. The purpose of taking this approach is to assess its feasibility and, hopefully, give an indication on how to balance flow optimisation functionality against compatibility with legacy routing protocols.

In this section we outline how the multi-commodity flow algorithm fits into a complete routing architecture. Figure 1 schematically illustrates its components. Flow measurements at all ingress nodes and the collection of the result are new components compared to legacy routing. The measurements continuously (at regular intervals) provide an estimate of the current demand matrix to the centralised flow optimisation. The demand matrix is aggregated at the level of all traffic from an ingress node destined for a certain egress node.

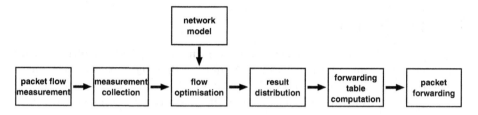

Fig. 1. Routing architecture with flow optimisation.

If a more fine-grained control over the traffic flows are desired, for instance to provide differentiated quality of service, a more fine-grained aggregation level can be chosen. This results in more commodities in the optimisation, which can be potential performance problem. One approach is to introduce two levels in the optimisation, one with a longer time-scale for quality of service flows.

The demand matrix is input to the flow optimiser together with a model of the network. The result of the optimisation is a set of values y_{ij}^t, which encode how traffic arriving at a certain node (i), destined for a certain egress node (t) should be divided between the set of next hops (j). These values are used at each node together with a mapping between destination addresses and egress nodes to construct forwarding tables. Finally, the packet forwarding mechanism is modified to be able to distinguish packets destined for a certain egress node, and to forward along multiple paths toward those egresses.

The computation of the multi-commodity flow optimisation algorithm is inherently centralised. In this paper we also think of the computation as implemented in a central server. If a so-called bandwidth broker is needed or desired for providing a guaranteed quality of service, it is natural to co-locate it with optimisation. We however see the design of a distributed mechanism implementing flow optimisation as an important future work item.

The timescale of operation is important in an optimising routing architecture. There are several performance issues that put lower bounds on the cycle flow measurement–optimisation–new forwarding tables. The flow measurement need to be averaged over a long enough time to get sufficiently stable values. Our current research as well as others [5] indicate that the needed stability exists in real networks at the timescale of a few, maybe five to ten, minutes. Other performance issues are the collection of the flow measurements, the computation of the optimisation algorithm, and the distribution of the optimisation result.

Our initial experiments indicate that a new optimisation cycle can be started in approximately each five minutes for typical intra-domain sizes.

An issue that we have identified is how to handle multiple egresses for a destination injected into the domain by BGP, the border gateway protocol. A straightforward way to solve this is to introduce additional virtual nodes in the network to represent a common destination behind both egresses. This approach may however introduce a large number of additional nodes. This will need to be more carefully considered in the future.

3   Optimisation

The routing problem in a network consists in finding a path or multiple paths that send traffic through the network without exceeding the capacity of the links. When using optimisation to find such (multiple) paths, it is natural to model the traffic problem as a (linear) multi-commodity network flow problem (see, e.g., Ahuja et al. [1]), as many authors have done.

First, the network is modelled as a directed graph (this gives the topology, i.e., the static information of the traffic problem), and then the actual traffic situation (i.e., the dynamic part of the problem, consisting of the current traffic demand and link capacity) as a linear program. In modelling the network as a graph, a node is associated to each router and a directed edge to each directional link physically connecting the routers. Thus, we assume a given graph $G = (N, E)$, where N is a set of nodes and E is the set of (directed) edges. We will abuse language and make no distinction between graph and network, node and router, or edge and link.

Every edge $(i, j) \in E$ has an associated capacity k_{ij} reflecting the bandwidth available to the corresponding link. In addition, we assume a given *demand matrix* $D = D(s, t)$ expressing the traffic demand from node s to node t in the network. This information defines the routing problem. In order to formulate it as a multi-commodity flow (MCF) problem we must decide how to model commodities. In the usual approach [1, 8, 11] commodities are modelled as source-destination pairs that are interpreted as "all traffic from source to destination". Thus, the set of commodities is a subset of the Cartesian product $N \times N$; consequently, the number of commodities is bounded by the square of the number of nodes. To reduce the size of the problem and speed-up computations, we model instead commodities as (only destination) nodes, i.e., a commodity t is to be interpreted as "all traffic to t". Thus, our set of commodities is a subset of N and, hence, there are at most as many commodities as nodes. The corresponding MCF problem can be formulated as follows:

$$\min \{ f(y) \,|\, y \in P_{12} \} \qquad\qquad (MCF_{12})$$

where $y = (y_{ij}^t)$, for $t \in N, (i, j) \in E$, and P_{12} is the polyhedron defined by the equations:

$$\sum_{\{j \,|\, (i,j) \in E\}} y_{ij}^t - \sum_{\{j \,|\, (j,i) \in E\}} y_{ji}^t = d(i, t) \qquad \forall i, t \in N \qquad (1)$$

$$\sum_{t \in N} y_{ij}^t \leq k_{ij} \qquad \forall (i,j) \in E \qquad (2)$$

where

$$d(i,t) = \begin{cases} -\sum_{s \in N} D(s,t) & \text{if } i = t \\ D(i,t) & \text{if } i \neq t \end{cases}.$$

The variables y_{ij}^t denote the amount of traffic to t routed through the link (i,j). The equation set (1) state the condition that, at intermediate nodes i (i.e., at nodes different from t), the outgoing traffic equals the incoming traffic plus traffic created at i and destined to t, while at t the incoming traffic equals all traffic destined to t. The equation set (2) state the condition that the total traffic routed over a link cannot exceed the link's capacity.

It will also be of interest to consider the corresponding problem *without* requiring the presence of the equation set (2). We denote this problem (MCF_1). Notice that every point $y = (y_{ij}^t)$ in P_{12} or P_1 represents a possible solution to the routing problem: it gives a way to route traffic over the network so that the demand is met and capacity limits are respected (when it belongs to P_{12}), or the demand is met but capacity limits are not necessarily respected (when it belongs to P_1). Observe that $y = (0)$ is in P_{12} or in P_1 only in the trivial case when the demand matrix is zero.

A general linear objective function for either problem has the form $f(y) = \sum_{t,(i,j)} b_{ij}^t y_{ij}^t$. We will, however, consider only the case when all $b_{ij}^t = 1$ which corresponds to the case where all commodities have the same cost on all links. We will later use different objective functions (including non-linear ones) in order to find solutions with desired properties.

3.1   Desirable Solutions

In short, the solutions we consider to be desirable are those which are *efficient* and *balanced*. We make these notions precise as follows.

We use the objective function considered above, $f(y) = \sum_{t,(i,j)} y_{ij}^t$, as a measure of efficiency. Thus, given y_1, y_2 in P_{12} or P_1, we say that y_1 is *more efficient* than y_2 if $f(y_1) \leq f(y_2)$. To motivate this definition, note that whenever traffic between two nodes can be routed over two different paths of unequal length, f will choose the shortest one. In case the capacity of the shortest path is not sufficient to send the requested traffic, f will utilise the shortest path to 100% of its capacity and send the remaining traffic over the longer path.

Given a point $y = (y_{ij}^t)$ as above, we let $Y_{i,j} = \sum_{t \in N} y_{ij}^t$ denote the total traffic sent through (i,j) by y. Every such y defines a *utilisation* of edges by the formula $u(y, i, j) = Y_{ij}/k_{ij}$, and $u(y, i, j) = 0$ when $k_{ij} = 0$. Let $u(y)$ denote the maximum value of $u(y, i, j)$ where (i,j) runs over all edges. Given an $\ell > 0$, we say that $y \in P_{12}$ (or $y \in P_1$) is ℓ-*balanced* if $u(y) \leq \ell$. For instance, a solution is (0.7)-balanced if it never uses any link to more than 70 % of its capacity.

3.2   How to Obtain Desirable Solutions

Poppe et al. [11] have proposed using different linear objective functions in order to obtain traffic solutions that are desirable with respect to several criteria (including balance, in the form of minimising the maximum utilisation of edges). Fortz and Thorup [8, 9], on the other hand, considers a fixed piece-wise linear objective function (consisting of six linear portions for each edge) which makes the cost of sending traffic along an edge depend on the utilisation of the edge. By making the cost increase drastically as the utilisation approaches 100 %, the function favours balanced solutions over congested ones. As the authors express it, their objective function "provides a general best effort measure".

Our contribution is related to the above mentioned work in that we use different objective functions to obtain desirable solutions, and the functions are piece-wise linear and depend on the utilisation. In contrast, our work defines different levels of balance (namely, ℓ-balance). For each such level, a simple piece-wise linear objective function consisting of two linear portions for each edge is *guaranteed* to find ℓ-balanced solutions provided, of course, that such solutions exist. Moreover, the solution found is guaranteed to be more efficient than any other ℓ-balanced solution.

Another distinctive feature of our functions is that they are defined through a uniform, theoretical "recipe" which is valid for every network. We thus eliminate the need to use experiments to adapt our definitions and results to each particular network. Finally, the fact that our functions consist of only two linear portions, shorten the execution time of the optimisation.

3.3   The Result

To formulate our result we need to introduce some notation. Let $y = (y_{ij}^t)$ be a point of P_{12} or P_1, and suppose given real numbers $\lambda > 1$ and $\ell > 0$. We define the link cost function (illustrated in Fig. 2)

$$C^{\ell,\lambda}(U) = \begin{cases} U & \text{if } U \leq \ell \\ \lambda\,U + (1-\lambda)\,\ell & \text{if } U \geq \ell \end{cases}.$$

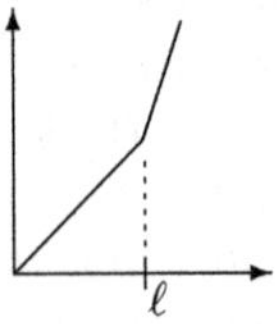

Fig. 2. The link cost function $C^{\ell,\lambda}$.

We use this function in the definition of the following objective function:

$$f^{\ell,\lambda}(y) = \sum_{(i,j)\in E} k_{ij}\,C^{\ell,\lambda}(u(y,i,j))$$

We also need to define the following constants:

$$v = \min \{f(y) \,|\, y \in P_{12}\} \qquad \text{and} \qquad V = \max \{f(y) \,|\, y \in P_{12}\}$$

Notice that $v > 0$ since $D(s,t) > 0$, and $V < \infty$ since the network is finite and we are enforcing the (finite) capacity conditions. At a more practical level, v can be computed by simply feeding the linear problem $\min \{f(y) \,|\, y \in P_{12}\}$ into CPLEX and solving it. Then, to compute V, one changes the same linear problem to a max problem (by replacing "min" by "max") and solves it.

Finally, let $\delta > 0$ denote the minimum capacity of the edges of positive capacity. We can now state the following theorem whose proof is given in a technical report [2]:

Theorem 1. *Let ℓ, ϵ be real numbers satisfying $0 < \ell < 1$ and $0 < \epsilon < 1 - \ell$. Suppose that $y \in P_1$ is ℓ-balanced, and let $\lambda > 1 + \frac{V^2}{v\delta\epsilon}$. Then any solution x of MCF_1 with objective function $f^{\ell,\lambda}$ is $(\ell + \epsilon)$-balanced. Moreover, x is more efficient than any other $(\ell + \epsilon)$-balanced point of P_1.*

Observe that, since $\ell < 1$ and $y \in P_1$ is ℓ-balanced, we can use MCF_1 instead of MCF_{12}. Informally, the theorem says that if there are ℓ-balanced solutions, then $f^{\ell,\lambda}$ will find one. The number $\epsilon > 0$ is a technicality needed in the proof. Notice that it can be chosen arbitrarily small.

Theorem 1 can be used as follows. Given a target utilisation ℓ, say $\ell = 0.7$, compute $\frac{V^2}{v\delta\epsilon}$, choose a λ as in Theorem 1, and choose $\epsilon > 0$, say $\epsilon = 0.01$. Finally, compute a solution, say x, of MCF_1 with objective function $f^{\ell,\lambda}$. Then there are two exclusive possibilities: either x is 0.71-balanced or there is no such solution. In the last case, x can be thought of as a "best effort" solution since we have penalised all utilisation above 0.7 (which forces traffic using edges to more than 70 % of capacity to try to balance) but no 0.71-balanced solution exists. At this point we can either accept this best effort solution or iterate, this time setting the balance target to, say, 0.85, etc. After a few iterations we arrive at a solution which is "sufficiently" balanced or we know that there is no solution that is ℓ-balanced for the current value of ℓ which, we may decide, is so close to 1 that it is not worthwhile to continue iterating.

3.4 A Generalisation

Theorem 1 has a useful generalisation that can be described as follows. Partition the set of edges E into a family (E_i) of subsets, and choose a target utilisation ℓ_i for each E_i. The generalised theorem says that for small $\epsilon > 0$ we can define a function corresponding to $f^{\ell,\lambda}$ in Theorem 1, such that solving MCF_1 with this objective function will result in efficient solutions that are $(\ell_i + \epsilon)$-balanced on E_i provided, of course, that such solutions exist. The generalised theorem is more flexible in that it allows us to seek solutions with different utilisation in different parts of the network.

3.5   Quantitative Results

We have used CPLEX 7.1[1] on a Pentium laptop to conduct numerical experiments with a graph representing a simplified version of a real projected network. The graph has approximately 200 nodes and 720 directed edges. If we had modelled MCF with source-destination pairs as commodities, the linear problem corresponding to MCF_{12} would consist of some 8 million equations and 30 million variables. Modelling commodities as traffic to a node, MCF_{12} contains, in contrast, "only" about 40 000 constraints and 140 000 variables. Solving MCF_1 with objective function $f^{\ell,\lambda}$ takes approximately 35 seconds.

Solving the same problem with the objective function considered by Fortz and Thorup [8, 9] takes approximately 65 seconds. Our experiments suggest that this function picks solutions that minimise balance. In contrast, with $f^{\ell,\lambda}$ we can choose any desired level of balance (above the minimum, of course).

4   Multi-path Forwarding

By modelling the routing problem as "all traffic to t", as described in the previous section, we get an output from the optimisation that is well suited for packet forwarding in the routers. The result from the optimisation, the y_{ij}^t values, tells how packets at a certain node (i) to a certain egress node (t) in the network should be divided between the set of next hops (j). We thus need a forwarding mechanism that can distinguish packets destined for a certain egress, and that can forward along multiple paths.

To enable forwarding along multiple paths, we introduce one more step in the usual forwarding process. An egress data structure is inserted in the address lookup tree just above the next hop data structure as illustrated in Fig. 3. A longest prefix match is done in the same manner as in a standard forwarding table, except that it results in the destination egress node. The egress data structure stores references to the set of next hops to which traffic for that egress should be forwarded, as well as the desired ratios (the y_{ij}^t for all js) between the next hops.

In order to populate the forwarding tables a mapping has to be created between destination addresses and egress nodes. The needed information is the same as a regular intra-domain routing protocol needs, and is obtained in much the same way. For destinations in networks run by other operators (i.e., in other routing domains), the mapping is obtained from the BGP routing protocol. For intra-domain destinations, the destination prefix is directly connected to the egress node.

Mechanisms for distributing traffic between multiple links have been thoroughly evaluated by Cao et al. [6]. We propose to use a table based hashing mechanism with adaptation, because it can distribute the load according to unequal ratios, is simple to compute, and adapts to the properties of the actual traffic.

Similar mechanisms already exist in commercial routers in order to handle the equal cost multi-path extension to OSPF and similar protocols.

---

[1] ILOG CPLEX 7.1 http://www.ilog.com

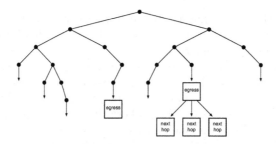

Fig. 3. Address lookup data structure for multiple path forwarding.

5   Related Work

With the prospect of better utilising available network resources and optimising traffic performance, a lot of research activity is currently going on in the area of traffic engineering. The general principles and requirements for traffic engineering are described in the RFC 3272 [3] produced by the IETF Internet Traffic Engineering working group. The requirements for traffic engineering over MPLS are described in RFC 2702 [4].

Several researchers use multi-commodity flow models in the context of traffic engineering. Fortz and Thorup [8, 9] use a local search heuristics for optimising the weight setting in OSPF. They use the result of multi-commodity flow optimisation as a benchmark to see how close to optimal the OSPF routing can get using different sets of weights. Mitra and Ramakrishnan [10] describes techniques for optimisation subject to QoS constraints in MPLS-supported IP networks. Poppe et al. [11] investigate models with different objectives for calculating explicit routes for MPLS traffic trunks. Multi-commodity flow and network flow models in general have numerous application areas. A comprehensive introduction to network flows can be found in Ahuja et al. [1].

A somewhat controversial assumption when using multi-commodity flow optimisation is that an estimate of the demand matrix is available. The problem of deriving the demand matrix for operational IP networks is considered by Feldmann et al. [7]. The demand matrix only describes the current traffic situation but, for an optimisation to work well, it must also be a good prediction of the near future. Current research in traffic analysis by Bhattacharyya et al. [5] and Feldmann et al. [7] indicate that sufficient long term flow stability exists on backbone links in timescales of minutes and hours and in manageable aggregation levels to make optimisation feasible.

6   Conclusions

We have taken the first steps to introduce flow optimisation as a routing mechanism for an intra-domain routing protocol. We have presented a routing algorithm based on multi-commodity flow optimisation which we claim is computationally tractable for on-line routing decisions and also only require a small modification to the legacy packet forwarding mechanism. More work is however needed on other components in order to design and implement a complete routing protocol using our algorithm.

The key issue, and our main contribution, is the mathematical modelling of commodities. Traffic destined for a certain egress node is aggregated into a single commodity. This results in computational requirements an order of magnitude smaller than in the traditional models where the problem is modelled with one commodity for each flow from one ingress to one egress node.

Multi-path forwarding of the aggregates produced by the optimiser is then handled by a hash based forwarding mechanism very similar to what is needed for OSPF with ECMP.

Another contribution is the design of a generic objective function for the optimisation which allows the network operator to choose a desired limit on link utilisation. The optimisation mechanism then computes a most efficient solution given this requirement, when possible, and produces a best effort solution in other cases. The process can be iterated with, e.g., binary search to find a feasible level of load balance for a given network load.

References

[1] R. K. Ahuja, T. L. Magnanti, and J. B. Orlin. *Network flows*. Prentice-Hall, 1993.

[2] J. Alonso, H. Abrahamsson, B. Ahlgren, A. Andersson, and P. Kreuger. Objective functions for balance in traffic engineering. Technical Report T2002:05, SICS – Swedish Institute of Computer Science, May 2002.

[3] D. Awduche, A. Chiu, A. Elwalid, I. Widjaja, and X. Xiao. Overview and principles of Internet traffic engineering. Internet RFC 3272, May 2002.

[4] D. Awduche, J. Malcolm, J. Agogbua, M. O'Dell, and J. McManus. Requirements for traffic engineering over MPLS. Internet RFC 2702, September 1999.

[5] S. Bhattacharyya, C. Diot, J. Jetcheva, and N. Taft. Pop-level and access-link-level traffic dynamics in a tier-1 pop. In *ACM SIGCOMM Internet Measurement Workshop*, San Francisco, USA, November 2001.

[6] Z. Cao, Z. Wang, and E. Zegura. Performance of hashing-based schemes for internet load balancing. In *Proc. of IEEE INFOCOM 2000*, Israel, March 2000.

[7] A. Feldmann, A. Greenberg, C. Lund, N. Reingold, J. Rexford, and F. True. Deriving traffic demands for operational IP networks: Methodology and experience. In *Proceedings of ACM SIGCOMM'00*, Stockholm, Sweden, August 2000.

[8] B. Fortz and M. Thorup. Internet traffic engineering by optimizing OSPF weights. In *Proceedings IEEE INFOCOM 2000*, pages 519–528, Israel, March 2000.

[9] B. Fortz and M. Thorup. Optimizing OSPF/IS-IS weights in a changing world. *IEEE Journal on Selected Areas in Communications*, 20(4):756–767, May 2002.

[10] D. Mitra and K. G. Ramakrishnan. A case study of multiservice, multipriority traffic engineering design for data networks. In *Proc. of Globecom'99*, Brazil, 1999.

[11] F. Poppe, S. van den Bosch, P. de la Vallée-Poussin, H. van Hove, H. de Neve, and G. Petit. Choosing the objectives for traffic engineering in IP backbone networks based on Quality-of-Service requirements. In *Proceedings of First COST 263 International Workshop, QofIS*, pages 129–140, Berlin, Germany, Sept. 2000.

[12] D. Yuan. *Optimization Models and Methods for Communication Network Design and Routing*. PhD thesis, Linköpings Universitet, 2001.

Proactive Multi-path Routing[1]

Jing Shen[*], Jiaoying Shi[*], Jon Crowcroft[+]

[*]State Key Lab of CAD&CG, ZheJiang University, P.R.China
{jshen,jyshi}@cad.zju.edu.cn
[+]Computer Lab, University of Cambridge, UK
Jon.Crowcroft@cl.cam.ac.uk

Abstract. Internet service provider faces a daunting challenge in provisioning network efficiently. We introduce a proactive multipath routing scheme that tries to route traffic according to its built-in properties. Based on mathematical analysis, our approach disperses incoming traffic flows onto multiple paths according to path qualities. Long-lived flows are detected and migrated to the shortest path if their QoS could be guaranteed there. Suggesting non-disjoint path set, four types of dispersion policies are analyzed, and flow classification policy which relates flow trigger with link state update period is investigated. Simulation experiments show that our approach outperforms traditional single path routing significantly.

1. Introduction

Provisioning ISP's network efficiently is a challenge for current technology. The difficulties root in two aspects: the rapid growth of network and the exponential growth of network usage. Although people have been working with network routing for years and overprovision are widely used by ISPs, there are times that a part of network resource is insufficient to deal with its load while adjacent part is idled.

To make full use of network resource, traffic dispersion has been an active research area in both circuit switching and packet switching network [2][3][4][5][6][7][8]. It is shown that network performance could be improved by spreading traffic equally onto multiple paths. In other related work, alternative routing has been investigated extensively[9][10][11][12][13], which proposes to improve QoS by rerouting traffic over a precomputed alternative path if the shortest path is congested.

Based on the research mentioned above, this paper addresses one of key issues in traffic transmission, i.e., how could traffic be routed proactively over network according to their built-in properties. We focus on networks with non-uniformed traffic pattern and those traffic exhibits correlation with a high variance over long time period. We propose routing scheme that routes traffic according to its built-in properties while making full use of network performance. The idea is to disperse incoming traffic flows proactively onto multiple paths and to optimize routing paths for long-lived flows according to QoS and load balancing requirements.

---

[1] This work was supported by China National Science Foundation for Innovative Colony under Contract No. 60021201.

B. Stiller et al. (Eds.): QofIS/ICQT 2002, LNCS 2511, pp. 145-156, 2002.

Section 2 outlines the proactive multipath routing scheme on the basis of theoretical analysis. Section 3 follows with policies for path establishment, traffic dispersion and rerouting of long-lived flows. The benefit of our proactive approach is illustrated in section 4 through detailed simulation. Section 5 concludes the paper with summary of our work.

2. Routing with Traffic Characteristics

In this section, we argue that routing mechanism should be built with careful consideration on traffic characteristics, we shows that proactive multipath routing can improve network performance effectively.

2.1 The Challenge for Internet Traffic Dispersion

People have been working with network routing for years. Most effort was expanded on path finding and on distributed routing algorithm. If we take routing system development as practice of complex system design, such research will not lead to the optimal solution for Internet traffic transmission, because properties of possible load is overlooked. Given an existing network infrastructure, routing mechanism should be designed with careful consideration on traffic characteristics, network connectivity and administration requirements.

It has been found that, internet traffic exhibits long-term correlations with high variance in packet arrivals, and connection size has a so-called "heavy-tail" distribution: most flows are short and small but the number which are very long tend to contribute the majority of traffic[17]. Defining burst as flows whose lifetime is shorter than the time scale of link state advertisements, Xun[8] invoked a theoretical analysis to find best way of bursty traffic transmission. In this model, a pair of ingress and egress nodes are interconnected via a set of n links with the same capacity c, bursty flows arrive at source node with λ_{sd} and average duration time is μ^{-1}. It is demonstrated that, minimum of network congestion could be reached by spreading traffic equally onto k paths according to one of the following equations:

$$\alpha^* = \min\left\{\sqrt{\frac{\lambda}{\mu c}\frac{(1-e^{-\mu t})}{e^{-\mu t}}}, 1\right\} \tag{1}$$

$$y^* = \min\left\{\sqrt{\frac{\lambda c}{\mu}\frac{(1-e^{-\mu t})}{e^{-\mu t}}}, c\right\} \tag{2}$$

Where, $\lambda = \lambda_{sd}/k$, $k = \alpha^* \cdot n$ or k indicates a set of links whose load are below the threshold y^*. Although Xun's model is not to the real situation in Internet, it gives a useful starting point. That is, network node should route its connection requests actively over a large set of paths if these requests are mostly short and arrive frequently, or direct its connection requests to the least loaded path if these requests are mostly long and arriving infrequently. To Internet traffic, this means: the large amount of short flows should be spread onto multiple paths, while those long-lived flows should be transmitted along the least burdened path. As most path computing algorithms relate path length with path load, the least burdened path usually means the "shortest" path.

In fact, the multiple paths for traffic dispersion need not to be totally disjoint. There are three reasons: 1) the goal for multipath routing is to improve network performance by making full use of network resource, when a new loopless path is calculated, it is ensured that new resource will be introduced even though it is not totally disjoint from existing paths; 2) if traffic is dispersed onto several disjoint paths between source and destination, it cannot be guaranteed that there is no overlapping between paths for different source destination pairs; 3) if we constraint multipath routing to disjoint paths, nodes with only one outgoing link will always fail in finding multiple paths even when there is idle resource.

Some other problems exist. Firstly, how could the path set keep up with time varying network state? Since link state propagates by relaying between network nodes, staleness in link state information is unavoidable, but the frequency of link state advertisement do have effect on path computation. To reflect network dynamics, path computing needs to be scheduled periodically or triggered by some conditions. On the other hand, the path set should not be modified frequently to avoid frequent route alternation and unreasonable signaling overhead.

Secondly, how should traffic be dispersed onto multiple paths with different quality? Since it's unavoidable that overlapping happens between logical paths, load on some links may accumulate in hot spots. On the other hand, great difference between traffic flows may derive to sharp variation in link state. To optimize network performance, the path assignment policy should be designed with full consideration of path quality, service quality and load balancing requirements.

2.2 Dispersing Traffic Proactively

To address these requirements, we propose a proactive mutlipath routing scheme that tries to optimize routes for long-lived flows on the basis of dispersity routing. Our approach improves network performance in three critical ways:

- **More resource for traffic transmission**: The amount of resource for traffic transmission is increased by establishing multiple paths between source and destination, dispersity routing restraints the transient overload caused by bursty flows. Thus, the requirement for router's buffer size may be reduced as well.
- **Flow distribution is optimized**: Migration of long-lived flows focuses the "quasi-static" part of network burden onto the shortest path, while bursty traffic is spread onto a large set of transmission links. This way, load distribution and resource utilization are optimized.
- **Fewer link state update message**: Load division lowers the additive effects of bursty flows, and the relative targeting of long-lived flows restrains the link state variation along the shortest path. The result is, links are kept in relatively stable state and link state advertisements can be scheduled with long period.

To exploit these potential benefits, two key design decisions were adopted:

- **Spreading traffic flows according to path quality**
  Considering both path overlapping and self-similarity in internet traffic, it cannot be expected that each path provides service with the same quality or that the resource consumed by traffic flows coincides with logical paths. To achieve a bal-

ance between QoS and traffic engineering, flow level traffic is spread across a set of paths according to their quality which is evaluated periodically.

- **Migrating long-lived flows only if their quality-of-service could be guaranteed**

   There are two possible methods to improve transmission quality of long-lived flows: 1.) reserving resource for long-lived flows; 2.) optimizing routes for long-lived flows on the basis of traffic dispersion. Although trunk reservation for long-lived flows could improve successful rate of flow migration, it is not adopted by our scheme. Three reasons contribute to this decision: 1.) if a part of network resource is reserved (hard or soft) for either long-lived flows or short flows, dispersity routing should be done according to the state of resource allocated to bursty traffic, it's difficult to predict how much resource should be reserved and the overhead of state maintenance will be high; 2.) if trunk reservation is proposed along the shortest path, it is difficult to predict whether a link carries a shortest-path; 3.) by choosing path quality metric and dispersing policy carefully, most of long-lived flows could be assigned to the shortest path from the start. Considering the requirement of QoS provisioning,long-lived flows are migrated only if their bandwidth requirement could be guaranteed.

Another critical problem is the number of paths. Xun[8] suggested a dynamic path set whose length is limited by $(1+\beta^*) \cdot l_{shortest}$, where β^* is a parameter varying with flow arrival rate, flow duration time and link update period. Great variation in Internet traffic makes it very difficult to implement this scheme, while random selection may affect the effectiveness of our scheme. Considering the fact that most of current networks have a diameter less than 10 and node degree is usually below 4, two or three recommended value may be reasonable.

3. Traffic Dispersion and Rerouting of Long-Lived Flows

This section describes details of proactive multipath routing, including k paths maintenance, traffic dispersion policy and rerouting of long-lived flows.

3.1 Computing and Establishing k Paths

According to discussion in section two, routing paths should be computed on the state of link capacity. There exist many types of link metrics, including bandwidth, delay, loss rate etc. In our approach, we choose the bandwidth related metric because it contains the fundamental information of throughput. In addition, it has been shown[1] that some other quality-of-service requirements can be converted to bandwidth requirement, and extensions to current routing protocols have been proposed to carry bandwidth information in LSAs. Getting link state information, routing paths could be computed easily by using an extended Dijkstra algorithm or a labeling algorithm at ingress node[16].

The k paths could be established by creating label switching paths in MPLS networks, which populates the forwarding tables in routers along the path. In this scenario, edge routers invoke explicit signaling along the path computed by using CR-LDP or RSVP-TE protocol. If network consists of ATM switches, the k paths can be established by connection signaling similar to MPOA. The path set is maintained in switch routing tables until refresh or recomputing happens.

In order to keep up with time varying network status, it's valuable to recompute the k paths periodically, but frequent computing and signaling may introduce unreasonable overhead and lead to frequent route alternation. To balance these factors, path computing should be scheduled with periods which are at least several times of long-lived flows' maximum lifetime, e.g. one or two days.

3.2 Path Evaluation and Traffic Dispersion

There may be great difference between the k paths for traffic dispersion. Evaluation result of the same path may also differ greatly if different metric is applied. As our approach focuses on improving network performance, only those metrics emphasizing TE&QoS requirements are considered.

With the first type of metric, each path is considered to have the same quality. This may be reasonable if the load on each path equals and their end-to-end transmission capacity is the same. Spreading traffic with this metric, each path should be assigned a same number of flows. Particularly, the outgoing path for an incoming flow could be chosen from the path set by random selection, hash based selection or round robin. We do not recommend hash selection based on header information, because it may introduce load accumulation on a particular path if most flows are between the same subnets. The most outstanding advantage of this metric is, no computing load is introduced. We denote this dispersion policy "random".

With the second type of metric, we focus on end-to-end transmission capacity. As throughput is determined by bottleneck bandwidth, it is chosen as path quality metric. That is, the wider the bottleneck is, the more traffic should be transferred along the path. In order to reflect difference between k paths, weight is assigned to each path by dividing its quality value with the minimum of those sharing the same source and destination. Ingress nodes use weighted round robin as path assignment algorithm. We denote this type of method "wrr-bw".

Using the third and the forth metric, we focus on path quality in end-to-end operational regime. Considering traffic routing as the multiplexing of packets over some part of network resource, hop-count and bottleneck bandwidth indicate the amount of resource consumed when packet stream is transferred from source to destination. In this way, path quality on resource occupation could be identified by "bottleneck_bandwidth/hop_count", which is the third metric we use. On the other hand, as we use bandwidth related link metric in route computation, the path cost reflects the end-to-end transmission delay, then the metric on QoS provisioning could be defined as "bottleneck_bandwidth/path_cost". As with metric two, weighted round robin is used as scheduling algorithm for metric three and four. We denote dispersion policy using metric three and four "wrr-bh" and "wrr-bp" respectfully.

To adapt to time varying network state, path quality evaluation has to be scheduled periodically or triggered by some threshold conditions. As with link state adver-

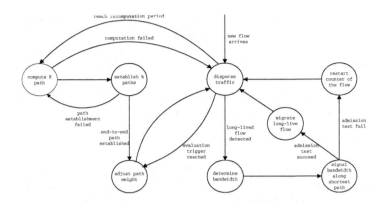

Figure 1: State Machine of Proactive Multipath Routing

tisement, the frequency of path weight adjustment is a tradeoff between performance and overhead. As all information for quality evaluation is learnt from link state advertisement, the evaluation process should be invoked when the number of links announcing new state reaches some threshold value. Based on research on link state advertisement policy, threshold between 0.3 and 0.5 is recommended.

3.3 Detecting Long-Lived Flows and Rerouting

Traffic flow can be defined by fields in the IP and TCP/UDP headers. Depending on the exact definition, flow could be of the same TCP connection, UDP stream or the same host end-points or subnets. As most of current routing equipments have been implemented with hardware monitoring and classifying incoming traffic, it possible to identify long-lived flows from a large amount of incoming traffic. By default, routers forward arriving flows according to the traffic dispersion policy discussed in the previous section. Once the accumulated size or duration time of a flow exceeds some threshold (in terms of bytes, packet number or seconds) a long-lived flow is identified. As has been found[17] that single feature classification could not achieve consistency simultaneously in the fraction of total load due to elephants and the number of elephants, two or more feature classification method is preferred by our approach, e.g. long-lived flows are those who have been active for more than a special time period and transfer more than N bytes of payload. Once a long-lived flow is identified on a non-shortest path, the router will try to determine it's bandwidth requirement, and signal that value along the shortest path to find out whether there is enough room for migration. If the admission test succeeds, long-lived flow will be migrated to the shortest path by modifying label-binding information, otherwise it will be left on the original path. If it is desired, a counter for the long-lived flow, which is not migrated, could be started over to give another chance of migration. It must be emphasized that the signaling proce- dure does not reserve any resource for long-lived flows. The path quality does not need not to be reevaluated immediately after flow migration, because "long-lived flow" does not necessarily equate to "high bandwidth flow". Figure 1 gives the state machine for our approach.

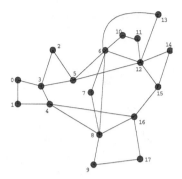

Ingress	Egress	Hop distance	Arrival rate
1	12	4	10.6
0	14	4	10.6
2	17	4	10.6
9	11	4	10.6
6	17	3	10.6
4	13	3	10.6
10	8	2	10.6

Figure 2: MCI Topology and Non-uniform Traffic Matrix

4.   Performance Evaluation

This section evaluates the performance of our approach to proactive multipath routing based on detailed simulation experiments. After a brief description of simulation methodology, we compare our scheme to traditional single path routing under a wide range of load levels and network topology. Experiment results show that, our approach outperforms traditional single path routing in various situations.

4.1 Simulation Setup

We use an event driven simulator in our experiments. The simulator operates at flow level to evaluate the essentials of multiple routing policies and network configuration. Figure 2 shows the topology used in our experiments. For simplicity all links are assumed to be bi-directional and of equal capacity in each direction. During the simulation, flows arrive to the network according to a Poisson process, and the flow holding time is Pareto distributed. The ingress and the egress node of the flows are selected according to Figure 2, which are set up to model a typical WAN traffic pattern. Other parameters are set as follows: link capacity is set to 155 units, and bandwidth request is uniformly distributed between 1.14 units and 2.46 units. Traffic load is increased by scaling the arrival rate as shown in Figure 3. Unless explicitly stated, we use dynamic link metric and the routers exchange link states periodically with a period of 10 seconds.

4.2 Network Throughput

From figure 3(a), it's evident that network performance is improved by our scheme when traffic is not uniformly distributed. Specifically, when traffic is not uniformly distributed the relative congestion status is decreased by a factor varying from 90% (arrival rate 10) to 3%(arrival rate 70). This means our approach can improve net work performance effectively, especially when the idled resource is plentiful or load on transmission path is not so heavy. This conclusion is confirmed by Figure 3(b) which plots the result of a 26-node waxman topology with uniformed traffic pattern. It could be found that, in a network with uniformed traffic pattern our approach outperforms single path routing only when the network is lightly burdened.

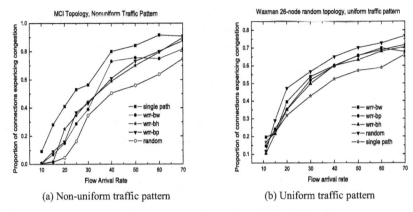

(a) Non-uniform traffic pattern   (b) Uniform traffic pattern

Figure 3: Throughput of Proactive Multipath Routing

These results reflect the essentials of our scheme: absorbing traffic by multiplexing them onto more network resource. When there is a plenty of idle resource, our method can improve network performance effectively; but if there is little idled resource, traffic dispersion may deteriorate network performance.

4.3 Stale Link-State Information

Figure 4(a) plots network performance as a function of link-state update period for dynamic single path routing (SPT) and our approach. It may be found that, although both methods succeed in path finding performance of single path routing is much more sensitive to inaccuracies in link-state information. Measuring either with absolute method or with relative method, performance of our approach is more stable than that of single path routing. In particular, when link-state update period increases from 10 seconds to 120 seconds proportion of flows experiencing congestion rises from 0.014 to 0.159 in single path routing network, while the value with "wrr-bp" is 0.008 to 0.032. Among the four quality metrics, "wrr-bp" and "wrr-bh" are more resistant to stale link state information than "wrr-bw" and "random".

Such a result is directly related to the policy behind each method. In single path routing, all traffic follows the unique path computed from link state information, the staleness of link state information affects network performance directly. With our approach, traffic is spread onto a set of paths, the increase of usable resource weakens the importance of computing accuracy. When a "random" policy is used, the great variation in Internet traffic makes it nearly impossible to reach an optimal flow distribution, so accuracy of path computing plays an important role in network performance. The situation improves with "wrr-bp" and "wrr-bh", whose robustness is tightened by dispersing traffic on basis of path quality evaluation, while properties of "wrr-bw" lies between "random" and "wrr-bp"("wrr-bh").

We also consider the effect of link-state staleness when traffic is aggregated. Figure 4(b) plots the situation when average duration time is improved by 1.8 times. It is found, as aggregation increases advantage of proactive multipath routing over single path routing dismisses somewhat but still significant.

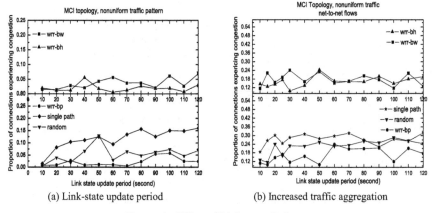

(a) Link-state update period   (b) Increased traffic aggregation

Figure 4: Effect of Link-state Staleness

To further evaluate the efficiency of our approach, we vary the flow arrivals to the network so that certain "hot-spot" present. Specifically, keeping total ejection rate the same, node 1 sends a part of its traffic to node 17 at a rate of 3.6 flows per second, and so does node 0 to node 13. Figure 5(a) shows that the improvement of network performance achieved by our approach over dynamic single path routing is still significant (20% to 90% of congestion removed). We note that, under this situation there is crossover between "random" and single path routing. This reflects the fact that random selection of outgoing path may worsen the performance under some situation. To be reliable for load balancing, the "random" policy should be avoided in network with hot spots. Unlike the previous experiments, "wrr-bw" outperforms others in hot-spotted network. The reason is, when hot spot appears some nodes receive more traffic than others, traffic to these destinations may compete for the same part of transmission resource. The result is that the bottleneck bandwidth becomes the key factor determining congestion status.

4.4 Doing with Long-Lived Flows

To make it simple, we focus on single feature classification in simulation experiments. Shaikh[9] showed that, by relating flow-trigger with time scale of link state advertisement network performance of load-sensitive routing could be improved. We investigate this policy with our approach by experimenting with "wrr-bp". Experiments varying the flow trigger illustrate that computing load decreases with increase of flow trigger. This stems directly from the fact that only a small number of flows will be identified and rerouted under large trigger time. Link state update period has little effect on flow detection, while the number of migrated flow decreases if link state update period is lengthened. In particular, with flow trigger of 20s, the percentage of flows migrated decreases from 3.4% to 0.03% when link state update period changes from 20 seconds to 50 seconds, while proportion of flows detected just decreases from 3.95% to 3.58%. The reason is that, when the staleness of link state information improves the warp of path computing increases; when a new long-lived flow is detected,

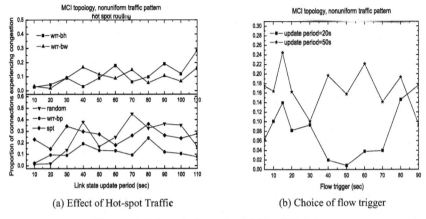

(a) Effect of Hot-spot Traffic   (b) Choice of flow trigger

Figure 5: Trigger in Proactive Multipath Routing

signaling of its bandwidth requirement may be carried along a busy path. Figure 5(b) plots network congestion as a function of flow trigger time. Similar to Shaikh results, congestion curve shows a cup-like shape, while the reason is different: when long-lived flows are triggered at a small threshold,. the bload of admission test improves, but performance of the shortest path decreases for more traffic and more bursty flows; if flow trigger is high a few of flows will be identified as long-lived, some of long-lived flows will be transferred along non-shortest paths, this improves the possibility of network congestion and leads to higher link state variation level. According to definition of bursty flow, we suggest relating flow trigger with link-state update period, and we emphasize that the optimal value of flow trigger should be decided to the state of implementation environment.

Although above results are gained from single feature classification, we argue that it reflects some of the essentials of two or more feature classification. As two or more features are adopted in flow classification to balance requirements between network performance, computing overhead and routing stability, one feature may be chosen as the fundamental feature for multi-criterion classification. According to discussion in previous sections, active time could be the basic metric in flow classification, and performance characteristics under different trigger time forms the fundamental part of performance characteristics in multi-criterion classification.

4.5 Tuning for Performance

In order to find way of performance optimization, we experiment with some of key parameters.

As the "random" policy takes any path with no preference, figure 6(a) plots relationship between path quality evaluation trigger and network performance of the other three polices. It is illustrated that network congestion decreases with increase of evaluation frequency. All three policies show good performance with trigger 0.1(the number of links announcing new state reaches 10% of the number of links), but we do observe that the computation overhead increases very quickly when evaluation trigger decreases. Based on our simulation results, we recommend 0.3 for evaluation trigger.

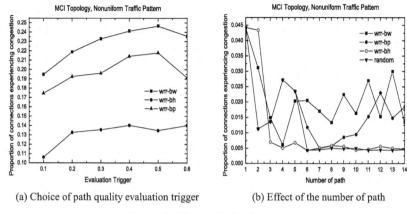

(a) Choice of path quality evaluation trigger   (b) Effect of the number of path

Figure 6: Tuning for Performance

Figure 6(b) plots how path number affects the performance of our approach with non-uniformed traffic pattern. It's evident that network performance varies with both number of path and path quality metric: using "random" and "wrr-bh", congestion decreases to a stable level after path number is larger than 6, while congestion curve of "wrr-bw" or "wrr-bp" has a cup-like shape. The reason is that, if traffic is not uniformly distributed there may be a lot of resource idled. Starting from the shortest path, more and more resource is introduced into traffic transmission and degree of overlapping improves. With "random" policy, bursty traffic will be spread onto k paths and most of long-lived flows will focus onto the shortest path gradually; once the amount of transmission capacity exceeds what is needed, network congestion will be lowered quickly to its minimum value which is caused by bursty flows and could not be removed by additional path. "wrr-bh" achieves this result more quickly by spreading traffic according to resource status on each path. By contrast, "wrr-bw" and "wrr-bp" spread traffic according to end-to-end path quality; when a new path does not share many links with existing paths, network performance will be improved; but, when the number of path exceeds some threshold overlapping will make some of the paths share the same bottleneck link, means constructive interference between flows will be prick up and network performance degrades. Referring to our experiments, if adaptive ability is required discrete value 1, 4 and 7 can be the candidates for the number of path.

5. Conclusion

Focusing on routing Internet traffic according to its built-in properties, this paper proposed a multipath routing scheme which routes incoming traffic proactively over network. In this scheme, network traffic is dispersed at flow level onto multiple paths connecting source and destination. The routing paths are established as LSPs that are not required to be totally disjoint. Path quality is used as reference in traffic dispersion and weighted round robin is used as path selection algorithm. Four types of path quality evaluation methods are proposed and experiments show that "wrr-bp" may be the best policy fitting requirements of TE&QoS. To optimize network performance, long-lived flows on non-shortest path are detected and migrated to the shortest path if their

QoS could be guaranteed there. Two or more feature flow classification method is preferred while policy relating flow trigger with link state update period is investigated. Simulation results demonstrate that, our approach can improve network performance significantly under various network environments, and its performance can be optimized by tuning of key parameters.

References

1. R. Guerin, A. Orda and D. Williams, "QoS Routing Mechanisms and OSPF extensions (QOSPF)", Proceedings of 2nd Global Internet Miniconference (joint with Globecom'97), November 1997
2. E. Gustafsson and G.Karlsson, "A literature survey on traffic dispersion", IEEE Network, pp28-36, 1997
3. N.F. Maxemchuk, "Dispersity routing", Proceeding of ICC'75, pp.41.10-41.13, 1975
4. N.F. Maxemchuk, "Dispersity routing on ATM networks", Proceeding of IEEE Infocom, Vol.1 pp347-357, 1993
5. C. Villamizar, "OSPF Optimized Multipath (OSPF-OMP)", Internet Draft, 1998
6. C. Villamizar, "MPLS Optimized Multipath (MPLS-OMP)", Internet Draft, 1998
7. S. Vutukury and J. J. Garcia-Luna-Aceves, "A simple approximation to minimum-delay routing", Proceeding of ACM Sigcomm'99, pp227-238, 1999
8. X. Sun and G. Veciana, "Dynamic multi-path routing: asymptotic approximation and simulations", Proceeding of Sigmetrics, 2001
9. A. Shaikh, Jennifer Rexford and Kang G. Shin, "Load-senstive routing of long-lived IP flows", Proceeding of ACM Sigcomm'99, 1999
10. D.Sidhu, S.Abdul and R.Haiy, "Congestion control in high speed networks via alternative path routing",High Speed Networks, Vol.1 ,No.2, 1992
11. S.Bahk and M.E.Zaiki, "A dynamic multi-path routing algorithm for ATM networks", High Speed Networks, Vol.1, No.3, 1992
12. S.A.Lippman, "Applying a new device in the optimization of exponential queuing systems", Operations Research, Vol.23, No.4, 1975
13. S.Sibal and A. Desimone, "Controlling alternate routing in general-mesh packet flow networks", Proceeding of Sigcomm'94, 1994
14. V. Nguyen, "On the optimality of trunk reservation in overflow processes", Probability in the Engineering and Informational Sciences, Vol.5, 1991
15. S. Nelakuditi and Zhi-Li Zhang, "On selection of paths for multipath routing", In *Proc. IWQoS'01*, June 2001
16. E. Qusiros etc., "The K shortest paths problem", 35th IEEE Symp. Foundations of Comp. Sci., Santa Fe, 1994
17. M.Crovella and A.Bestavros, "Self-similarity in world wide web traffic: evidenceand possible causes", IEEE/ACM Trans. On Networking, Vol.5, Dec. 1997
18. K.Papagiannaki, N.Taft etc., "On the feasibility of identifying elephants in internet backbone traffic", Sprint ATL Technical Report TR01-ATL-110918, 2001

Panel

Premium IP: On the Road to Ambient Networking

Organizer: Paulo de Sousa

European Commission, Belgium

Abstract

This panel will contribute to the understanding of the role of QoS in networking; specifically what are emerging topics in Premium IP networking. This topic relates closely to the vision of Ambient Networking as a research area that links network interoperability with reconfiguration capabilities to meet challenges such as mobility, QoS, multi-homing, and multimedia.

Premium IP networking is currently in its infancy, and applications that can exploit Premium IP features are rare today. In order to bring the excellent initial achievements of Europe towards full-scale Premium IP deployment, a concerted effort is needed by key participants to develop, validate and demonstrate on a large scale improved features for the provision of enhanced IP communications services.

Proposed trends for Premium IP include:

- Shifting the focus from QoS mechanisms and SLAs to a more holistic view of the "end-to-end performance problem":

- Building expertise across several areas that are traditionally separate, but whose interaction determines the end-to-end performance: performance in the individual parts of the network, computer/OS architecture, transport and higher-layer protocols, etc.

- Investigating end-to-end Premium IP services in an environment consisting of a multiplicity of heterogeneous domains.

The ability to support new services (such as Premium services) requires the deployment of new hardware and/or software features in the network.

B. Stiller et al. (Eds.): QofIS/ICQT 2002, LNCS 2511, pp. 157-158, 2002.
© Springer-Verlag Berlin Heidelberg 2002

The list of panelists shows distinguished experts in their ares of research as well as their current or most recent European projects they worked on:

- Martin Potts, Martel, Switzerland (CADENUS)
- Rui Aguiar, University of Aveiro, Portugal (MOBY DICK)
- Bert Koch, Siemens, Germany (AQUILA)
- Bob Briscoe, BT Research, U.K. (M3I)
- Simon Leinen, Switch, Switzerland (SEQUIN, GEANT)

Chair and Organizer: Paulo de Sousa, European Commission, Belgium

Service Differentiation and Guarantees for TCP-based Elastic Traffic

Nidhi Hegde and Konstantin E. Avrachenkov

INRIA, 2004 route des Lucioles, B.P.93
06902 Sophia Antipolis Cedex, France
{Nidhi.Hegde, K.Avrachenkov}@sophia.inria.fr

Abstract. We compare buffer management policies that offer differentiated services to TCP traffic, with the goal of providing some sort of performance guarantees to a premium class of traffic. Specifically, we study the effectiveness of a scheduling policy combined with various buffer management policies on the performance of competing TCP connections. In this work we consider a stochastic model for a class-based weighted-fair-queueing scheduling policy where packets of different classes are scheduled according to their pre-assigned static weights. We consider two buffer management policies: complete partitioning, and complete sharing with pushout at various thresholds. We consider two classes of TCP traffic in our model. Our goal is to propose mechanisms with easily adjustable parameters to achieve service differentiation as required. In the numerical results we show how the scheduling and buffer policies can be used to provide some sort of performance guarantee to the higher class.

1   Introduction

It is expected that the next generation IP networks will be able to provide service differentiation [1,2]. With many different types of applications on such networks, users may also expect a performance guarantee. Our goal is to study the possibility of time priority and space priority schemes at the packet level to provide such service differentiation, and some sort of performance guarantees. The DiffServ Architecture[1] proposes packet-level mechanisms to achieve service differentiation. Traffic is classified upon entry into the network and packets are forwarded in the core network based on such pre-classified per-hop behaviours. Two main classes of per-hop behaviour are the Expedited Forwarding(EF) class for stream traffic and the Assured Forwarding(AF) class for elastic traffic. Here we assume that for stream traffic, the EF class receives a guaranteed fixed bandwidth[3]. Thus, we concentrate on the remaining bandwidth being shared by elastic traffic. Users might want to have the possibility of further traffic differentiation for the elastic traffic. This can be achieved by using Assured Forwarding (AF) scheme [4]. In principle, the Assured Forwarding scheme might have up to four classes of traffic and up to three drop precedence levels within each traffic class. We restrict ourselves to the simplest case of two AF classes in our model, which is

B. Stiller et al. (Eds.): QofIS/ICQT 2002, LNCS 2511, pp. 159–168, 2002.

still tractable for analysis. Our model can also be viewed as a model for two different drop precedence levels within one AF class.

We consider mechanisms that do not include adjustments to the underlying architecture and transport protocols such as TCP. More desirable are simple mechanisms that can be easily implemented and easily tunable as required. To differentiate between classes of traffic we propose to use either time priority (scheduling) or space priority (buffer management) schemes or both. We will show that the very same policies may also be used towards the goal of performance guarantees. As the time priority scheme, we use stochastic Weighted Fair Queueing(WFQ) which is a packet level approximation of Generalized Processor Sharing(GPS) [5,6]. For buffer management we consider both a partitioned buffer and a shared buffer space. For the shared buffer space, we propose a Pushout with Threshold (POT) policy to achieve further service differentiation. We suppose that the traffic is sent by greedy TCP sources [7,8,9]. Therefore the TCP loss-based control mechanism has to be taken into account.

To the best of our knowledge, the DiffServ scheme with stochastic weighted fair queueing and the buffer management schemes described above along with the TCP congestion control mechanism is analyzed for the first time here. Most of the related work concentrates on either scheduling priority or buffer priority, not both. In [10,11], the authors considered RED with colours as a tool for differentiation. Such a scheme uses packet marking and various drop thresholds at the buffer. RED with colours can attain various levels of differentiation by adjusting several parameters. The advantage of our scheme is that there are only two tunable parameters, allowing enough freedom to alter differentiation levels without being too complex. In [12], pushout without threshold and threshold-based drop mechanisms have been proposed as means to provide priority. A pushout scheme is proposed where an arriving higher priority packet may pushout one of lower priority if the buffer is full. This corresponds to a special case of our pushout with threshold scheme where the threshold is set to 0. We will show in Section 4 that setting a pushout threshold of at least one still achieves service differentiation without starvation of class 2 traffic. In addition, this threshold parameter may be adjusted to achieve varying levels of differentiation. For the threshold mechanism in [12], arriving lower priority packets are dropped if they number more than some given threshold. In [13], an alternate threshold policy is proposed where the drop threshold for any particular class of packets is based on the total number of all packets in the queue. Both such threshold-based drop schemes may be easy to implement, but suffer from wasted resources if the arrival rate of higher priority packets is not high and possible starvation of lower priority packets.

In this paper we consider a network where a bandwidth guarantee for a premium class of traffic may be required. For instance, consider an enterprise with multiple sites that may be connected through an IP network managed by some Internet Service Provider (ISP). Such a company might expect its traffic to receive service similar to that of a leased line[14]. The ISP then might classify this company's traffic as high priority, and offer a higher level of service. In order to achieve some desired level of service for adaptive connections (such as TCP traffic), it is important that scheduling be combined with a proper buffer management scheme[14]. In this work, we first consider a partitioned buffer space with priority given at the scheduler. However, because such a buffer allocation wastes resources, we recommend a shared buffer space with the POT policy to provide service differentiation in addition to that achieved through scheduling. We use our numerical results to show how an ISP might adjust the scheduling and buffer parameters to achieve service differentiation as well as some sort of service guarantees.

The paper is organized as follows. Section 2 outlines the priority policies we consider and how the TCP performance measures are obtained. Section 3 details the performance analysis for each priority scheme proposed. Numerical results from our analysis are shown in Sect. 4, and we conclude with Sect. 5.

2   Model

Consider a bottleneck link with greedy TCP connections of two AF classes with different service level requirements compete. We assume that the aggregate EF class traffic receive a fixed capacity, leaving a fixed capacity for AF classes. Our goal is to study how some sort of bandwidth guarantee may be offered to a high priority class of TCP traffic. We model this bottleneck link as a single server queue with two priority classes. Differentiation between the two service classes can be realized through priority at the server and/or priority in the buffer space. We combine this queueing system with a TCP model in order to calculate performance measures of the types of traffic in the system. The "guarantees" are therefore on the average TCP throughput achieved.

2.1   Scheduling Priority

We provide priority in scheduling through Stochastic WFQ which can be seen as a practical packet-level implementation of GPS. Each class i is given a weight of α_i where $\sum_i \alpha_i = 1$. At the end of each service time, a packet of class i is served with probability $\alpha_i / \sum_{j:N_j>0} \alpha_j$, where N_i is the number of packets of class i in the buffer. In our case of two classes, class 1 packets have a weight of $\alpha_1 = \alpha$ and class 2 packets a weight of $\alpha_2 = 1 - \alpha$. Such a priority policy may be combined with either a partitioned or shared buffer space. In what follows we explain buffer management schemes for both types of buffers, that allow for greater service differentiation or better resource utilization.

2.2   Buffer Priority

Complete Partitioning(CP). This system consists of a single server that serves two separate buffers. Each class i of packets has a buffer of size K_i. Our model is general enough to offer priority to a particular class by increasing its buffer size relative to the other. However,we assume that that in a partitioned buffer space, the size of each buffer is not easily tunable, and thus in Sect. 4 we present results for the case of equal buffer size.

Complete Sharing(CS). Packets of both type share a single buffer of size K. Neither class has priority at the buffer and may be differentiated only at the server. For such a buffer space, we can offer priority to the higher class, of type 1, with the Pushout with Threshold(POT) scheme in the following way. A type 1 arriving packet that finds the shared buffer full may enter the buffer by pushing out a type 2 packet if the number of type 2 packets in the buffer is above a given threshold, K_2. Priority to type 1 class may be changed by adjusting the threshold K_2. For instance when $K_2 = K$, a type 2 packet is never pushed out by a type 1 packet, and no priority is given to class 1. When $K_2 = 0$, a type 1 packet arriving to a full buffer always pushes out a type 2 packet if any, and type 1 has maximal priority at the buffer. Note that for $K_2 = 0$, the lower priority traffic may experience starvation.

CP is the simplest policy for implementation. However, it leads to under-utilization of resources. The CS buffer without a threshold policy utilizes the buffer space well, but does not provide any differentiation. The POT policy on the other hand, not only utilizes the resources well, but also provides differentiation in buffer management.

3   Performance Analysis

3.1   TCP Model

Let m_i be the number of class i connections at the link, S_i be the average sending rate of each TCP source corresponding to class i, L_i be the loss probability of class i packets, and $\overline{W}_i$ be the average waiting time for class i packets. We use the standard square-root equation for calculating the TCP sending rate of each source [8]:

$$S_i(L_i, \overline{W}_i) = \text{MSS} \cdot \min\left(\frac{1}{d_i + \overline{W}_i}\sqrt{\frac{k}{L_i}}, \frac{M}{d_i + \overline{W}_i}\right), \tag{1}$$

where MSS is the maximum segment size, d_i is the two-way propagation delay, and M is the maximum window size of the receiver. The constant k represents the TCP model considered. For instance, in our results in Section 4, we consider $k = 2$ where the inter-loss times are assumed to be exponentially distributed and the delay ACK mechanism is disabled[7]. We assume fair link sharing within each class, and define the load for each class of traffic by: $\rho_i = m_i\frac{S_i}{C}$ where C is the capacity of the link.

As a measure of quality of service, we use the throughput of each connection, $T_i = S_i(1 - L_i)$.

In the next subsection, we show how the packet loss probabilities and the average waiting time can be calculated as functions of ρ_1 and ρ_2. Namely,

$$L_i = L_i(\rho_1, \rho_2), \qquad \overline{W}_i = \overline{W}_i(\rho_1, \rho_2). \tag{2}$$

One can see that Equations (1) and (2) form a system of four equations in four unknowns. This system can be solved using a fixed-point approach.

3.2   Performance Measures

The loss probabilities and waiting time averages for each priority policy are computed by solving the respective queueing system. We assume that packets of class i arrive at the link according to an independent Poisson process with rate $\lambda_i = m_i S_i$. The service time for each packet is assumed to be exponential with rate $\mu_i = C$. Let N_i be the number of class i packets in the buffer. For each of the priority policies, the vector (N_1, N_2) will be a Markov Chain. The generators corresponding to different buffer management policies are given in the Appendix.

Loss Probability. For the system with partitioned buffers (CP), the loss probability for each class is the probability that its buffer is full:

$$L_1 = \sum_{n_2=0}^{K_2} P(K_1, n_2), \qquad L_2 = \sum_{n_1=0}^{K_1} P(n_1, K_2).$$

With the shared buffer policy (CS), the loss probabilities of both classes are equal. This can be determined from the solution to a M/M/1/K+1 system with an arrival rate of $\sum_i S_i$ and a service rate of C. Namely, we have:

$$L_i = \frac{(1-\rho)\rho^{K+1}}{1 - \rho^{K+2}}, \qquad \rho = \sum_i \rho_i.$$

For the POT policy, the loss probability is calculated as follows. Let P_K denote the probability that the shared buffer is full, φ denote the probability that a class 1 arrival pushes out a class 2 packet, and ϕ denote the probability that a packet in the buffer is pushed out by a class 1 arrival. An arriving class 1 packet is lost only when $N_2 \leq K_2$:

$$L_1 = P_K - \varphi, \qquad P_K = \frac{(1-\rho)\rho^{K+1}}{1 - \rho^{K+2}}, \qquad \varphi = \sum_{n_2=K_2+1}^{K} P(K - n_2, n_2).$$

A packet of class 2 is lost either when the buffer is full upon its arrival, or when it enters the buffer and then is pushed out by a class 1 arrival:

$$L_2 = P_K + (1 - P_K)\phi.$$

We solve for ϕ by equating the rates of pushout loss: the rate of class 1 packets that push out a class 2 packet must equal the rate of accepted class 2 packets that are subsequently pushed by a class 1 arrival, as follows:

$$S_2(1 - P_K)\phi = S_1\varphi.$$

Average Delay. For both CP and CS without POT, the average delay is determined by invoking Little's Law: $\overline{W}_i = \frac{\overline{N}_i}{T_i}$, where $\overline{N}_i$ is the average number of class i packets in the buffer. We use Little's Law also for the POT scheme, noting that only those arrivals that are eventually served are considered in the determination of average waiting time. Therefore, for both the average waiting time and loss probability calculations, it is irrelevant which class 2 packet is pushed out: the first in queue, the last in queue, or a random class 2 packet. The waiting time averages are as follows:

$$\overline{W}_1 = \frac{\overline{N}_1}{T_1} \qquad \overline{W}_2 = \frac{\overline{N}_2}{S_2(1 - P_K)(1 - \phi)} = \frac{\overline{N}_2}{T_2}.$$

4   Numerical Results

In this section we will show how the network operator may use our priority schemes to provide type 1 connections with a throughput guarantee. We consider two cases of core routers, those with a partitioned buffer space and those with a common buffer for both types of traffic. For the first type with partitioned buffer space, we use only stochastic WFQ to provide priority. For the case of a shared buffer, we use the pushout with threshold buffer management scheme in addition to stochastic WFQ at the server.

We consider a 100Mbps bottleneck link. For all our numerical results we have assumed the maximum window size at the sender $M = 20$, the average maximum segment size MSS of 512 Bytes, and a buffer space of 40 segments in total for both type 1 and type 2 packets. We assume that the high priority class is a premium class and therefore the number of this type of connections is very small relative to the number of low priority connections. We vary the number of high priority connections through this link, m_1, from two to five, and the number of low priority connections $m_2 = 20, \ldots, 50$. Our model is general to allow any values of m_1 and m_2, however we use these values for illustration purposes.

4.1   Partitioned Buffer

Let us consider the case where the total throughput of all high priority connections require some portion of the total bandwidth, and a partitioned buffer space. We consider this case as a form of bandwidth reservation, but only loosely so because since all connections are of greedy TCP sources, if the "reserved bandwidth" is not utilized, it is shared among the remaining low priority connections. We use Fig. 1 and 2 to show how the network operator may choose the scheduling parameter α to provide type 1 connections at least 20% of the total bandwidth. We note that in all cases, such performance *guarantees* are for the average TCP throughput T_1. For example, let us assume that type 1 traffic request 20% of total bandwidth and there will be three type 1 connections. Then, given the number of type connections, Fig. 1 can be used to set α. We note that in general, because of the stochastic nature of the arrival traffic, in order to guarantee some portion x of the total bandwidth to type 1 traffic, it is not sufficient to set $\alpha = x$. The value of α depends not only on the arrival rates, or number of connections of both types of traffic, but also on the type of buffer management scheme.

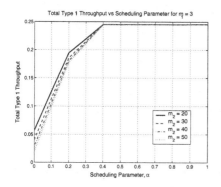

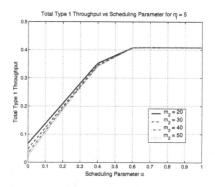

Fig. 1. Providing Bandwidth Guarantee with WFQ, $m_1 = 3$

Fig. 2. Providing Bandwidth Guarantee with WFQ, $m_1 = 5$

We note that ideally, a TCP connection with no queueing delay and no packet losses will have a maximum sending rate of $S_i^{\max} = \frac{M \cdot \text{MSS}}{d_i} = 8.19\text{Mbps}$. Note also that in Figs. 1 and 2 as α is increased to 1, the total bandwidth occupied by type 1 traffic does not go beyond $m_1 S^{\max_1}$.

The remaining bandwidth which is not used by type 1 connections is shared equally by type 2 connections. For instance, when $m_1 = 2$, $m_2 = 40$, and $\alpha = 1$, type 1 connections get a total throughput of about 16.4 Mbps. Type 2 connections share the rest of the total overall throughput, and get a total throughput of 80.4 Mbps.

4.2   Shared Buffer

We now consider a shared buffer space with the pushout with threshold policy. We first offer priority only at the buffer, by setting $\alpha = 0.5$. Figure 3 shows how the pushout threshold K_2 may be set to achieve some guarantee for type 1 connections, when $m_1 = 5$. For instance, for a 20% total bandwidth guarantee for type 1 connections, may be set to a value between 40 and 37, depending on the number of type 2 connections. A relatively small change in K_2 is required to achieve this performance guarantee.

We may offer priority both at the buffer and at the scheduler. The network operator may find it more convenient to change one paramater at smaller time scales and the other at longer time scales. We show examples of when one parameter is set to a fixed value, the other can be changed to achieve the performance guarantee. In Fig. 4 we set $\alpha = 0.2$ for $m_1 = 5$, and show the change in total

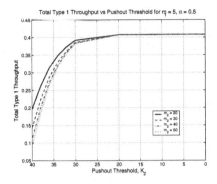

Fig. 3. Providing Bandwidth Guarantee with POT, $m_1 = 5$

type 1 bandwidth as K_2 is varied, for various values of m_2. For instance, if type 1 connections require a total throughput of 0.2, K_2 can be set to a value between 37 and 28 depending on the number of type 2 connections.

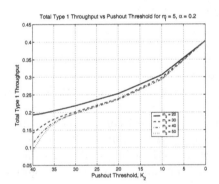

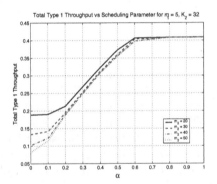

Fig. 4. Providing Bandwidth Guaran-   **Fig. 5.** Providing Bandwidth Guarantee with WFQ and POT, $m_1 = 5$ tee with WFQ and POT, $m_1 = 5$

In Fig. 5 we set $K_2 = 32$ and vary α for $m_1 = 5$. $K_2 = 32$ corresponds to 20% of the buffer space "reserved" for type 1 packets. If we now want to give type 1 connections a total throughput of 0.2, α can be set to a value between 0.14 and 0.22 depending on m_2.

5   Conclusion

We have proposed scheduling and buffer management schemes in order to achieve performance guarantees in IP networks in the form of average TCP throughput. For a partitioned buffer space, we have used a stochastic weighted-fair queueing policy at the scheduler. For a shared buffer, we have used a pushout with threshold policy at the buffer along with stochastic weighted-fair queueing at the server. We have combined analysis of such priority policies with the TCP model to study guarantees in the achieved throughputs of the TCP sources.

References

1. S. Blake, D. Black, M. Carlson, E. Davies, Z. Wang, and W. Weiss, "An architecture for differentiated service", RFC 2475, December 1998.
2. Kakevi Kilkki, *Differentiated Services for the Internet*, Macmillan Technical Publishing, 1999.
3. V. Jacobson, K. Nichols, and K. Poduri, "An expedited forwarding PHB group", RFC 2598, June 1999.
4. J. Heinanen, F. Baker, W. Weiss, and J. Wroclawski, "Assured forwarding PHB group", RFC 2597, June 1999.
5. Srinivasan Keshav, *An Engineering Approach to Computer Networking, ATM Networks, the Internet, and the Telephone Networks*, Addison-Wesley, 1997.
6. Abhay Parekh and Robert Gallager, "A generalized processor sharing approach to flow control in integrated services networks: The single-node case", *IEEE/ACM Transactions on Networking*, vol. 1, no. 3, pp. 344–57, June 1993.
7. Eitan Altman, Konstantin E. Avrachenkov, and Chadi Barakat, "A stochastic model of TCP/IP with stationary random losses", in *Proceedings of ACM SIGCOMM'00*, Stockholm, Sweden, Aug. 28 - Sept. 1 2000.
8. Mathew Mathis, Jeffrey Semke, Jamshid Mahdavi, and Teunis Ott, "The macroscopic behavior of the TCP congestion avoidance algorithm", *Computer Communications Review*, vol. 27, no. 3, July 1997.
9. Jitendra Padhye, Victor Firoiu, Don Towsley, and Jim Kurose, "Modeling TCP throughput: A simple model and its empirical validation", in *Proceedings of ACM SIGCOMM'98*, August 1998.
10. Chadi Barakat and Eitan Altman, "A markovian model for TCP analysis in a differentiated services network", in *Proceedings of the First International Workshop on Quality of future Internet Services (QofIS)*, Berlin, Germany, September 2000.
11. Sambit Sahu, Philippe Nain, Don Towsley, Christophe Diot, and Victor Firiou, "On achievable service differentiation with token bucket marking for TCP", in *Proceedings of ACM SIGMETRICS'00*, Santa Clara, CA, USA, June 18-21 2000.
12. Martin May, , Jean-Chrysostome Bolot, Alain Jean-Marie, and Christophe Diot, "Simple performance models of differentiated services schemes for the internet", in *INFOCOM (3)*, 1999, pp. 1385–94.
13. Eeva Nyberg, Samuli Aalto, and Jorma Virtamo, "Relating flow level requirements to diffserv packet level mechanisms", Tech. Rep., COST279, 2001, available at http://tct.hut.fi/tutkimus/cost279/.
14. Vijay P. Kumar, T. V. Lakshman, and Dimitrios Stiliadis, "Beyond best effort: Router architectures for the differentiated services of tomorrow's internet", *IEEE Communications Magazine*, pp. 152–64, May 1998.
15. Marcel Neuts, *Matrix-geometric solutions in stochastic models : an algorithmic approach*, John Hopkins University Press, 1981.
16. D. P. Gaver, P. A. Jacobs, and G. Latouche, "Finte birth-and-death models in randomly changing environments", *Advances in Applied Probability*, vol. 16, pp. 715–31, 1984.

Appendix A: Solution to Queueing Models

Complete Partitioning. The state space of the Markov chain is $\{(n_1, n_2) : n_1 = 0 \ldots K_1, n_2 = 0 \ldots K_2\}$ and the generator is given by:

$$
Q = \begin{pmatrix}
-B_1 & B_1 & & & \\
B_2 & E_0 & A & & \\
& D & E & \ddots & \\
& & \ddots & \ddots & A \\
& & & D & E+A
\end{pmatrix}
$$

where $\quad B_1 = [\lambda_1 \lambda_2], \quad B_2 = \begin{bmatrix} \mu_1 \\ \mu_2 \end{bmatrix}, A = \mathrm{diag}\{\lambda_2\},$

$$
E_0 = \begin{pmatrix}
\mathbf{a} & \Lambda_1 & & & \\
\mathbf{d} & \mathbf{a} & \Lambda_1 & & \\
& \ddots & \ddots & \ddots & \\
& & \mathbf{d} & \mathbf{a} & \Lambda_1 \\
& & & \mathbf{d} & \mathbf{a}+\Lambda_1
\end{pmatrix}, \quad
E = \begin{pmatrix}
\mathbf{a} & \Lambda_1 & & \\
\mathbf{b}_1 & \ddots & \ddots & \\
& \ddots & \mathbf{a} & \Lambda_1 \\
& & \mathbf{b}_1 & \mathbf{a}+\Lambda_1
\end{pmatrix}, \quad
D = \begin{pmatrix}
0 & \mu_1 & & \\
0 & \mu_2 & & \\
& & \mathbf{b}_2 & \\
& & & \ddots \\
& & & & \mathbf{b}_2
\end{pmatrix},
$$

$\Lambda_i = \mathrm{diag}\{\lambda_i\}_{2\times 2}, \mathbf{a} = \mathrm{diag}\{-(\lambda + \mu_i)\}_{i=1,2}, \mathbf{d} = \begin{bmatrix} \mu_1 & 0 \\ \mu_2 & 0 \end{bmatrix}, \mathbf{b}_1 = \alpha_1 \mathbf{d}, \mathbf{b}_2 = \begin{bmatrix} 0 & \alpha_2\mu_1 \\ 0 & \alpha_2\mu_2 \end{bmatrix}$

B_1 and B_2 are of size (1×2) and (2×1) respectively, and D, E, E_0, and A are of size $(2(K_1 + 1) \times 2(K_1 + 1))$. The generator matrix Q is of QBD-type with complex boundary behaviour. We use matrix-geometric methods [15] to solve for the steady-state probabilities $P(n_1, n_2) = \Pr[N_1 = n_1, N_2 = n_2]$.

Pushout with Threshold. The state space of the Markov chain $\{(n_1, n_2) : n_1 + n_2 \leq K\}$. The generator is given by:

$$
Q = \begin{pmatrix}
-B_1 & B_1 & & & & \\
B_2 & E_0 & A_0 & & & \\
& D_1 & E_1 & A_1 & & \\
& & \ddots & \ddots & \ddots & \\
& & & D_{K-1} & E_{K-1} & A_{K-1} \\
& & & & D_K & E_K + \Lambda_2
\end{pmatrix}
$$

where $\quad A_j = \begin{pmatrix}
\Lambda_2 & & \\
& \ddots & \\
& & \Lambda_2 \\
0 & \cdots & 0 \\
0 & \cdots & 0
\end{pmatrix}, \quad
E_j = \begin{pmatrix}
\mathbf{a} & \Lambda_1 & & \\
\mathbf{b}_1 & \ddots & \ddots & \\
& \ddots & \mathbf{a} & \Lambda_1 \\
& & \mathbf{b}_1 & \tilde{\mathbf{a}}_j
\end{pmatrix}, \quad
D_j = \begin{pmatrix}
0 & \mu_1 & & \\
0 & \mu_2 & & \\
& & \mathbf{b}_2 & \\
& & & \ddots \\
& & & & \mathbf{b}_2 \mathbf{v}_j
\end{pmatrix}
$$

$\tilde{\mathbf{a}}_k = \mathbf{a} + \Lambda_1 + \Lambda_2$ and $\mathbf{v}_k = \mathbf{0}$ for $k \leq K_2$, $\tilde{\mathbf{a}}_k = \mathbf{a} + \Lambda_2$ and $\mathbf{v}_k = \Lambda_1$ for $k > K_2$. The matrices A_{K-j}, E_{K-j}, and D_{K-j} are of sizes $2j \times 2(j+1)$, $2(j+1) \times 2(j+1)$, and $2(j+1) \times 2(j+2)$ respectively. This generator is of type level-dependent QBD. We use the algorithm described in [16] to solve for the steady-state probabilities.

Service Differentiation in Third Generation Mobile Networks

Vasilios A. Siris[1], Bob Briscoe[2], and Dave Songhurst[2]

[1] Institute of Computer Science, FORTH, P.O. Box 1385,
GR 711 10, Heraklion, Crete, Greece
vsiris@ics.forth.gr
[2] BT Research, B54/130, Adastral Park, Ipswich, IP5 3RE, UK
bob.briscoe@bt.com, dsonghurst@jungle.bt.co.uk

Abstract. We present and analyse an approach to service differentiation in third generation mobile networks based on Wideband CDMA, that exposes a new weight parameter designed to reflect allocation of the congestible resource. The approach naturally takes into account the difference in resource scarcity for the uplink and downlink, because it is grounded on fundamental economic models for efficient utilization of resources in WCDMA. Discrete values of the weight parameter can be presented as different service classes. Finally, we present numerical experiments demonstrating the effectiveness of our approach, and investigate its performance and transient behaviour under power control and signal quality estimation errors.

1 Introduction

The percentage of users accessing packet switched networks through wireless access networks is increasing at a very large pace. Hence, the ability to support quality of service (QoS) differentiation in wireless systems is becoming increasingly important. Indeed, the UMTS (Universal Mobile Telecommunication System) third generation mobile telecommunication system allows user negotiation of bearer service characteristics, such as throughput, error rate, and delay [1,2]. WCDMA (Wideband Code Division Multiple Access) is the main air interface for UMTS. With WCDMA all mobile users can simultaneously transmit, utilizing the whole radio spectrum, and unique digital codes are used to differentiate the signal from different mobiles. Variable bit rates are achieved using variable spreading factors, where the spreading factor determines how much a data bit is spread in time, and multiple codes. The signal quality (error rate) is determined by the signal-to-interference ratio, SIR, which is the ratio of the signal's received power over the total interference, the latter given by the sum of the noise and the interference due to signals from other mobiles.

In this paper we propose models and procedures for service differentiation in WCDMA networks, and present numerical investigations that demonstrate the effectiveness of our approach, and show how various characteristics of the wireless system, such as power control and SIR estimation errors, and discrete

B. Stiller et al. (Eds.): QofIS/ICQT 2002, LNCS 2511, pp. 169–178, 2002.

transmission rates, affect service differentiation and the system's transient behaviour. By considering actual resource usage in the uplink and the downlink, our procedures are fair and efficient, and are robust to varying demand for wireless resources. In the uplink, resource usage is radio spectrum limited, being an increasing function of the product of the transmission rate and *SIR*. In the downlink, resource usage is constrained by the base station's total transmission power. Our approach modifies outer loop power control and load control, which runs on the radio network controller (RNC), and does not affect fast closed-loop power control, which operates on a much faster timescale.

In related work, [3] presents a class-based quality of service framework. The performance for a particular class depends on its elasticity, which specifies how the rate will decrease in periods of congestion. Our approach differs from the above in that allocation of resources is done proportional to weights, thus leading to fair, in terms of weights, allocations. [4] provides an overview of radio resource allocation techniques, focusing on power and rate adaptation in overload periods, and [5] discusses rate adaptation for different wireless technologies.

The rest of the paper is organized as follows. In Sect. 2 we discuss resource management procedures in WCDMA. In Sect. 3 we first discuss resource usage in the uplink and downlink, and then propose procedures for service differentiation in each direction. In Sect. 4 we present and discuss numerical investigations demonstrating the effectiveness of our approach, and in Sect. 5 we conclude the paper identifying related and future research issues.

2   Resource Management in WCDMA

Resource management in WCDMA includes fast closed-loop power control, outer loop power control, and load control [2]. In the uplink, with fast closed-loop power control, Fig. 1(a), the base station (BS) continuously measures the received *SIR* for each mobile, and compares it with a target *SIR*. If the measured *SIR* is smaller (larger), then the BS instructs the mobile to increase (decrease) the transmission power. The above cycle has frequency 1500 Hz, which corresponds to one power update every 0.67 msec. In WCDMA, a similar fast closed-loop power control loop exists in the downlink direction, where now the *SIR* is measured by the mobile, which sends power update commands to the BS.

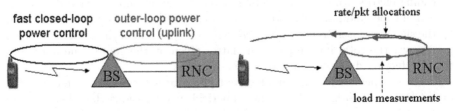

(a) Closed-loop & outer loop power ctrl   (b) Load control in uplink & downlink

Fig. 1. Resource management functions in WCDMA

In second and third generation systems based on CDMA, the main objective of fast closed-loop power control is to tackle the *near-far* problem: If all mobiles transmitted with the same power, then the signal from the mobile nearer to the base station would overwhelm all other signals. Fast closed-loop power control resolves this problem by maintaining the same *SIR* at the base station, for all mobiles. In third generation system, which will support applications with different quality of service requirements, the target *SIR* need not be the same, since a different target *SIR* yields a different signal quality, in terms of the frame error rate, *FER*. Typically, for non-real-time services the frame error rate is in the range $10 - 20\%$, whereas for real-time services it is close to 1% [2, p. 193].

Because there is no one-to-one correspondence between the target *SIR* and the achieved *FER*, outer loop power control between the base station and the RNC is required, Fig. 1(a); its objective is to adjust the target *SIR* in order to achieve some constant, predefined *FER*. Typically, outer loop power control operates in timescales slower than those of fast closed-loop power control.

In WCDMA data transmission occurs in fixed-size frames, which have minimum duration 10 msec; the rate is allowed to change between frames but remains constant within a single frame. Hence, the timescales of rate control are slower than those of fast closed-loop power control, where one power update occurs every 0.67 msec. Moreover, WCDMA supports discrete bit rates: Specifically, in the uplink the user data rate can obtain the values 7.5, 15, 30, 60, 120, 240, 480 Kbps, which correspond to a spreading factor of 256, 128, 64, 32, 8, 4, respectively [2]. Higher bit rates can be achieved with the use of multiple codes. In addition to code-division scheduling, WCDMA supports time-division scheduling that controls which mobiles can transmit in each frame.

Finally, load control decreases the target *SIR* that is used in fast closed-loop power control, or decreases the transmission rate and/or adjusts the time scheduling of data, for both the uplink and the downlink, during periods of overload, Fig. 1(b); such periods are detected by the RNC based on measurements it receives from the base station.

3   Models for Service Differentiation

In this section we first discuss resource usage in the CDMA uplink and downlink. Then we present our approach for service differentiation, which involves allocating resources according to weights. An important property of CDMA networks is that there are two control parameters that affect the quality of service: the transmission rate and the signal-to-interference ratio. However, resource usage in the two directions is different, leading to different models for service differentiation.

3.1   Resource Usage in CDMA

Resource Usage in the Uplink. Consider a single CDMA cell. In the uplink, the signal-to-interference ratio at the base station for the transmission from mobile i is given by [6,7]

$$SIR_i = \frac{W}{r_i} \frac{g_i p_i}{\sum_{j \neq i} g_j p_j + \eta}, \qquad (1)$$

where W is the chip rate, which is equal to 3.84 Mcps for WCDMA, r_i is the transmission rate, p_i is the transmission power, g_i is the path gain between the base station and mobile i, and η is the power of the background noise at the base station. The ratio W/r_i is the spreading factor or processing gain for mobile i.

The value of SIR corresponds to the signal quality, since it determines the bit error rate, BER [6,7], and hence the frame error rate, FER. If we assume perfect power control, then the value of SIR in (1) will be equal to the target signal-to-interference ratio that is used in fast closed-loop power control.

Solving the set of equations given by (1) for each mobile i, we get [7,8]

$$g_i p_i = \frac{\eta \alpha_i^{\text{UL}}}{1 - \sum_j \alpha_j^{\text{UL}}}, \quad \text{where} \quad \alpha_i^{\text{UL}} = \frac{1}{\left(\frac{W}{r_i SIR_i} + 1\right)}. \qquad (2)$$

Since the power p_i can take only positive values, from (2) we get

$$\sum_i \alpha_i^{\text{UL}} < 1. \qquad (3)$$

The last equation illustrates that the uplink is *interference-limited*: Even when they have no power constraints, mobile hosts cannot increase their power with no bound, due to the increased interference they would cause to the other mobiles. If (3) is violated, then the target SIR values cannot be met for all mobiles.

In practise, due to the limited transmission power of the mobile hosts, imperfect power control, shadowing, etc, the total load must be well below 1. Indeed, in radio network planning [2], all the above factors are used to determine an interference margin (or noise rise) I_{margin}, based on which (3) becomes

$$\sum_i \alpha_i^{\text{UL}} \leq \frac{I_{margin} - 1}{I_{margin}}. \qquad (4)$$

When each mobile user uses a small portion of the available resources, we have $\frac{W}{r_i SIR_i} \gg 1$, hence $\alpha_i^{\text{UL}} \approx \frac{r_i SIR_i}{W}$ and the constraint (4) can be approximated by

$$\sum_i r_i SIR_i \leq \rho^{\text{UL}} W, \quad \text{where} \quad \rho^{\text{UL}} = \frac{I_{margin} - 1}{I_{margin}}. \qquad (5)$$

The above results can be generalized for multiple cells by considering the intercell interference coefficient $f =$(other cell interference)/(intracell interference) [6,2], in which case ρ^{UL} is multiplied by $1/(1 + f)$.

Resource Usage in the Downlink. In the downlink, for the case of a single cell the signal-to-interference ratio at mobile i is

$$SIR_i = \frac{W}{r_i} \frac{g_i p_i}{\theta_i g_i \sum_{j \neq i} p_j + \eta_i}, \qquad (6)$$

where r_i is the transmission rate, p_i is the transmission power, g_i is the path gain between the base station and mobile i, θ_i is the orthogonality factor for the codes used in the downlink, and η_i is the power of the background noise at mobile i. The orthogonality factor θ_i depends on multipath effects, hence can be different for different mobiles. In the case of multiple cells, (6) can be generalized by multiplying the term $\theta_i g_i \sum_{j \neq i} p_j$ with $(1 + f_i)$, where f_i is the intercell interference coefficient, which in the downlink can be different for different mobiles.

The total transmission power at the base station has an upper bound, say P. Hence, the downlink is *power-limited* and resource usage is determined by the transmission power. As with the uplink, in the downlink the utilization in practise cannot reach 100%. Hence, the resource constraint in the downlink is

$$\sum_i p_i \leq \rho^{\mathrm{DL}} P . \tag{7}$$

3.2   Service Differentiation in the Uplink

In this section we discuss service differentiation in the uplink. Assume that each mobile user has an associated weight; this weight can correspond to a service class selected by the mobile user. To achieve fair resource allocation, wireless resources should be allocated in proportion weights. Due to such proportional allocation, and since resource usage in the uplink is given by the product of the transmission rate and signal-to-interference ratio, from (5) we have

$$r_i SIR_i = \frac{w_i}{\sum_j w_j} \rho^{\mathrm{UL}} W . \tag{8}$$

Recall that users can potentially control both the transmission rate and the signal-to-interference ratio. How this selection is done depends on what the user values. Indeed, for users that value only the average throughput, i.e., the product of the transmission rate and frame success rate, one can prove that the optimal signal-to-interference ratio depends solely on the frame error rate as a function of the signal-to-interference ratio, and is independent of the transmission rate [9]. In this case, (8) can be used to compute the transmission rate as follows

$$r_i = \frac{1}{SIR_i} \frac{w_i}{\sum_j w_j} \rho^{\mathrm{UL}} W . \tag{9}$$

The application of the above equation would be part of load control. Equation (9) can also be applied for traffic that is rate adaptive, but has fixed quality of service requirements in terms of the frame error rate, *FER*. In this case, outer loop power control would be responsible for adjusting the target *SIR* in order to achieve the predetermined *FER*. Note that for traffic with fixed quality of service requirements, there is no requirement that *FER* is a function of only *SIR*; *FER* can also depend on the transmission rate r, which is the case in practise.

The application of (9) allows two alternatives regarding the value for the signal-to-interference ratio that appears in the right-hand side: SIR_i can be either the target *SIR* for mobile i, or it can be the actual *SIR* for mobile i, which is estimated from (1). We investigate these two alternatives in Sect. 4.

In the case of traffic with fixed-rate requirements that is adaptive to the signal quality, based on (8), *SIR* values can be allocated according to

$$SIR_i = \frac{1}{r_i} \frac{w_i}{\sum_j w_j} \rho W \,.$$

Application of this equation would involve outer loop power control.

Although the models and procedures discussed above lead to simple (proportional) allocation of resources, they can be theoretically justified in terms of economically efficient resource usage [9]. Indeed, in the case of rate adaptive traffic with fixed quality of service requirements, mobile users having a fixed weight correspond to users with a logarithmic utility function of the form $w_i \log(r_i)$, where a user's utility function represents his value for a particular level of service. If each user is charged in proportion to his weight, equivalently in proportion to the amount of resources he is receiving, then the resulting allocations in the equilibrium maximize the aggregate utility of all users (social welfare). If the utility of a user i has a more general form $U_i(r_i)$, then the user's weight can be modified slowly in order to satisfy $w_i(t) = U_i'(r_i(t))r_i(t)$.

3.3   Service Differentiation in the Downlink

In the downlink the resource constraint is related to the total transmission power, (7). Based on this equation we can allocate instantaneous power levels in proportion to weights. However, due to multipath fading, this approach has the disadvantage that the received signal quality at a mobile host would fluctuate. Moreover, it requires modification of the fast closed-loop power control procedure, which is implemented in the physical layer of CDMA systems. Another alternative is to determine *average* power levels, which are then used to compute the transmission rate. According to this approach the average power for user i would be

$$\bar{p}_i = \frac{w_i}{\sum_j w_j} \rho^{\mathrm{DL}} P \,.$$

As was the case in the uplink, if users value only the average throughput of data transmission, then the optimal signal-to-interference ratio is independent of the transmission rate. Hence, from (6), the transmission rate for user i will be

$$r_i = \frac{W}{SIR_i} \frac{1}{l_i I_i} \frac{w_i}{\sum_j w_j} \rho^{\mathrm{DL}} P \,, \tag{10}$$

where $l_i = 1/g_i$ and $I_i = \theta_i g_i \sum_{j \neq i} p_j + \eta_i$ is the path loss and interference, respectively, for user i. If a mobile is moving, it is appropriate to take average values for its path loss and interference. In the last equation observe that, as was the case for the power, the transmission rate is proportional to the weight.

Equation (10) requires estimation of the path loss from the base station to the mobile, which can be done using the pilot bits in the downlink control channel. There are two alternatives as to where the selection of r_i using (10) is performed: the mobile host or the radio network controller (RNC). The first alternative results in more complexity at mobile hosts. Moreover, it requires communicating the ratio $\rho^{\mathrm{DL}} P / \sum_j w_j$ from the RNC to the mobiles. On the other hand, if the RNC performs the selection, then there would be increased signalling overhead between the mobile and the RNC, since the values of the path loss and the interference would need to be communicated to the RNC; how often these parameters change depends on the mobile's movement.

It is interesting to observe from (10) that, for the same weight, a higher path loss gives a smaller transmission rate. To avoid such differentiation due to a mobile's position, the parameters in (10) can be replaced with their corresponding averages *over all mobile hosts*. Hence, if $\bar{l}$ is the average path loss and $\bar{I}$ is the average interference over all mobiles, rates can be allocated using

$$r_i = \frac{W}{SIR_i} \frac{1}{\bar{l}\bar{I}} \frac{w_i}{\sum_j w_j} \rho^{\mathrm{DL}} P \,.$$

Whether to allocate resources in the downlink with or without dependence on the mobile's position will be determined by a wireless operator's policy.

4   Numerical Investigations

In this section we present simulation investigations that demonstrate the effectiveness of our approach for supporting service differentiation, and investigate how various characteristics of the wireless system, such as power control and SIR estimation errors, and discrete transmission rates, affect service differentiation and the system's transient behaviour. Due to space limitations, we present results only for the uplink, where rates are allocated using (9).

The accuracy of approximation (5), on which (9) is based, depends on the number of mobile users and the utilization. Indeed, for the parameters considered in our experiments, the ratio of throughput for mobile users with different weights using (4) differs from the ratio of throughput using the approximation (5) by less than 5%. Moreover, observe from (9) that the rate is inversely proportional to the signal-to-interference ratio, and proportional to the utilization; recall that such an allocation corresponds to a logarithmic utility function. Results for general forms of user utilities, including comparison of resource sharing in the uplink and downlink, and how it is affected by a mobile's distance from the base station and the wireless network's load appear in [9].

We consider a single cell. The simulation parameters are shown in Table 1. Both the power control and SIR estimation errors are assumed to be lognormally distributed. We assume that the start time for each mobile is randomly distributed in the interval $[0, \lceil N/2 \rceil]$, where N is the total number of mobiles. The simulation experiments were performed in MATLAB, and utilized some functionality of the RUNE library [10].

There are two alternatives for applying (9), which refer to the value of the signal-to-interference ratio that appears on the right-hand side: SIR_i can be

Table 1. Simulation parameters. d is distance in Km

parameter	values
mobile power, $\bar{p}$	250 mW
interference margin, I_{margin}	3 dB
noise, η	10^{-13} Watt
path gain, $g(d) = kd^{-u}$	$u = 3.52, k = 1.82 \cdot 10^{-14}$
target SIR	5
power control error, PCE	0 or 1 dB
SIR estimation error, SIR_{err}	0 or 1 dB
# of mobiles, N	11

either the target SIR for mobile i, or it can be the actual SIR for mobile i, which is estimated from (1). The latter leads to more robust behaviour in cases when the mobile does not obey power control commands from the base station.

Figs. 2(a) and 2(b) show the transmission rate as a function of frame number, for continuous and discrete rate values, when the target SIR is used in (9). Each graph displays the rate for two mobiles with weights 1 and 2. Figs. 3(a) and 3(b) show the same results when the actual SIR is used. Observe that when the target SIR is used, Figs. 2(a) and 2(b), convergence is reached very fast, as soon as the last mobile has entered the system. On the other hand, when the actual SIR is used, Figs. 3(a) and 3(b), convergence takes longer.

Figs. 4(a) and 4(b) show the rate as a function of frame number, in the case of imperfect power control. Observe in Fig. 4(a) that imperfect power control has no effect when the target SIR is used in the rate allocation. On the other hand, Fig. 4(b) shows that imperfect power control has an effect when the actual SIR is used in the rate allocation. This occurs because the transmission powers appear in (1). Despite the rate variations in Fig. 4(b), observe that average service differentiation is still achieved.

Figs. 5(a) and 5(b) show the rate as a function of frame number, when SIR estimation errors occur. These errors affect the estimation of the target SIR, hence they affect rate allocations when the target SIR is used, Fig. 5(a). Moreover, errors in the target SIR result in fluctuation of the transmitting powers, hence they also affect rate allocations when the actual SIR is used, Fig. 5(b).

5   Conclusions

We have presented models and procedures for fair and efficient service differentiation in third generation mobile networks based on WCDMA. With simulation experiments, we demonstrated the effectiveness of our approach, and investigated the effects of power control and SIR estimation errors, and discrete transmission rates on service differentiation and the system's transient behaviour.

Ongoing work includes extensions for traffic that is adaptive to both the transmission rate and the signal quality and, unlike best-effort traffic, has a utility that depends on the loss rate in addition to the average throughput.

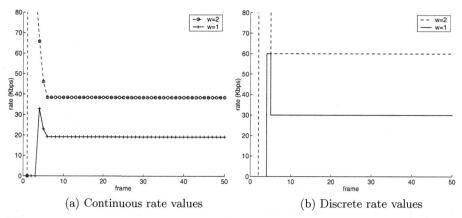

(a) Continuous rate values   (b) Discrete rate values

Fig. 2. Rate as function of frame number, when rate allocation is based on the target SIR. $PCE = 0, SIR_{err} = 0$

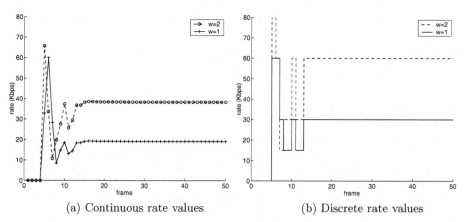

(a) Continuous rate values   (b) Discrete rate values

Fig. 3. Rate as function of frame number, when rate allocation is based on the actual SIR. $PCE = 0, SIR_{err} = 0$

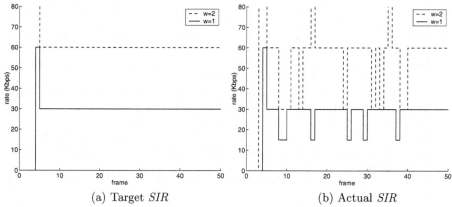

(a) Target SIR (b) Actual SIR

Fig. 4. Rate (discrete values) as function of frame number. $PCE = 1$ dB, $SIR_{err} = 0$

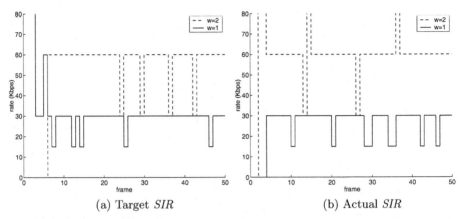

(a) Target *SIR*   (b) Actual *SIR*

Fig. 5. Rate (discrete values) as function of frame number. $PCE = 0, SIR_{err} = 1$ dB

Related work involves the investigation of models and procedures for service differentiation and seamless congestion control in wireless and wired networks.

Acknowledgments. The authors would like to thank the anonymous reviewers for their helpful comments and suggestions.

References

1. Dixit, S., Guo, Y., Antoniou, Z.: Resource management and quality of service in third-generation wireless networks. IEEE Commun. Mag. (2001) 125–133
2. Holma, H., Toskala, A.: WCDMA for UMTS (revised edition). Wiley, New York (2001)
3. Guo, Y., Chaskar, H.: Class-based quality of service over air interfaces in 4G mobile networks. IEEE Commun. Mag. (2002) 132–137
4. Jorguseski, L., Fledderus, E., Farserotu, J., Prasad, R.: Radio resource allocation in third-generation mobile communication systems. IEEE Commun. Mag. (2001) 117–123
5. Nanda, S., Balachandran, K., Kumar, S.: Adaptation techniques in wireless packet data services. IEEE Commun. Mag. (2000) 54–64
6. Gilhousen, K.S., *et al*: On the capacity of a cellular CDMA system. IEEE Trans. on Vehicular Technology **40** (1991) 303–312
7. Yun, L.C., Messerschmitt, D.G.: Power control for variable QoS on a CDMA channel. In: Proc. of IEEE MILCOM'94, NJ, USA (1994)
8. Sampath, A., Kumar, P.S., Holtzman, J.M.: Power control and resource management for a multimedia CDMA wireless system. In: Proc. of IEEE Int. Symp. Personal, Indoor, Mobile Radio Commun. (PIMRC). (1995)
9. Siris, V.A.: Resource control for elastic traffic in CDMA networks. In: Proc. of ACM International Conference on Mobile Computing and Networking (MOBICOM). (2002)
10. Zander, J., Kim, S.L.: Radio Resource Management for Wireless Networks. Artech House (2001)

Policy-Driven Traffic Engineering for Intra-domain Quality of Service Provisioning

Panos Trimintzios, Paris Flegkas, and George Pavlou

Centre for Communication Systems Research
School of Electronics and Physical Sciences
University of Surrey, Guildford, Surrey, GU2 7XH, U.K.
{P.Trimintzios, P.Flegkas, G.Pavlou}@eim.surrey.ac.uk

Abstract. Given the emergence of IP networks and the Internet as the multi-service network of the future, it is plausible to consider its use for transporting demanding traffic with high bandwidth and low delay and packet loss requirements. Emerging technologies for scalable quality of service such as Differentiated Services and MPLS can be used for premium quality traffic. We are looking at the problem of intra-domain provisioning in an automated manner from an Internet Service Provider's (ISPs) point of view, i.e. we want to satisfy the contracts with our customers while optimising the use of the network resources. We need to be able to dynamically guide the behaviour of such an automated provisioning system in order to be able to meet the high-level business objectives. The emerging policy-based management paradigm is the means to achieve this requirement. In this paper we devise first a non-linear programming formulation of the traffic engineering problem and show that we can achieve the objectives and meet the requirements of demanding customer traffic through the means of an automated provisioning system. We extend the functionality of the automated system through policies. We define resource provisioning policies, and we present example scenarios of their enforcement.

1 Introduction

Differentiated Services (DiffServ) [1] is seen as the emerging technology to support Quality of Service (QoS) in IP backbone networks in a scalable fashion. Multi-Protocol Label Switching (MPLS) [2] can be used as the underlying technology to support traffic engineering. It is possible to use these technologies to support premium traffic with stringent QoS requirements. This can be done through careful traffic forecasting based on contracted premium services with customers and subsequent network provisioning in terms of routing and resource management strategies. In this paper we show that this is a feasible solution for guaranteeing QoS for demanding premium traffic. In order to provide adequate quality guarantees for demanding traffic over an IP Autonomous System (AS), we propose to use the DiffServ framework together with MPLS for Traffic Engineering (TE). Customers have contractual Service Level Agreements (SLAs). ISPs on the other hand want to meet the customers' demands as these are described in the Service Level Specification (SLS) [3], which is

B. Stiller et al. (Eds.): QofIS/ICQT 2002, LNCS 2511, pp. 179-193, 2002.
© Springer-Verlag Berlin Heidelberg 2002

technical part of an SLA, while at the same time optimising the use of network resources.

Policy-based Management has been the subject of extensive research over the last decade [4]. Policies are seen as a way to guide the behaviour of a network or distributed system through high-level, declarative directives. We view policy-based management as a means of extending the functionality of management systems dynamically, in conjunction with pre-existing "hard-wired" logic [5]. Policies are defined in a high-level declarative manner and are mapped to low-level system parameters and functions, while the system intelligence can be dynamically modified added and removed by manipulating policies.

The rest of the paper is organised as follows. Section 2 presents the Traffic Engineering and resource provisioning system architecture together with the policy-based extensions. In section 0 we present our network dimensioning algorithm and the corresponding simulation results. In section 4 we enlist potential policies related to network dimensioning and we present policy enforcement examples. Section 5 presents the related work and finally section 6, concludes and suggests extensions of this work.

2   Architecture

In [6] we have designed a system for supporting QoS in IP DiffServ Networks. This architecture can be seen as a detailed decomposition of the concept of an extended Bandwidth Broker (BB) realized as a hierarchical, logically and physically distributed system. A detailed description can be found in [6].

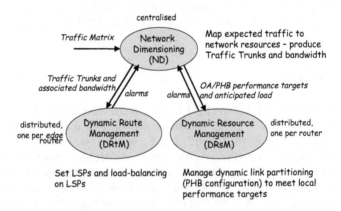

Fig. 1 The components of the Traffic Engineering system

The Traffic Engineering (TE) aspects of this architecture are shown in Fig. 1 The Network Dimensioning (ND) component is responsible for mapping traffic requirements to the physical network resources and for providing Network Dimensioning directives in order to accommodate the predicted traffic demands. The lower level of the traffic engineering part intends to dynamically manage the resources allocated by Network Dimensioning during the system operation in real-time, in order to react to statistical traffic fluctuations and special arising conditions. This part is realized by the Dynamic Route (DRtM) and Dynamic Resource Management (DRsM), which both monitor the network resources and act to medium to short term fluctuations. DRtM operates at the edge nodes and is responsible for managing the routing processes in the network. It mainly influences the parameters based on which the selection of one of the established MPLS Labelled Switched Paths (LSPs) is effected at an edge node with the purpose of load balancing. An instance of DRsM operates at each router and aims to ensure that link capacity is appropriately distributed among the PHBs in that link. It does so by managing the buffer and scheduling parameters according to the guidelines provided by ND. Thus, the provisioning of the network is effectively achieved by both taking into account the long-term service level subscriptions in a time dependent manner (ND) and the dynamic network state.

We extended the traffic engineering system to be able to drive its behaviour through policies. The resulting extended system architecture is depicted in Fig. 2. The Policy extensions include components such as the Policy Management Tool, Policy Repository, and the Policy Consumers. A single Policy Management Tool exists for providing a policy creation environment to the administrator where policies are defined in a high-level declarative language and after validation and static conflict detection tests, they are translated into object-oriented representation (information objects) and stored in a repository. The Policy Repository may be physically distributed since the technology for implementing this component is the LDAP (Lightweight Directory Access Protocol) Directory. After policies are stored, activation information is passed to the responsible Policy Consumer in order to retrieve and enforce them.

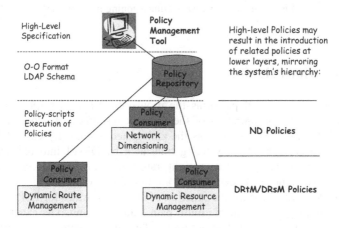

Fig. 2 Policy-driven Traffic Engineering Architecture

The methodology for applying policies to a hierarchically distributed system, like our architecture, is described in detail in [5]. Our model assumes many instances of Policy Consumers, every instance attached to the component that is influenced by the policies this Policy Consumer is responsible to enforce, complementing the static management intelligence of the above layer of hierarchy. Policies may be introduced at every layer of our system but higher-level policies may possibly result in the introduction of related policies at lower levels, mirroring the system hierarchy.

Input:

Network topology, link properties (capacity, propagation delay, supported PHBs)

Pre-processing:

- Request traffic forecast, i.e. the potential traffic trunks (TT)
- Obtain statistics for the performance of each PHB at each link
- Determine maximum allowable hop count K per TT according to PHB statistics

Optimisation phase:

- Start with an initial allocation (e.g. using the shortest path for each TT)
- Iteratively improve the solution such that for each TT find a set of paths for which:
- The minimum bandwidth requirements of the TT are met
- The hop-count constraints K is met (delay and loss requirements are met)
- The overall cost function is minimized

Post-processing:

- Allocate extra capacity to paths of each OA according to resource allocation policies
- Sum all the path requirements per link per OA, give minimum (optimisation phase) and maximum (post-processing phase) allocation directives to DRsM
- Give the appropriate multiple paths calculated in the optimisation phase to DRtM
- Store the configuration into the Network Repository

Fig. 3 The basic Network Dimensioning functionality

3   Network Dimensioning

Network Dimensioning (ND) performs the provisioning and is responsible for the long to medium term configuration of the network resources. By configuration we mean the definition of LSPs as well as the anticipated loading for each PHB on all interfaces, which are subsequently being translated by DRsM into the appropriate scheduling parameters (e.g. priority, weight, rate limits) of the underlying PHB implementation. The values provided by ND are not absolute but are in the form of a range; constituting directives for the function of the PHBs, while for LSPs they are in the form of multiple paths to enable multi-path load balancing. The exact PHB configuration values and the load distribution on the multiple paths are determined by DRsM and DRtM respectively, based on the state of the network, but should always adhere to the ND directives.

ND runs periodically, getting the expected traffic per Ordered Aggregate [7] (OA) in order to be able to compute the provisioning directives. The dimensioning period is in the time scale of a week while the forecasting period is in the time scale of hours. The latter is a period in which we have considerably different predictions as a result of the time schedule of the subscribed SLSs. For example, ND might run every Sunday evening and provide multiple configurations i.e. one for each period of each day of the week (morning, evening, night). So, effectively the resource provisioning cycle is at the same time scale of the forecasting period.

The objectives are both traffic and resource-oriented. The former relate to the obligation towards customers, through the SLSs. These obligations induce a number of restrictions about the treatment of traffic. The resource-oriented objectives are related to the network operation, more specifically they are results of the high-level business policy that dictates the network should be used in an optimally. The basic Network Dimensioning functionality is summarised in Fig. 3.

3.1   Network Dimensioning Algorithm

The network is modelled as a directed graph $G = (V, E)$, where V is a set of nodes and E a set of links. With each link $l \in E$ we associate the following parameters: the link physical capacity C_l, the link propagation delay d_l^{prop}, the set of the physical queues K, i.e. Ordered Aggregates (OAs), supported by the link. For each OA, $k \in K$ we associate a bound d_l^k (deterministic or probabilistic depending on the OA) on the maximum delay incurred by traffic entering link l and belonging to the $k \in K$, and a loss probability p_l^k of the same traffic.

The basic traffic model of ND is the traffic trunk (TT). A traffic trunk is an aggregation of a set of traffic flows characterized by similar edge-to-edge performance requirements [8]. Also, each traffic trunk is associated with one ingress node and one egress node, and is unidirectional. The set of all traffic trunks is denoted by T. Each trunk $t \in T$, from ingress node v_i to egress node v_e ($v_i, v_e \in V$), is associated with bandwidth requirements in the form of a minimum B_t^{min} and a maximum B_t^{max}, where the minimum represents the requirement of the currently subscribed SLSs aggregation, and the maximum reflects the over-provisioning policies. We view TTs as the abstract representation of traffic with specific characteristics.

The *primary objective* of such an allocation is to ensure that the requirements of each traffic trunk are met as long as the traffic carried by each trunk is at its specified minimum bandwidth. However, with the possible exception of heavily loaded conditions, there will generally be multiple feasible solutions. The design objectives are further refined to incorporate other requirements such as: (a) avoid overloading parts of the network while other parts are under loaded, and (b) provide overall low network load (cost).

The last two requirements do not lead to the same optimisation objective. In any case, in order to make the last two requirements more concrete, the notion of "load" has to be quantified. In general, the load (or cost) on a given link must be an increasing function of the amount of traffic the link carries. This function may refer to link utilization or may express an average delay, or loss probability on the link. Let x_l^k denote the capacity demand for OA $k \in K$ satisfied by link l. Then the link cost induced by the load on OA $k \in K$ is a convex function, $f_l^k(x_l^k)$, increasing in x_l^k. The total link cost per link is defined as: $F_l(\overline{x}_l) = \sum_{k \in K} f_l^k(x_l^k)$, where $\overline{x}_l = \{x_l^k\}_{k \in K}$ is the vector of demands for all OAs of link l. In order to take into account that link capacities may be different, the cost may be modified to reflect equivalent utilization by normalizing with the link capacities C_l.

Provided that appropriate buffers have been provided at each router and the scheduling policy has been defined, then $f_l^k(x_l^k)$ may specify the *equivalent capacity* needed by PHB $k \in K$ on link l in order to satisfy the loss probability associated with that PHB. Hence, the total cost per link is the total equivalent capacity allocated on link l. This has the following drawbacks: the cost definition depends on the PHB implementation at the routers, and the cost functions may not be known, or may be too complex. Hence we are using approximate functions, e.g. $f_l^k(x_l^k) = a_l^k x_l^k$.

We can now formally define the objectives (a) and (b) described above as follows: *Avoid overloading parts of the network*:

$$\text{minimize} \ \max_{l \in E} F_l(\overline{x}_l) = \ \max_{l \in E} \left\{ \sum_{k \in K} f_l^k(x_l^k) \right\} \tag{1}$$

Minimize overall network cost:

$$\text{minimize} \ \sum_{l \in E} F_l(\overline{x}_l) = \sum_{l \in E} \left(\sum_{k \in K} f_l^k(x_l^k) \right) \tag{2}$$

We provide an objective that compromises between the previous two:

$$\text{minimize} \ \sum_{l \in E} (F_l(\overline{x}_l))^n = \sum_{l \in E} \left(\sum_{k \in K} f_l^k(x_l^k) \right)^n, \ n \geq 1 \tag{3}$$

When $n = 1$, the objective (3) reduces to (2), while when $n \to \infty$ it reduces to (1). Because of (3), even if the cost function $f_l^k(x_l^k)$ is linear function of load, the problem remains still a non-linear optimisation problem.

Handling the delay and loss constraints. Each traffic trunk is associated with an end-to-end delay and loss probability constraint of the traffic belonging to the trunk. Hence, the trunk routes must be designed so that these two constraints are satisfied. Both the constraints above are constraints on additive path costs under specific link costs (d_l^k and p_l^k respectively). However the problem of finding routes satisfying these constraints is, in general, NP-complete [9]. Given that this is only part of the problem we need to address, the problem in its generality is rather complex.

Usually, loss probabilities and delay for the same PHB on different nodes are of similar order. We simplify the optimisation problem by transforming the delay and loss requirements into constraints for the maximum hop count for each traffic trunk (TT). This transformation is possible by keeping statistics for the delay and loss rate of the PHBs per link, and by using the maximum, average or n-th quantile in order to derive the maximum hop count constraint. By using the maximum we are too conservative (appropriate for EF traffic), while by using an average we possibly underestimate the QoS requirements e.g. for AF traffic we may use the 80-th percentile. The accuracy of the statistics is determined by the period used to obtain them, methods like exponential weighted moving average over long period must be used.

Optimisation problem. For each traffic trunk $t \in T$ we denote as R_t the set of (explicit) routes defined to serve this trunk. For each $r_t \in R_t$ we denote as b_{r_t} the capacity we have assigned to this explicit route. We seek to optimise (3), such that the hop-count constraints are met, the explicit routes per traffic trunk should be equal to the trunks' capacity requirements.

This is a network flow problem and considering the non-linear formulation described above, for the solution we make use of the general gradient projection method [10]. This is an iterative method, where we start from an initial feasible solution, and at each step we find the minimum first derivative of the cost function path and we shift part of the flow from the other paths to the new path, so that we improve our objective function (3). If the path flow becomes negative, the path flow simply becomes zero. This method is based on the classic unconstraint non-linear optimisation theory, and the general point is that we try to decrease the cost function through incremental changes in the path flows.

The optimisation variables are the capacity variables b_{r_i} assigned to each route of each trunk, i.e. $\mathbf{b} = \{b_{r_i} : r_i \in R_t, t \in T\}$. In order to apply the gradient projection method we need to handle all the constraints. The non-negativity constraint, is handled by defining a cost function which increases very fast after $x_l \geq C_l$. At each iteration i and for each of the trunks $t \in T$, one of the variables b_{r_i} $r_i \in R_t$, say $b_{\bar{r_i}}$, is substituted by $b_{\bar{r_i}} = B_t - \sum_{r_i \in R_t - \{\bar{r_i}\}} b_{r_i}$, in order to equal the .capacity assigned to each variable b_{r_i} to the trunks' capacity requirements.

The hop-count constraint is handled as follows. At each step of the algorithm we are required to find a minimum weight path for each trunk $t \in T$ with the weights being the first derivative of the cost function (3). The minimum weight path computation algorithm has to check whether the path found satisfies the hop-count constraint. If not, then we need to find another path (not necessarily with the minimum weight but with a total weight less than at least one of the paths in R_t) that meets the hop-count constraint. This can be achieved by using a *k-shortest path* algorithm [11].

We control the iterative procedure described above by ending the iterations when the relative improvement in a step $i + 1$ from step i, is less than a parameter ε. More specifically the iterative procedure terminates when

$$\left| \frac{F(\mathbf{b}^{i+1}) - F(\mathbf{b}^i)}{F(\mathbf{b}^{i+1})} \right| < \varepsilon. \tag{4}$$

3.2   Simulation Results

The topologies used for experimentation were random, according to the models for random topology generation presented in [12]. We opted for a 90% confidence level and achieved confidence interval of about 8-10% of the corresponding values. The initial solution (step 0) of the iterative procedure of the ND algorithm is as if the traffic trunks were to be routed with a shortest path first (SPF) algorithm. This corresponds to the case where all traffic of a particular class from ingress to an egress is routed through the shortest path. The routing metric used for the SPF algorithm was set to be inversely proportional to the physical link capacity. The parameter ε was set to 0.001.

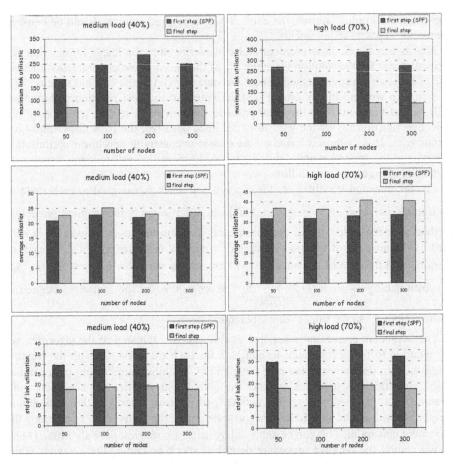

Fig. 4 Maximum, average and standard deviation of link load utilisation for different network sizes and network loading profiles

The edge nodes were 40-60% of the total network nodes. We defined as the *total throughput* of a network the sum of the capacities of the first-hop links emanating from all edge nodes. This is the upper bound of the throughput, and in reality it is a much greater than the real total throughput a network can handle, since the sum of the first-hop link capacity imposes the limit, but the rest of the backbone might not be able to handle that traffic. In our experiments we used 70% load of the *total throughput*, as the highly loaded condition, and a 40% for medium load.

Fig. 4 shows the maximum, average and standard deviation of the link load distribution for the different topology and traffic loading profiles. We show the results after the first step and the final step algorithm. It is clear that at step 0 solution, which corresponds to the SPF, parts of the network are over-utilized while others have no traffic at all. After the final step, which corresponds to the final output of our dimensioning algorithm, the traffic is balanced over the network. We can see that the algorithm manages to reduce the maximum link load below 100% for all the cases, while the SPF algorithm gives solutions with more than 300% maximum link load utilisation.

The average link utilisation increases slightly, since the algorithm uses paths with more number of links than the shortest path, and therefore the same load is accounted in more links than in the SPF case. We can also see that the traffic load is balanced over the network since the standard deviation of the link load utilisation from the average reduces to more than half of that in the case of SPF.

We run those experiments with the exponent in equation (3) being $n = 2$. This value compromises between minimizing the total (sum) of link costs and minimizing the maximum link load. In section 4.2 we are going to look at the effect of exponent n of the cost function.

Finally, in Table 1 we show the average running time of the various experiments conducted. We can see that even for quite large networks the running times are relatively low. For example for 300 node networks, for medium load the running time is about 17 minutes, and for high load about 25 minutes. These times are perfectly acceptable taking into account the timescales of the ND system operation.

Table 1 Average running time in seconds for the various network sizes

Network Size	Medium load	High load
10	0.055	0.061
50	9.761	10.164
100	123.989	302.079
200	529.532	1002.245
300	981.175	1541.937

4 Policy-Driven Network Dimensioning

In the architecture shown in Fig. 1, ND besides providing long-term guidelines for sharing the network resources, it can also be policy influenced so that its behaviour can be modified dynamically at run-time reflecting high-level, business objectives. The critical issue for designing a policy capable resource management component is to specify the parameters influenced by the enforcement of a policy that will result in different allocation of resources in terms of business decisions. These policies that are in fact management logic, are not hard-wired in the component but are downloaded on the fly while the system is operating.

4.1 Network Dimensioning Policies

We identify two categories of policies, *initialisation policies*, which concern policies that result in providing initial values to variables, which are essential for the functionality of ND and do not depend on any state but just reflect decisions of the policy administrator. The second category, *resource provisioning policies*, concerns those that influence the way it calculates the capacity allocation and the path creation configuration of the network. Such policies are those that their execution is based on the input from the traffic forecast module and on the resulting configuration of the network.

Since dimensioning is triggered mainly periodically, the policy administrator should specify this period. The priority of this policy should be specified in order not to cause any inconsistencies when re-dimensioning is triggered by notifications sent from the dynamic control parts of the system, that is when DRtM and DRsM are unable to perform an adaptation of the network with the current configuration. Another parameter that should be defined by policies is the cost-function used by ND. The administrator should be able either to choose between a number of pre-specified cost functions and/or setting values to parameters in the desired function. For example, if the approximate cost function used by ND is linear to the bandwidth allocated to a PHB, that is $f_l^k(x_l^k) = a_l^k x_l^k$ where x_l^k is the bandwidth allocated to PHB k on link l and a_l^k is a constant, the value of this constant could be specified by the policy administrator depending on the cost, i.e. importance, of a particular PHB. Another constraint imposed by policies is the maximum number of alternative paths that ND defines for every traffic trunk for the purpose of load balancing. Finally, the exponent n, as defined in equation (3), is another parameter that is specified by policies allowing the administrator to choose the relative merit of low overall cost and avoiding overload parts of the network.

The policy administrator should be able to specify the amount of network resources (giving a minimum, maximum or a range) that should be allocated to each OA. This will cause ND to take into account this policy when calculating the new configuration for this OA. More specifically, ND should allocate resources in a way that does not violate the policy and then calculate the configuration taking into account the remaining resources. A more flexible option should be for the policy administrator to indicate how the resources should be shared in specific (critical) links. After the optimisation phase ends, ND enters a post-processing stage where it will try to assign the residual physical capacity to the various OAs. This distribution of spare capacity is left to be defined by policies that indicate whether it should be done proportionally to the way resources are already allocated or it can be explicitly defined for every traffic class. A related policy is to specify the way the capacity allocated to each OA should be reduced because the link capacity is not enough to satisfy the predicted traffic requirements. ND actually translates the delay and loss requirements on an upper bound on the number of hops per route, the way this translation is done can also be influenced by policy rules. For example, the safest approach to satisfy the TT requirements would be to assume that every link and node belonging to the route induces a delay equal to the maximum delay caused by a link and node along the route. So, this policy rule will allow the administrator to decide if the maximum, average or minimum delay or loss induced by a node or link along the route should be used to derive the hop count constraint. Policies that allow the administrator for a particular reason to explicitly specify an LSP that a TT should follow can also be defined. Of course, this should override the algorithm's decision about the creation of the LSP for this TT.

4.2   Policy Enforcement Examples

In order to demonstrate the results of the enforcement of policies we used the topology depicted in the Fig. 5 as input to the ND component and a traffic load of 70 % of the total throughput of the network as defined in section 3.2.

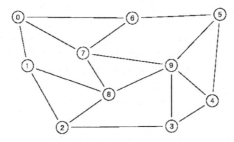

Fig. 5 Topology used for the policy examples

Our first example (P1) is a policy that wants to create an explicit LSP following the nodes 4, 9, 7, 6 with the bandwidth of the TT being 2 Mbps that is associated with this LSP. The administrator enters the policy in the Policy Management Tool using our proprietary policy language, which is then translated in LDAP objects according to an LDAP schema based on the Policy Core LDAP Schema [13] and stored in the Policy Repository. The syntax of our language as well as the extension to the Policy Core Information Model [14] with specific classes that reflect the policies described in the previous section are presented in [5]. The policy is entered with the following syntax:

If OA=<u>EF</u> and Ingress=<u>4</u> and Egress=<u>6</u> **then** SetupLSP <u>4-9-7-6</u> <u>2</u>Mbps   (P1)

After this rule is correctly translated and stored in the repository, the Policy Management Tool notifies the Policy Consumer associated with ND that a new policy rule is added in the repository, which then retrieves all the associated objects with this policy rule. From these objects the consumer generates code that is interpreted and executed representing the logic added in our system by the new policy rule. The pseudo-code produced by the Policy Consumer for (P1) is shown in Fig. 6. It searches for a TT in the traffic matrix that matches the criteria specified in the conditions of the policy rule regarding the OA, the ingress and egress node. If a TT is found then it executes the action that creates an LSP with the parameters specified and subtracts the bandwidth requirement of the new LSP from the TT. Note that if the administrator had in mind a particular customer then this policy should be refined into a lower level policy enforced on DRtM, mapping the address of this customer onto the LSP.

```
TT_OA: the set of TTs belonging an OA
v_i, v_e: ingress, egress nodes
b(tt): bandwidth requirement of tt

for each tt e TT_EF do
   if ((v_i == 4) and (v_e == 6))
      add_lsp ('4-9-7-6', 2000)
      b(tt) = b(tt) - 2000
   else
      policy failed - TT not found

              (P1)
```

```
maxLinkLoad: maximum link load
utilisation after optimisation
n=1: cost function exponent


optimisation_algorithm n
while (maxLinkLoad > 80 )
   n = n+1
   optimisation_algorithm n

              (P2)
```

Fig. 6 Pseudo-code produced for enforcing policy (P1) and policy (P2)

The second example (P2) of a policy rule concerns the effect of the cost function exponent in the capacity allocation of the network. As we mentioned earlier by increasing the cost function exponent the optimisation objective that avoids overloading parts of the network is favoured. If the administrator would like to keep the load of every link below a certain point then he/she should enter the following policy rule in our system by using our policy notation as:

$$\text{\textbf{If} maxLinkLoad > \underline{80\%} \textbf{then} Increase Exponent by \underline{1}} \tag{P2}$$

The same procedure explained in the previous example is followed again and the policy consumer enforces this policy by generating a script, which is shown in Fig. 6.

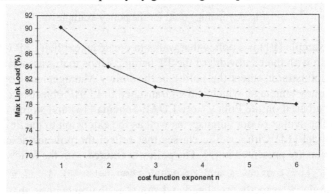

Fig. 7 Effect of the cost function exponent on the maximum link load utilisation

As it can be observed from Fig. 7 the enforcement of the policy rule caused the optimisation algorithm to run for 4 times until the maximum link load utilisation at the final step drops below 80%. The policy objective is achieved when $n = 4$.

5 Related Work

Next we describe some of the related works. This should not be considered an exhaustive survey of all related works. We have to mention that none of the previous works used a non-linear formulation and solution, and none of them considered and used the policy framework implications, as we do in this work.

The Traffic Engineering working group of the Internet Engineering Task Force (IETF) has chartered to define, develop, and specify principles, techniques, and mechanisms for traffic engineering in the Internet. Their main output up to now, is the definition of the basic principles for traffic engineering [15], and the requirements to support the interoperation of MPLS and Diffserv for traffic engineering [16].

Two similar works with the work presented in this paper are Netscope [17] and RATES [18]. Both of them try to automate the configuration of the network in order to maximise the network utilisation. The first one uses measurements to derive the traffic demands and then by employing the offline algorithm described in [19] tries to offload overloaded links. The latter uses a semi-online algorithm described in [20] to find the critical links which if they are chosen for routing would cause the greatest interference (i.e. reduce the maximum flow) of the other egress-ingress pairs of the network. Both of these works do not take into account any QoS requirements of the traffic and try only to minimize the maximum load of certain links.

The work described in this paper can be categorised as a time-dependent offline traffic engineering [15] Such problems can modelled as multi-commodity network flow optimisation problems [21]. The related works use such optimisation formulations, focusing on the use of linear cost functions, usually the sum of bandwidth requirements, and in most of the cases try to optimise a single criterion, minimize total network cost, or combine multiple criteria in a linear formula.

In Mitra et al [22] the traffic-engineering problem is seen as a multi-priority problem, which is formulated as a multi criterion optimisation problem on a predefined traffic matrix. This approach uses the notion of predefined admissible routes, and the objective is to maximise the carried bandwidth. The main objective of [23], is to design Virtual Private networks (VPNs), which will have allocated bandwidth on the links such that, when the traffic of a customer is optimally routed, a weighted aggregate measure over the service provider's infrastructure is maximized, subject to the constraint that each VPN carries a specified minimum. The weighted measure is the network revenue is the sum of the linear capacity costs for all links.

In [24] a model is proposed for off-line centralized traffic engineering over MPLS. The traffic engineering uses the following objectives: resource-oriented or traffic-oriented traffic engineering. The resource-oriented problem targets to load balance and minimise the resource usage. The objective function is a linear combination of capacity usage and load balancing, subject to link capacity constraints. The traffic-oriented model suggests an objective function that is a linear combination of fairness and throughput, where throughput is the total bandwidth guaranteed by the network and fairness as the minimum weighted capacity allocated to a traffic trunk. In [25] the authors propose an algorithm, which has a pre-processing phase and an on-line phase. In the pre-processing the goal is to find paths in the network to accommodate as much traffic of the traffic classes as possible from the source to the destination node. The algorithm minimizes a link cost function of the bandwidth assigned to each link for a traffic class. The second phase performs the on-line path selection for LSPs requests by using the pre-computed output of the multi-commodity pre-processing phase.

Works like [19], [26], and [27] try to achieve the optimal routing behaviour by appropriately configuring the shortest path routing metric. Wang et al. in [26] proved theoretically that any routing configuration, including the optimal one, could be achieved by the appropriate setting of the shortest path routing metric.

Finally, as far as online traffic engineering algorithms is concerned, they are mainly extensions of the QoS-routing paradigm [28]. These approaches are heuristics, known as Constraint-Shortest Path (CSPF). They utilise information kept in traffic engineering databases, which are populated from the routing flooding mechanisms, about link capacities, unreserved capacity, etc. Other online traffic engineering approaches, e.g. [29], [30], focus on load balancing on multiple equal or non-equal cost paths. These works are complementary to this work, since they can be used in conjunction (e.g. as parts of DRtM or DRsM functionality).

6   Conclusions and Further Work

Supporting demanding services requires dedicated networks with high switching capacity. In this paper we investigate the possibility of using common IP packet networks, with DiffServ and MPLS, as the key QoS technologies, in order to provision the network for such traffic. We proposed an automated provisioning system, targeting to support demanding SLSs while at the same time optimising the use of network resources. We seek to place the traffic demands to the network in such a way as to avoid overloading parts of the networks and minimize the overall network cost. We devised a non-linear programming formulation and we proved though simulation that we achieve our objectives.

We showed how this system can be policy-driven and described the components of necessary policy-based system extensions that need to be deployed in order to enhance or modify the functionality of policy influenced components reflecting high-level business decisions. We enlisted the policy parameters that influence the behaviour of dimensioning and presented enforcement of two example policies. As a continuation of this work, we will be focusing on defining policies for the rest of the components of the TE system and explore the issue of the refinement of policies to lower level policies forming a policy hierarchy. Also we intend to look at conflict detection and resolution mechanisms specific to our problem domain.

Acknowledgements

The work presented in this paper was supported by the European Commission under the IST Project IST-1999-11253 "Traffic Engineering for Quality of Service in the Internet at Large" (TEQUILA). The authors would like to thank their project partners for the fruitful discussions while forming the ideas presented in this paper.

References

[1]   S. Blake, et al., "An Architecture for Differentiated Services", IETF Informational RFC-2475, December 1998
[2]   E. Rosen, A. Viswanathan, R. Callon, "Multiprotocol Label Switching Architecture", IETF Standards Track RFC-3031, January 2001
[3]   D. Goderis, et al. "Service Level Specification Semantics and Parameters", IETF draft-tequila-sls-01.txt, December 2001 (available at: www.ist-tequia.org/sls)
[4]   M. Sloman, "Policy Driven Management For Distributed Systems", *Journal of Network and Systems Management*, vol. 2, no. 4, pp. 333-360, December 1994.
[5]   P. Flegkas, P. Trimintzios, G. Pavlou, "A Policy-based Quality of Service Management Architecture for IP DiffServ Networks", *IEEE Network Magazine*, vol. 16, no. 2, pp. 50-56, March/April 2002.
[6]   P. Trimintzios, et al., "A Management and Control Architecture for Providing IP Differentiated Services in MPLS-based Networks", *IEEE Communications Magazine*, vol. 39, no. 5, May 2001
[7]   D. Grossman, "New Terminology and Clarifications for DiffServ", IETF Informational RFC 3260, April, 2002

[8] T. Li, and Y. Rekhter, "Provider Architecture for Differentiated Services and Traffic Engineering (PASTE)" IETF Informational RFC-2430, October 1998

[9] Z. Wang, and J. Crowcroft, "Quality of Service Routing for Supporting Multimedia Applications", *IEEE JSAC*, vol. 14, no. 7, pp. 1228-1234, September 1996

[10] D. Bertsekas, *Nonlinear Programming*, (2nd ed.) Athena Scientific, 1999

[11] D. Eppstein, "Finding k-shortest paths", *SIAM J. on .Computing*, vol.28, no 2, pp.652-673, 1998

[12] E. W. Zegura, K. L. Calvert, and S. Bhattacharjee. "How to model an internetwork", In Proceedings of IEEE INFOCOM 96, vol.2, pp. 594-602, USA, March 1996

[13] J. Strassner, et al., "Policy Core LDAP Schema", IETF draft-ietf-policy-core-schema-14.txt, January 2002

[14] B. Moore et al., "Policy Core Information Model – Version 1 Specification", IETF Standards Track RFC-3060, February 2001

[15] D. Awduche, et al. "Overview and Principles of Internet Traffic Engineering", IETF Informational RFC-3272, May 2002

[16] F. Le Faucheur, et al. "Requirements for support of Diff-Serv-aware MPLS Traffic Engineering", IETF Internet draft, <draft-ietf-tewg-diff-te-reqts-05.txt>, work in progress, June 2002

[17] A. Feldmann and J. Rexford, "IP Network Configuration for Intradomain Traffic Engineering", *IEEE Network Magazine*, vol. 15, no. 5, pp. 46-57, September 2001

[18] P. Aukia, et al. "RATES: A Server for MPLS Traffic Engineering", *IEEE Network Magazine*, vol. 14, no. 2, pp. 34-41, March 2000

[19] B. Fortz, and M. Thorup, "Internet Traffic Engineering by Optimizing {OSPF} Weights", In Proc. of IEEE INFOCOM 2000, pp. 519-528, Israel, March 2000

[20] M. Kodialam, and T.V. Lakshman, "Minimum Interference Routing with Applications to Traffic Engineering", in Proc. IEEE INFOCOM00, pp. 884-893 March 2000

[21] R.K. Ahuja, T.L. Magnanti, and J.B. Orlin, *Network Flows: Theory, Algorithms and Applications*, Prentice Hall, 1993

[22] D. Mitra, and K. G. Ramakrishnan, "A Case Study of Multiservice, Multipriority Traffic Engineering Design for Data Networks", In Proc. IEEE GLOBECOM 99, pp. 1087-1093, Brazil, December 1999

[23] D. Mitra, J.A. Morrison and K.G. Ramakrishnan, "Virtual Private Networks: Joint Resource Allocation and Routing Design", In Proc. IEEE INFOCOM 99, USA, March 1999

[24] F. Poppe, et al. "Choosing the Objectives for Traffic Engineering in IP Backbone Networks Based on Quality-of-Service Requirements", In Proc. Workshop on Quality of future Internet Services (QofIS'00), pp. 129-140, Germany, September 2000

[25] S. Suri, et al. "Profile-based Routing: A New Framework for MPLS Traffic Engineering", In Proc. of the 2nd International Workshop on Quality of future Internet Services (QofIS'01), pp. 138-157, Portugal, September 2001

[26] Z. Wang, Y. Wang, and L. Zhang, "Internet Traffic Engineering without Full Mesh Overlaying", In Proc. of IEEE INFOCOM 2001, Alaska, April 2001

[27] Y. Breitbart, M. Garofalakis, A. Kumar and R. Rastogi, "Optimal Configuration of OSPF Aggregates", In Proc. of IEEE INFOCOM02, New York, USA, June 2002

[28] S. Chen, K. Nahrstedt, "An Overview of Quality-of-Service Routing for the Next Generation High-Speed Networks: Problems and Solutions", *IEEE Network Magazine*, vol. 12, no. 6, pp. 64-79, November 1998

[29] A. Elwalid, C. Jin, S H. Low, and I. Widjaja, "MATE: MPLS Adaptive Traffic Engineering", In Proc. of IEEE INFOCOM2001, pp. 1300-1309, Alaska, USA, April 2001

[30] Z. Cao, Z. Wang, and E. Zegura, "Performance of Hashing-based Schemes for Internet Load Balancing", In Proc. of IEEE INFOCOM 00, pp. 332-341, March 2000

A Congestion Control Scheme for Continuous Media Streaming Applications*

Pantelis Balaouras and Ioannis Stavrakakis

Department of Informatics and Telecommunications
University of Athens, Greece
{P.Balaouras, istavrak}@di.uoa.gr

Abstract. In this paper a novel congestion control scheme is proposed that aims at providing a smoother and reduced loss rate adaptation behavior compared to that under the basic Additive Increase Multiplicative Decrease (AI/MD) scheme; such a scheme is expected to be more appropriate for continuous media streaming applications. The basic features of the proposed scheme are that: (a) the rate increment is modulated by the distance between the flow's current rate and a targeted maximum rate determined by the streaming application; (b) the multiplicative decrease factor is shaped by the induced packet losses. These features are responsible for a dynamic and self-adjusting rate adaptation behavior that is shown to improve convergence to fairness, the oscillatory behavior and induced packet losses. Numerical results illustrate the good properties and intrinsic advantages of the proposed scheme.

1   Introduction

Internet's stability is due to the congestion control and avoidance algorithm [1, 2] implemented in the Transport Control Protocol (TCP), which belongs in the class of the Additive Increase / Multiplicative Decrease (AI/MD) algorithms [3]. TCP has been designed and successfully used for unicast reliable data transfer. Its window-based congestion control scheme and retransmission mechanism introduce typically end-to-end delays and delay variations, that make TCP unsuitable for the Continuous media (CM) streaming applications. For CM streaming services the User Datagram Protocol (UDP) is used as the base transport protocol, in conjunction with the Real Time Protocol (RTP) and Real Time Control Protocol (RTCP) [5]. Applications that use the UDP transport protocol should also implement end-to-end congestion control to retain the stability of the Internet otherwise all supported applications will suffer (UDP flows will get most of the bandwidth) and eventually the network will collapse [6]. The main requirements of the CM streaming applications are that their adaptation behavior should be as smooth as possible, and the induced losses be controlled and kept low. These requirements are taken into consideration in the design of the rate adaptation scheme proposed in this paper.

---

* Work supported in part by the Special Account for Research Grants of the National and Kapodistrian University of Athens and the General Secretariat for Research and Technology of Greece under the Joint Greek-Italian Scientific and Technological Cooperation Programme.

B. Stiller et al. (Eds.): QofIS/ICQT 2002, LNCS 2511, pp. 194–204, 2002.
© Springer-Verlag Berlin Heidelberg 2002

A rate adaptation scheme needs to be communicated the network state through a feedback mechanism. The feedback mechanisms that have been proposed are typically either RTCP-based [7]-[12] or ACK-based [14, 15]. In the first case the feedback delay is about 5 sec ($\pm$ 1.5 randomly determined), whereas in the second case it is equal to one round trip time. Most rate adaptation schemes for multimedia communication use the current packet loss rate to infer whether the network is congested or not, and thus adjust the rate in response. The increase function is either additive increase or multiplicative increase [15], whereas the decrease function is multiplicative decrease [7, 8, 14]. Alternatively, the rate may be set to a value calculated by an equation-based formula [4] which estimates the equivalent TCP throughput [10-14,16]. The latter schemes aim primarily at inducing a behavior that is TCP friendly.

Independently from whether a rate adaptation scheme is used for data or CM transferring, only a few of them [9, 16] exploit the packet loss rate to determine the new (decreased) rate. In the approach described in the present paper, the introduced loss dependent multiplicative decrease policy is simpler than that in [16] and more meaningful than that in [9], as will be shown later. Furthermore, the simplicity allows for a better understanding of the intrinsic properties of such a loss dependent decrease policy. In addition to this, an innovative aspect of the proposed congestion control scheme is that the associated increase policy takes into consideration the distance between the flow's current rate and a predetermined upper bound. The proposed scheme is compared against the basic AI/MD scheme in terms of smoothness and packet loss rates and its better behavior is demonstrated through a set of simulation results in an environment where only adaptable CM (RTP/UDP/IP) flows are transmitted (i.e., free of TCP flows), as it may be the case under the forthcoming differentiated services.

2 Description of the Distance Weighted Additive Increase and Loss Rate Dependent Multiplicative Decrease Scheme

In this section, a novel congestion control scheme - referred to as the DWAI/LDMD scheme - is introduced. The major characteristics of this scheme are: (a) the rate increment depends on the distance between the current rate and a maximum allowable rate (Distance Weighted Additive Increase (DWAI) policy); (b) the rate decrease depends on the actual value of the reported packet loss rate (Loss rate Dependent Multiplicative Decrease (LDMD) policy).

A discrete time network model that consists of n users (flows) is considered ; Let $\{\ldots, t-1, t, t+1, \ldots\}$ denote the discrete time instants which are assumed to coincide with the times at which all users receive feedback from their peer entities (synchronized feedback assumption). Let $f_i(t)$ denote the feedback received by user i at time instant t; $f_i(t)$ specifies the packet loss rate that the corresponding flow i experienced over the preceding time interval $(t - 1, t)$. The propagation delays between the peer entities are assumed to be zero. Under the assumption that the total network losses are distributed to all flows in proportion to their rate, it is implied that either all flows will experience packet losses or none (see Proposition 1). This implies - as determined by the proposed scheme described bellow - that all flows will react in the same manner: they will all either increase (under zero losses) or decrease (under non-zero losses) their rate. Such rate increases and decreases are subject to the constraint that the resulting rate remain in the range $[m, M]$, that is a minimum (m) and a maximum (M) allowed rate be respected.

Let $x_i(t)$ denote the transmission rate of user i over the interval $(t - 1, t)$. The proposed DWAI/LDMD scheme is determined by the following function describing the next rate $(x_i(t + 1))$ in terms of the current $(x_i(t))$ and the reported packet losses $f_i(t)$

$$x_i(t + 1) = \begin{cases} \min\{M, x_i(t) + \frac{M - x_i(t)}{M - m} I_{\text{DW}}\} & \text{if } f_i(t) = 0 \text{ (DWAI policy)}; \\ \max\{m, x_i(t)d(1 - f_i(t))\} & \text{if } f_i(t) > 0 \text{ (LDMD policy)}; \end{cases} \quad (1)$$

where, I_{DW} is the base increase step $(M - m > I_{\text{DW}} > 0)$, and d is a constant which is less than one $(d < 1)$. Notice that the normalized distance between the current rate and the maximum allowable rate M is taken into consideration when the rate is increased; the normalization quantity is equal to the distance between the maximum and the minimum allowable rates. The increment to the rate is maximized (equal to I_{DW}) when the rate is equal to the minimum allowable (m) while the increment to the rate is minimum (equal to 0) when the rate is equal to the maximum allowable (M). For a rate between the minimum and the maximum, the increment is a linear function of the distance of the current rate from the maximum allowable (equal to $I_{\text{DW}} \frac{M - x_i(t)}{M - m}$). Regarding the decrement of the rate when packet losses are reported, notice that the associated decrease factor $(d(1 - f_i(t))$ is not constant - as in the case of the basic AI/MD scheme - but depends on the reported packet loss rate $f_i(t)$, $(0 \leq f_i(t) \leq 1)$.

Let $x(t) = \{x_1(t), x_2(t), \ldots, x_n(t)\}$ denote the rate vector at time instant t associated with n supported users (flows) and let $X(t)$ denote the total network load associated to $x(t)$. That is, $X(t) = \sum_{i=1}^{n} x_i(t)$. Let X_{eff} denote the targeted maximum total load which is typically equal to the network capacity. Let $L(t)$ and $l_i(t)$ denote the number of packet losses for the network and user i, respectively, associated with time instant t (that is, occurred over $(t - 1, t)$). Since the length of the interval $(t - 1, t)$ is equal to 1, $L(t) = 1|X(t) - X_{\text{eff}}|^+$, where $|w|^+ = w$ if $w > 0$ and 0 otherwise. Let $f(t)$ and $f_i(t)$ denote the packet loss rate for the network and user (flow) i, respectively, associated with time instant t.

Proposition 1. *Under the assumption that the packet losses experienced by a flow i, are proportional to its rate, that is, $l_i(t) = L(t) \frac{x_i(t)}{X(t)}$, the resulting packet loss rates are identical for all flows and are given by*

$$f_i(t) = \frac{l_i(t)}{x_i(t)} = \frac{L(t)}{X(t)} = \frac{|X(t) - X_{\text{eff}}|^+}{X(t)} = f(t)$$

Since $f_i(t) = f(t)$, $f_i(t)$ will be used instead of $f(t)$ only when the association with flow i is to be emphasized.

The study of the properties of the proposed DWAI/LDMD scheme (presented in the next sections) is facilitated by re-writing the rate adaptation equation (1) as follows

$$x_i(t + 1) = \begin{cases} \min\{M, a_{\text{I}} + b_{\text{I}} x_i(t)\} & \text{if } f_i(t) = 0; \\ \max\{m, a_{\text{D}} + B_{\text{D}}(f(t)) x_i(t)\} & \text{if } f_i(t) > 0; \end{cases} \quad \text{where} \quad (2)$$

$$a_{\mathrm{I}} = cM, \quad c = \frac{I_{\mathrm{DW}}}{M - m}, \quad 0 < c < 1; \quad b_{\mathrm{I}} = 1 - c, \; b_{\mathrm{I}} < 1; \tag{3}$$

$$a_{\mathrm{D}} = 0; \quad B_{\mathrm{D}}(f(t)) = d(1 - f(t)), \quad B_{\mathrm{D}}(f(t)) < 1; \tag{4}$$

For comparison purposes (to be used occasionally in the next sections), the rate adaptation equation for the basic AI/MD scheme is given by

$$x_i(t+1) = \begin{cases} \min\{M, I_{\mathrm{AI}} + x_i(t)\} & \text{if } f(t) = 0; \text{ where } I_{\mathrm{AI}} > 0 \\ \max\{m, b_{\mathrm{D}} x_i(t)\} & \text{if } f(t) > 0; \text{ where } 0 < b_{\mathrm{D}} < 1. \end{cases} \tag{5}$$

Notice that the decrease factor $B_{\mathrm{D}}(f(t))$ (under the LDMD policy) is shaped by the induced packet losses, unlike the factor b_{D} (under the basic MD policy) which is unaffected by the reported packet losses. Thus, it is expected to apply a large decrease factor when losses are high (caused, for instance, by the initiation of a new high rate flow), which should help avoiding congestion soon and reduce the losses and a small decrease factor when losses are low, preserving thus the smoothness of the rate adaptation. Also notice that the rate increment $I_{\mathrm{DW}} \frac{M - x_i(t)}{M - m}$ under the DWAI policy is variable, takes into account the current rate $x_i(t)$ of the flow and is always equal to or less than the rate increment I_{AI} under the basic AI policy (assuming that $I_{\mathrm{DW}} = I_{\mathrm{AI}}$). Thus, it is expected that the total and per flow rate increment will be less than that under the basic AI policy, therefore the rate adaptation will be smoother and the induced packet losses lower under the DWAI policy, whereas the resulting load allocation among the flows will be more fair since the rate increment for a lower rate flow will be larger than that of a higher rate flow. These claims are also supported by the results of Section 4.

3   Analysis of the Behavior of the DWAI/LDMD Policy

It is easy to show, by employing approaches and definitions as in [3], that the proposed scheme converges to the efficiency line X_{eff} and fairness. Details as well as the proof of Proposition 2 may be found in [17].

In the sequel the responsiveness to BW availability (convergence time to efficiency) under the DWAI policy is quantified by determining the number k_{DW} of successive rate increases (steps) required in order for the network to reach the efficiency line X_{eff}, from an initial total load $X(t)$. BW may become available due to (a) initial startup, (b) flow termination(s), and (c) rate decrease actions after congestion detection.

Proposition 2. *Under the DWAI policy, the rate $x_i(t + k)$ of flow i after k successive increase steps, starting from some initial rate $x_i(t) \in (m, M)$, statifies the following: (a) $x_i(t + k) = x_i(t)(1 - c)^k + M(1 - (1 - c)^k) = M - (M - x_i(t))(1 - c)^k$ and (b) $x_i(t + k) < M$ for any $k \geq 1$.*

The following corollary is self-evident in view of Proposition 2.

Corollary 1 *Given an initial n-dimensional rate vector $\mathbf{x}(t)$ with total network load $X(t)$, the total network load $X(t + k)$, after k successive increases is given by*

$$X(t + k) = (1 - c)^k X(t) + nM(1 - (1 - c)^k) = nM - (nM - X(t))(1 - c)^k$$

Let $k_{\mathrm{DW}}(X(t))$ denote the number of successive increase steps required in order for the network load to reach X_{eff} starting from $X(t)$. Then

$$X_{\mathrm{eff}} = (1-c)^k X(t) + nM(1-(1-c)^k) \Leftrightarrow (1-c)^k = \tfrac{nM-X_{\mathrm{eff}}}{nM-X(t)} \Rightarrow$$

$$k_{\mathrm{DW}}(X(t)) = \left\lceil \log_{(1-c)} \tfrac{nM-X_{\mathrm{eff}}}{nM-X(t)} \right\rceil \qquad (6)$$

where $\lceil w \rceil$ is the smallest integer exceeding w. $k_{\mathrm{DW}}(X(t))$ determines the responsiveness to BW availability and is independent of $x(t)$.

In the sequel it is established that following the first application of the LDMD policy (that occurs when the total load exceeds X_{eff} for the first time and packet losses are reported) the rate adaptation behavior is periodic. That is the total load fluctuates between fixed levels below and above the efficiency line in a periodic fashion.

Let t_j denote the jth overload time instant, that is the jth time in which the total load $X(t)$ exceeds X_{eff} (that is, $X(t_j) > X_{\mathrm{eff}}$), $j \geq 1$. The following proposition determines the behavior of the total load in the next time instant $X(t_j + 1)$ and the resulting oscillation below the efficiency line, $S^-(t_j + 1)$, defined below.

Proposition 3. *If t_j is the jth overload time instant then: (a) $X(t_j + 1) = dX_{\mathrm{eff}}$ and, thus, the total load in the next time instant always falls below the efficiency line and this new total load is always the same, independent of $X(t_j)$. (b) $S^-(t_j + 1) = X_{\mathrm{eff}} - X(t_j + 1) = (1-d)X_{\mathrm{eff}}$ which is independent of $X(t_j)$ and, thus, the size of the oscillations below the efficiency line caused by decrease actions are constant.*

Proof. Since $X(t_j) > X_{\mathrm{eff}}$ and $f_i(t_j) = f(t_j) > 0$, the total load will be decreased according to the LDMD policy at time instant $t_j + 1$. The new total rate at $t_j + 1$ is given by (see Proposition 1)

$$X(t_j + 1) = \sum_{i=1}^{n} x_i(t_j + 1) = d(1 - \frac{X(t_j) - X_{\mathrm{eff}}}{X(t_j)})X(t_j) = dX_{\mathrm{eff}}$$

which proves (a) and (b). $\qquad\qquad\qquad\qquad\qquad\qquad\qquad\qquad\qquad\qquad\qquad\square$

It is obvious that the value of parameter d should be less that 1 to ensure that the total load will fall below the efficiency line. For video streaming applications it is suggested that d be set very close to 1 (e.g. 0.99) since this value preserves both the smooth adaptation behavior of the flows (small rate decrease) and the throughput of the network.

After a decrease step, the total load $X(t)$ will start increasing from an initial load dX_{eff} until it exceeds the efficiency line; let k^{T} denote the required number of increase steps. k^{T} is a metric of the responsiveness to small BW availability (due to a rate decrease action(s) as a reaction to congestion) in the sense that captures the required number of intervals (increase steps) to reach the efficiency line after such a rate decrease. In view of Proposition 3, after k^{T} increase steps the total load $X(t)$ will be reduced again to dX_{eff} in the next time instant and thus the total load adaptation behavior will be periodic: the total load will be at the overload level $X(t_j)$ at time instant t_j, $j > 1$, then will fall at the level dX_{eff} in the next time instant and then climb up to the same level $X(t_j)$ after k^{T} steps always in the same manner (reaching time instant $t_{j+1} = t_j + 1 + k^{\mathrm{T}}$ from which the process will be repeated). Thus, the function of the total load $X(t), t > t_1$ is periodic with period $T = k^{\mathrm{T}} + 1$. It should be noted that $X(t_j), j > 1$, is independent of the initial total load $X(0)$ and fixed. The size of the oscillations above X_{eff}, S^+, below X_{eff}, S^-, and the induced packet loss rates f^c (during the periodic adaptation period) are given by

$$S^+ = X(t_j) - X_{\text{eff}}, j > 1 \text{ fixed for all } j > 1 ; \quad S^- = (1-d)X_{\text{eff}} \tag{7}$$

$$f^c = \frac{S^+}{X(t_j)} = \frac{S^+}{X_{\text{eff}} + S^+} \tag{8}$$

Since f^c is fixed due to the periodic oscillatory behavior discussed above, as long as the number n of the flows remains unchanged, it is concluded that the decrease factor $B_D(f(t))$ will also be fixed. Notice that $X(t_1)$, that is the total load when the efficiency line is exceeded for the first time, will in general - be different than $X(t_j), j > 1$, since it would be determined by a sequence of increase steps starting from some initial value $X(0)$ and not dX_{eff}, and therefore depends on $X(0)$.

As discussed earlier, after a decrease step the total load $X(t)$ starts increasing from load dX_{eff} (see Proposition 3). The total load $X(t)$ will exceed the efficiency line after k_{DW}^{T} increase steps, where k_{DW}^{T} is given by (6) for $X(t) = dX_{\text{eff}}$. Since the base $(1-c)$ decreases (increases) as I_{DW} increases (decreases) (see (3), recall that $1 - c < 1$) it is concluded that k_{DW}^{T} decreases (increases) with I_{DW}.

Clearly, k_{DW}^{T} decreases with d, $0 < d < 1$ and $k_{\text{DW}}^{\text{T}} \to 1$ as $d \to 1$. Thus, d affects the number of increase steps required to reach the efficiency line from dX_{eff} and for a value sufficiently close to 1 it will require only one increase step to exceed the efficiency line. The total load $X(t_j), j > 1$ when the efficiency line is exceeded again is determined by the DWAI policy and it is given by (see Corollary 1)

$$X(t_j) = nM - (nM - dX_{\text{eff}})(1-c)^{k_{\text{DW}}^{\text{T}}}, j > 1 \tag{9}$$

In view of (9), S^+ and f^c (see (7) and (8)) become

$$S^+ = nM - (nM - dX_{\text{eff}})(1-c)^{k_{\text{DW}}^{\text{T}}} - X_{\text{eff}} \tag{10}$$

$$f^c = 1 - \frac{X_{\text{eff}}}{nM - (nM - dX_{\text{eff}})(1-c)^{k_{\text{DW}}^{\text{T}}}} \tag{11}$$

Proposition 4. *The DWAI/LDMD scheme presents a periodic adaptation behavior. After the first overload time instant t_1 the total load $X(t)$ exceeds the efficiency line every $k_{\text{DW}}^{\text{T}} + 1$ time instants, where k_{DW}^{T} is given by (6) for $X(t) = dX_{\text{eff}}$. Following t_1, the overshoot S^+, the undershoot S^- and the packet loss rate f^c are fixed, and given by (10), (7) and (11), respectively.*

4   Numerical Results and Discussion

In this section the better behavior of the DWAI/LDMD scheme compared to the basic AI/MD scheme (see (5)) in terms of smoothness and packet loss rates is discussed and demonstrated through a set of simulation results.

The LDMD policy ensures that after a decrease step the total load will be adapted to a level below the efficiency line after a single action, regardless of the number of flows in the network and the level of the packet loss rate (see Proposition 3). Contrary to the LDMD policy, the basic MD policy cannot ensure such an adaptation because of its fixed multiplicative decrease factor b_D. If b_D is selected to be relatively large, then the rate decrease under large losses would not be sufficient and thus, more losses would occur in the next time interval(s). On the other hand, if b_D is selected to be relatively small, then the rate decrease under small losses would be unnecessarily large with the obvious impact on smoothness and throughput. The DWAI policy induces a typically diverse (for diverse current rates) per flow rate increase, unlike the basic AI policy that induces the same rate increase for all flows independently from their current rate. As a consequence, the resulting rates will come closer after an application of the DWAI policy than of the basic AI policy, leading to a faster convergence to fairness.

In addition to the previous, it is expected that the packet losses induced under the DWAI policy would be lower than those under the basic AI policy (assuming that $I_{DW} = I_{AI}$) since more conservative increments are applied under the former policy when the per flow rate is away from its minimum, that is likely to be the case when the efficiency line is exceeded by the total load and losses occur. Finally, the DWAI policy results in an overall smoother rate increase compared to that under the basic AI policy (assuming that $I_{DW} = I_{AI}$). This is clearly the case since the increase step under the DWAI policy is (typically) lower than that under the basic AI policy.

In the sequel a set of simulation results concerning the rate adaptation behavior of the DWAI/LDMD and the basic AI/MD schemes under different configuration parameters are presented. The objective is to demonstrate the aforementioned relative advantages of the DWAI/LDMD scheme over the basic AI/MD scheme with respect to smoothness and packet losses, as well as point to some intrinsic characteristics of the DWAI/LDMD scheme, such as the dynamic nature of the increase/decrease step. The considered network model is implemented in Matlab. In all simulations the network capacity is set to 8 Mb/s and the number of flows n is equal to 12, 13 and 14 for the time periods [0,700), [700,900) and [900,1000], respectively. The initial rate vector for the first 12 flows is given by $x = \{m + 1 * (M - m)/n, m + 2(M - m)/n, \cdots, M\}$ where $m = 56$ kb/s, $M = 1.2$ Mb/s, $n = 12$ and, thus, the initial total load is $X(1) = 8.108.000$ bps. Flows 13 and 14 are initiated at time instants 700 and 900 with initial loads $x_{13}(700) = x_{14}(900) = 600$ kb/s. In the figures that follow the behavior of flows 1 and 12 (which have the furthest apart initial rates of rates of 151.3 kb/s and 1.2 Mb/s, respectively) are shown. Fig. 1 shows the results under the DWAI/LDMD scheme for parameters $I_{DW} = 30$ kb/s, $d = 0.99$, $m = 56$ kb/s, $M = 1.2$ Mb/s. Fig. 1 illustrates the convergence of two initially far away rates to a (common) fair rate, the oscillatory behavior both during the transition period ([0,~200]) as well as during the "steady state" period ([~200,700]), the response of flows 1 and 12 to the introduction of flow 13 ([700,900]) and to flow 14 ([900,1000]). It is clear that fairness is achieved and the oscillations are fairly low. The latter may be attributed to the dynamic and "self-adjusting" rate increment that assumes a small value when flows are away from their minimum rate m and to the dynamic and "self-adjusting" decrease factor $B_D(t) = d(1 - f(t))$, which assumes a large value (about $d = 0.99$) when losses are low (typical case) leading to a small decrement.

Fig. 2 shows the results under the basic AI/MD scheme with the "equivalent" fixed parameters. That is, $I_{AI} = I_{DW} = 30$ kb/s and $b_D = 0.99 = d$; that is, when losses are

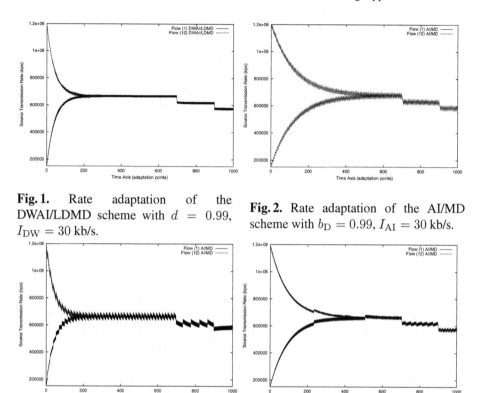

Fig. 1. Rate adaptation of the DWAI/LDMD scheme with $d = 0.99$, $I_{DW} = 30$ kb/s.

Fig. 2. Rate adaptation of the AI/MD scheme with $b_D = 0.99$, $I_{AI} = 30$ kb/s.

Fig. 3. Rate adaptation of the AI/MD scheme with $b_D = 0.95$, $I_{AI} = 30$ kb/s.

Fig. 4. Rate adaptation of the AI/MD scheme with $b_D = 0.979$, $I_{AI} = \frac{M - \frac{X_{eff}}{n}}{M - m} I_{DW}$.

very low $b_D \approx d(1 - f(t))$ and about the same rate decrement would be in effect for both schemes. The idea here is to show the behavior of a scheme that is equivalent to the above DWAI/LDMD scheme where the increments and decrements are not shaped by the rate distance from M and the induced losses $f(t)$, respectively.

From Fig. 1 and 2 it is observed that both the convergence speed to fairness and the oscillatory behavior is worst under the basic AI/MD scheme. The poor convergence speed to fairness is expected due to the application of the same increment to all flows independently from their rate as well as because of the typically larger decrease factor $b_D = d > d(1 - f(t)) = B_D(t)$. As a consequence, the rates would come closer under the DWAI/LDMD scheme as a smaller increment and a larger decrement would be applied to higher rate flows because of the rate dependent increment size and the smaller decrease factor ($B_D(t)$). The poorer oscillatory behavior under the basic AI/MD scheme during the times that fairness has been reached may be attributed to the larger increment under the basic AI policy, that leads to higher overshoots. These higher over-shoots - along with the smaller decrement under large packet losses - is responsible for

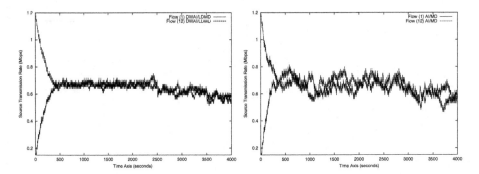

Fig. 5. Rate adaptation of the DWAI/LDMD scheme with $d = 0.99$, $I_{DW} = 30$ kb/s.

Fig. 6. Rate adaptation of the AI/MD scheme with $b_D = 0.979$, $I_{AI} = 30$ kb/s.

the induced higher mean conditional packet loss rate under the basic AI/MD scheme. Specifically, these loss rates are 2.22% (1.11%) , 2.61% (1.49%), 2.72% (1.88%) under the basic AI/MD (DWAI/LDMD) scheme, over the periods [0,700], [700,900] and [900,1000], respectively. The mean conditional packet loss rate is defined as the long term average of the following quantity defined only over periods with non zero packet losses: (number of packet lost over the current loss report period)/(number of packets sent over the current loss report period). The above simulation results point to the advantages of a scheme that adapts dynamically its key parameters in a meaningful way (DWAI/LDMD) over a scheme that does not (AI/MD). The remaining results emphasize some key trade-offs in the selected values of key parameters and the fact that a static parameter cannot effectively balance competing objectives.

Fig. 3 shows the results under the AI/MD scheme with a lower decrease factor $b_D = 0.95$ (and $I_{AI} = 30$ kb/s). This decrease factor would lead to a larger diversification of the decrements that would be applied to flows of different rates (larger decrement to higher rates) and thus improve the convergence speed to fairness (compared to $b_D = 0.99$). This indeed is observed to be the case. In fact $b_D = 0.95$ brings the convergence speed to fairness under the basic AI/MD scheme to almost match that under the DWAI/LDMD scheme for the parameters considered in Fig. 3 and 1. The price paid for this improvement (wrt $b_D = 0.99$) are larger oscillations (less smooth rates). The induced conditional packet loss rates are higher than those under the DWAI/LDMD scheme. Specifically the former are equal to 2.1%, 2.04% and 4.04% over the periods [0,700], [700,900] and [900,1000], respectively.

In Fig. 4 a lower fixed increment is selected that is about equal to the increment applied by the DWAI/LDMD scheme in Fig. 1 when the flows have reached the fairness line (period [∼200,700]). That is $I_{AI} = \frac{M - \frac{X_{eff}}{n}}{M - m} I_{DW}$. Also, the parameter b_D is selected to be equal to 0.979 which is the vaule of $B_D(t)$ that is applied by the DWAI/LDMD scheme in Fig. 1 during the period [0,700]. Therefore, the flows under the basic AI/MD scheme present the same smooth adaptation behavior as the flows under the DWAI/LDMD scheme as the former flows converge to the fairness line.

In the simulation results presented above it has been assumed that the packet loss rate is the same for all the flows; that is, $f_1(t) = f_2(t) = \cdots = f_n(t)$. This is a result of the assumption made in the study of the DWAI/LDMD scheme that packet losses are proportional to the flow's rate (see Proposition 1). As this may not be practically the case, further simulations have been carried out in a more realistic environment based on the ns-2 simulator. RTP/UDP/IP flows and drop tail routers that tend to distribute losses not necessarily in accordance with the flow's rates. The RTCP provides the packet loss rate feedback though the RTCP reports sent every 5 seconds. The round trip time is common for all flows, therefore the feedback is synchronized. The behavior of the DWAI/LDMD and the basic AI/MD schemes is examined in an environment where only adaptive CM (RTP/UDP/IP) flows are transmitted (i.e., free of TCP flows), as it may be the case under the forthcoming differentiated services. The results derived from the ns-2 simulations are in line with the theoretical as reported earlier. Flows under the DWAI/LDMD scheme (Fig. 5) present smoother adaptation behavior than those under the basic AI/MD scheme (Fig. 6) and lower packet losses. Details and comments - omitted here due to space limitations - may be found in [17].

References

[1] V. Jacobson, Congestion avoidance control. In Proceedings of the SIGCOMM '88 Conference on Communications Architectures and Protocols (1988).
[2] RFC-2001. TCP Congestion Avoidance, Fast Retransmit and Recovery Algorithms.
[3] D.M Chiu, R. Jain, Analysis of the Increase and Decrease Algorithms for Congestion Avoidance in Computer Networks, Computer Networks and ISDN Systems 17 (1989) 1-14.
[4] J. Padhye, V. Firoiu, D. Towsley, J.Kurose, Modeling TCP Throughput: A Simple Model and its Empirical Validation, ACM SIGCOMM'98.
[5] RFC 1889. RTP: A Transport Protocol for Real-Time Applications.
[6] Floyd, S., and Fall, K., Promoting the Use of End-to-End Congestion Control in the Internet. IEEE/ACM Transactions on Networking. August 1999.
[7] I. Busse, B. Defner, H. Schulzrinne, Dynamic QoS Control of Multimedia Application based on RTP. May 1995.
[8] J. Bolot,T. Turletti, Experience with Rate Control Mechanisms for Packet Video in the Internet. ACM SIGCOMM Computer Communication Review, Vol. 28, No 1, Jan. 1998.
[9] D. Sisalem, F. Emanuel, H. Schulzrinne, The Loss-Delay Based Adjustment Algorithm: A TCP-Friendly Adaptation Scheme. 1998.
[10] D. Sisalem and A. Wolisz, LDA+: Comparison with RAP, TFRCP. IEEE International Conference on Multimedia (ICME 2000), July 30 - August 2, 2000, New York.
[11] D. Sisalem and A. Wolisz, MLDA: A TCP-friendly congestion control framework for heterogeneous multicast environments. Eighth International Workshop on Quality of Service (IWQoS 2000), 5-7 June 2000, Pittsburgh.
[12] D. Sisalem and A. Wolisz, Constrained TCP-friendly congestion control for multimedia communication tech. rep., GMD Fokus, Berlin Germany, Feb. 2000.
[13] D. Sisalem, A. Wolisz, Towards TCP-Friendly Adaptive Multimedia Applications Based on RTP. Fourth IEEE Symposium on Computers and Communications (ISCC'1999).
[14] R. Rejaie, M. Handley, D. Estrin, An End-to-end Rate-based Congestion Control Mechanism for Realtime Streams in the Internet. Proc. INFOCOM 99, 1999.
[15] J. Padhye, J. Kurose, D. Towsley, R. Koodli, A Model Based TCP-Friendly Rate Control Protocol. Proc. IEEE NOSSDAV'99 (Basking Ridge, NJ, June 1999).

[16] T. Kim, S. Lu and V. Bharghavan,   Improving Congestion Control Performance Through Loss Differentiation.   International Conference on Computers and Communications Networks '99, Boston, MA. October 1999.

[17] P. Balaouras, I. Stavrakakis, A Self-adjusting Rate Adaptation Scheme with Good Fairness and Smoothness Properties, June 2002. http://cgi.di.uoa.gr/ istavrak/. Submitted in Journal Publication.

A New Path Selection Algorithm for MPLS Networks Based on Available Bandwidth Estimation*

Tricha Anjali[1][**], Caterina Scoglio[1], Jaudelice C. de Oliveira[1], Leonardo C. Chen[1],
Ian F. Akyildiz[1], Jeff A. Smith[2], George Uhl[2], and Agatino Sciuto[2]

[1] Broadband and Wireless Networking Laboratory,
School of Electrical and Computer Engineering,
Georgia Institute of Technology, Atlanta, GA 30332 USA
{tricha, caterina, jau, leochen, ian}@ece.gatech.edu
[2] NASA Goddard Space Flight Center,
Greenbelt, MD 20771 USA
{jsmith, uhl, asciuto}@rattler.gsfc.nasa.gov

Abstract. A network should deploy QoS-aware path selection algorithms for efficient routing. To this end, measurements of the various characteristics of the
network can provide insight into the state and performance of the network. In
this paper, we present a new QoS-aware path selection algorithm for flows requiring bandwidth guarantees based on an estimation algorithm for the available
bandwidth on the links of the network. The estimation algorithm for the available
bandwidth predicts the available bandwidth and also tells the duration for which
the estimate is valid with a high degree of confidence. Thus, the path selection
algorithm is adaptive and not very computationally intensive.

Keywords: QoS-aware path selection, Passive measurement, Available bandwidth.

1 Introduction

In recent years, there has been a tremendous growth in the Internet. New applications
present new traffic patterns and new Quality of Service (QoS) requirements. A QoS-
aware path selection algorithm is needed to improve the satisfaction of the user QoS
as well as for improved Traffic Engineering (TE). An efficient path selection procedure
enables a network operator to identify, for a new flow, a path with sufficient resources to
meet the flow's QoS requirements. These requirements are typically specified in terms
of bandwidth guarantees. Thus, the path identification involves some knowledge of the
availability of resources throughout the network. Thus, understanding the composition
and dynamics of the Internet traffic is of great importance. But observability of Internet traffic is difficult because of the network size, large traffic volumes, and distributed
administration. The main tools towards ensuring appropriate QoS satisfaction for all

---

* This work was supported by NASA Goddard and Swales Aerospace under contract number
S11201 (NAS5-01090).
** Corresponding Author. Research Assistant, 250 14th Street, Suite 556, Atlanta, GA 30318,
USA Phone: +1-404-894-6616 Fax: +1-404-894-7883

B. Stiller et al. (Eds.): QofIS/ICQT 2002, LNCS 2511, pp. 205–214, 2002.
© Springer-Verlag Berlin Heidelberg 2002

applications will be measurements of the network and appropriate path selection procedures.

Without loss of generality, we assume in this paper that a source node is presented with a request to reserve bandwidth resources for a new flow with specific QoS requirements, and is responsible for finding a suitable path to the destination. This is similar to the source routing model. Also, we assume the link state information model, where a network topology database is available for state information about nodes and links in the network. The state information is obtained by measurements from the network elements. This information can then be used by the path selection process.

A network can be monitored either in an *active* or *passive* manner. First gives a measure of the performance of the network whereas the latter of the workload on the network. Both have their merits and should be regarded as complementary. The active approach relies on the capability to inject packets into the network and then measure the services obtained from the network. It introduces extra traffic into the network. But the active approach has the advantage of measuring the desired quantities at the desired time. Passive measurements are carried out by observing normal network traffic, without the extra load. The passive approach measures the real traffic. But the amount of data accumulated can be substantial because the network will be polled often for information.

The rest of the paper is organized as follows. In Sect. 2, we present the importance of available bandwidth and a description of various bandwidth measurement techniques. This is followed by a description of our path selection algorithm in Sect. 3. In Sect. 4, we present our algorithm for available bandwidth estimation. In Sect. 5, the results of the experiments with the available bandwidth estimation algorithm are presented. Finally, we conclude in Sect. 6.

2   Measuring Available Bandwidth

There are various quantities of interest that can be insightful about the state of the network. Available bandwidth (together with latency, loss etc.) can predict the performance of the network. Based on the bandwidth available, the network operator can obtain information about the congestion in the network, decide the admission control, perform routing etc. For MPLS networks, the available bandwidth information can be used to decide LSP setup [1], routing (Shortest Widest Path [2], Widest Shortest Path [3]), LSP preemption [4], etc. Several applications could benefit from knowing the available bandwidth of an LSP. One such application is congestion control. Available bandwidth information can also be used to build multicast routing trees more efficiently and dynamically.

It is desirable to obtain the available bandwidth information by measurements from the actual LSPs because they give more realistic information about the available bandwidth. The available bandwidth information can also be obtained by subtracting the nominal reservation for the tunnels from the link capacity which gives a lower bound. In this paper, we will use the terms LSP and link interchangeably because an LSP is equivalent to links in an MPLS network. The path between two LSRs can be composed of multiple LSPs.

The available bandwidth on a link is indicative of the amount of additional load that can be routed over the link. Obtaining an accurate measurement of the available bandwidth can be crucial to effective deployment of QoS services in a network. Available bandwidth can be measured using both active and passive approaches. Various tools to measure available bandwidth of a link in the network are available. The bottleneck bandwidth algorithms can be split into two families: those based on pathchar [5] algorithm and those based on Packet Pair [6] algorithm. The pathchar algorithm is an active approach which leads to the associated disadvantages of consumption of significant amount of network bandwidth etc. The packet pair algorithm measures the bottleneck bandwidth of a route. It can have both active and passive implementations. In [7], the authors have proposed another tool to measure bottleneck link bandwidth based on packet pair technique. None of these tools measures the available bandwidth or utilization of a desired link of a network. In [8], the authors have proposed an active approach to measure the available bandwidth of a route which is the minimum available bandwidth along all links of the path. Another active approach to measure the throughput of a path is Iperf [9] from NLANR that sends streams of TCP/UDP flows. Cisco has introduced the NetFlow technology that provides IP flow information for a network. But in a DiffServ environment, the core of a network is interested in aggregate rather than per-flow statistics, due to the scalability issues.

All the tools, except NetFlow, give path measurements based on an active approach. A network operator, on the other hand, would be interested in finding the available bandwidth on a certain link of the network. He does not need the end-to-end tools that utilize the active approach of measurement. One solution is to use Simple Network Management Protocol (SNMP) which is a short-term protocol to manage nodes in the network. Its operation revolves around two key components: managed devices and management systems. Managed devices store management information in Management Information Bases (MIBs) and make this information available to the management systems. Thus SNMP can be used as a passive technique to monitor a specific device. MRTG [10] is a tool based on SNMP to monitor the network links. It has a highly portable SNMP implementation and can run on most operating systems.

Thus, to obtain the available bandwidth on a certain link of the network in a passive manner whenever he desires, the network manager can use MRTG. But MRTG has the limitation that it gives only 5 minute averages of link utilization. For applications like path selection, this large interval averaging may not be enough. We have modified MRTG to MRTG++, to obtain averages over 10 second durations. This gives the flexibility to obtain very fine measurements of link utilization. We utilize a linear regression-based algorithm to predict the utilization of a link. The algorithm is adaptive because a varying number of past samples can be used in the regression depending on the traffic profile.

3   Path Selection Algorithm

In addition to just finding a path with sufficient resources, other criteria are also important to consider during the path selection. They include optimization of network utilization, carried load, etc. In the following, we describe the related work followed by the description of the proposed path selection scheme.

3.1 Related Work

Various QoS routing algorithms [3, 11, 12] have been extensively studied in literature. All these schemes utilize the nominal available bandwidth information of the links during the path selection. As most of the traffic flows do not follow the SLA agreement very closely, the nominal utilization of a link is highly overestimated than the actual instantaneous utilization. This leads to non-efficient network resource utilization.

A few measurement based methods are proposed in literature for resource allocation, admission control etc. In [13], a resource allocation method is proposed that estimates the entropy of the system resources based on measurements. A connection admission control algorithm for ATM networks based on the entropy of ATM traffic (obtained from measurements) is given in [14]. Another measurement based admission control framework is given in [15]. A measurement-based routing algorithm is given in [16]. The algorithm tries to minimize the end-to-end delay.

3.2 QoS-aware Path Selection

we propose a path selection scheme that is based on obtaining more accurate link utilization information by actually measuring the same from the network elements involved. However, it is not efficient to update the link state database with the available bandwidth information with very fine granularity, due to scalability concerns. Also, due to the varying nature of the available bandwidth, a single sample of measured available bandwidth does not have much significance. This is because the actual traffic profile may be variable and path selection decisions based on a single sample are more than likely to be wrong. Thus, a method has to be devised that performs an estimation of the link utilization in the future. We have proposed an algorithm for available bandwidth estimation that is explained in next sections.

Once, we obtain the available bandwidth estimates, we can then use these values as the input to a shortest widest path routing algorithm to find an efficient path. Even though there is some inaccuracy involved in path selection based on estimates of available bandwidth, the results in [17] demonstrate that the impact of the inaccuracy in bandwidth information on the path selection procedure is minimal. Our path selection procedure is detailed in Fig. 1. As shown, the available bandwidth estimation algorithm is used only for the links where the estimate has not been calculated before. The parameter $threshold$ in the figure is used as a benchmark for path selection. If the bandwidth request is more than a certain fraction of the bottlenect link bandwidth, the request is rejected. This is done to limit the congestion in the network. Once the estimate is obtained, the available bandwidth information can be used in the shortest widest path routing algorithm as weights of the links. The new flow will be admitted into the network if the flow's bandwidth requirement is less than a certain threshold, which is calculated based on the available bandwidth on the path. Optionally, the arrival of the flow can trigger the re-dimensioning/setup of an LSP between the ingress and the egress LSRs. An optimal policy for this procedure is given in [1].

Path Selection Algorithm:
1. At time instant k, a bandwidth request r arrives between nodes i and j.
2. Run the available bandwidth estimation algorithm links with no bandwidth estimation available.
3. Compute the best path using the shortest widest path algorithm with weights as calculated in step 2.
4. Obtain the available bandwidth A on the bottleneck link of the path.
5. If $r > A * threshold$, reject this path and return to step 3. Else, path is selected for the request.
6. If no path available, request rejected and network is congested.

Fig. 1. The path selection algorithm

4   The Available Bandwidth Estimator for MPLS Networks

Our approach for available bandwidth estimation is based on the use of MRTG. A network operator can enquire each router in the domain through SNMP and obtain the information about the available bandwidth on each of its interfaces. The most accurate approach will be to collect information from all possible sources at the highest possible frequency allowed by the MIB update interval constraints. However, this approach can be very expensive in terms of signaling and data storage. Furthermore, it can be redundant to have so much information. Thus we present an approach that minimizes the redundancies, the memory requirements for data storage and the signaling effort for data retrieval.

We can set the measurement interval of MRTG and measure the average link utilization statistic for that interval. We define for a link between two nodes i and j:

- C: Capacity of link in bits per sec,
- $A(t)$: Available capacity at time t in bits per sec ,
- $L(t)$: Traffic load at time t in bits per sec,
- τ: Length of the averaging interval of MRTG,
- $L_\tau[k]$, $k \in \mathbf{N}$: Average load in $[(k-1)\tau, k\tau]$.

The available capacity can be obtained as $A(t) = C - L(t)$. So, it would be sufficient to measure the load on a link to obtain available bandwidth. Note that we have not explicitly shown the $i - j$ dependence of the the defined variables. This is because the analysis holds for any node pair independent of others. We also define

- p is the number of past measurements in prediction,
- h is the number of future samples reliably predicted,
- $A_h[k]$: the estimate at $k\tau$ valid in $[(k+1)\tau, (k+h)\tau]$.

Our problem can be formulated as linear prediction:

$$L_\tau[k + a] = \sum_{n=0}^{p-1} L_\tau[k - n]\, w_a[n] \qquad \text{for } a \in [1, h] \tag{1}$$

Available Bandwidth Estimation Algorithm:
1. At time instant k, available bandwidth measurement is desired.
2. Find the vectors $\boldsymbol{w_a}, a \in [1, h]$ using Wiener-Hopf equations given p and the previous measurements.
3. Find $\left[\hat{L}_\tau[k+1], \ldots, \hat{L}_\tau[k+h]\right]^T$ from (1) and $[L_\tau[k-p+1], \ldots, L_\tau[k]]$.
4. Predict $A_h[k]$ for $[(k+1)\tau, (k+h)\tau]$.
5. At time $(k+h)\tau$, get $[L_\tau[k+1], \ldots, L_\tau[k+h]]^T$.
6. Find the error vector $[e_\tau[k+1], \ldots, e_\tau[k+h]]^T$.
7. Set $k = k + h$.
8. Obtain new values for p and h.
9. Go to step 1.

Fig. 2. The Available Bandwidth Estimation Algorithm

where on the right side are the past samples and the prediction coefficients $w_a[n]$ and on the left side, the predicted values. The problem can be solved in an optimal manner using Wiener-Hopf equations. We dynamically change the values of p and h based on the traffic dynamics.

The available bandwidth estimation algorithm is given in Fig. 2. We define p_0 and h_0 as the initial values for p and h. In step 2 of the algorithm, we need to solve the Wiener-Hopf equations. They are given in a matrix form as $\boldsymbol{R_L w_a} = \boldsymbol{r_a}$, $a = 1, \ldots, h$, i.e.,

$$\begin{bmatrix} r_L(0) & \cdots & r_L(p-1) \\ \vdots & \ddots & \vdots \\ r_L(p-1) & \cdots & r_L(0) \end{bmatrix} \begin{bmatrix} w_a(0) \\ \vdots \\ w_a(p-1) \end{bmatrix} = \begin{bmatrix} r_L(a) \\ \vdots \\ r_L(a+p-1) \end{bmatrix}$$

In order to derive the autocorrelation of the sequence from the available measurements, we estimate it as

$$r_L(n) = \frac{1}{N} \sum_{i=0}^{N} L_\tau[k-i] L_\tau[k-i-n]$$

where N affects the accuracy of the estimation, i.e., more samples we consider, more precise the estimation is. The number of samples needed for a given n and N is $(n + N)$. Since the assumption about stationarity of the measurement sequence may not be accurate, we update the values of the autocorrelation every time we change the value of p in step 8 of the algorithm. The solution of the Wiener-Hopf equations will provide w_a that can be used for predicting $\hat{L}_\tau(k+a)$, $a = 1, \ldots, h$. Since the matrix $\boldsymbol{R_L}$ is symmetric Toeplitz, Levinson recursion can be used to solve for $\boldsymbol{w_a}$ efficiently. For a size p matrix, it involves $(2p^2 - 2)$ multiplications and $(2p^2 - 3p + 1)$ additions.

From the w_a's, we predict $\left[\hat{L}_\tau[k+1], \ldots, \hat{L}_\tau[k+h]\right]^T$ using (1). Next step is to obtain an estimate of the available bandwidth for the interval $[(k+1)t, (k+h)t]$. The available bandwidth $A_h[k]$ is given as $A_h[k] = C - max\left\{\hat{L}_\tau[k+1], \ldots, \hat{L}_\tau[k+h]\right\}$ for a conservative estimate. As the estimate is valid for a duration longer than the individual sampling times, the path selection scheme is feasible for applications requiring

Interval Variation Algorithm:
  1. If $\sigma/\mu > Th_1$, decrease h till h_{min} and increase p till p_{max} multiplicatively.
  2. If $Th_1 > \sigma/\mu > Th_2$, decrease h till h_{min} and increase p till p_{max} additively.
  3. If $\sigma/\mu < Th_2$, then:
      (a) If $\mu > Th_3 * M_E^2$, decrease h till h_{min} and increase p till p_{max} additively.
      (b) If $Th_3 * M_E^2 > \mu > Th_4 * M_E^2$, keep h and p constant.
      (c) If $\mu < Th_4 * M_E^2$, increase h and decrease p till p_{min} additively.

Fig. 3. Algorithm for h and p

estimates of available bandwidth valid over long durations. Thus, even though LSPs are long-living, our path selection procedure is still valid.

After obtaining the actual load $[L_\tau[k+1], \ldots, L_\tau[k+h]]^T$ at time $(k+h)t$, we find the prediction error vector $[e_\tau[k+1], \ldots, e_\tau[k+h]]^T$ where each element is given as

$$e_\tau[k+a] = \left(L_\tau[k+a] - \widehat{L}_\tau[k+a]\right)^2 \quad \text{for } a = 1, \ldots, h.$$

Next, we estimate new values for p and h based on a metric derived from the mean (μ) and standard deviation (σ) of error e_τ. The algorithm is given in Fig. 3.

In the algorithm, M_E is the maximum error value and we have introduced h_{min} and p_{max} because small values of h imply frequent re-computation of the regression coefficients and large values of p increase the computational cost of the regression. Also, we have introduced the thresholds Th_1 to Th_4 to decide when to change the values of the parameters p and h. They are determined based on the traffic characteristics and the conservatism requirements of the network domain. The values of these parameters are held constant throughout the network. In other words, they are not decided on a link-by-link basis.

In this section, we have presented an algorithm for estimating the available bandwidth of a link by dynamically changing the number of past samples for prediction and the number of future samples predicted with high confidence. The objective of the algorithm is to minimize the computational effort while providing a reliable estimate of available bandwidth of a link. It provides a balance of the processing load and accuracy. The algorithm is based on the dynamics of the traffic, *i.e.*, it adapts itself.

5   Experimental Results

In this section, we describe the experiments we used to validate and quantify the utility of the available bandwidth estimation algorithm, and thus effectively demonstrating the performance of the proposed path selection algorithm. First, we describe a few implementation details about the estimator.

A manager can retrieve specific information from the device using the Object Identifiers (OIDs) stored in the devices. The in-bound and out-bound traffic counters on the router interfaces can be periodically fetched to calculate the traffic rate. We have modified the MRTG software to reduce its minimum sampling down to 10 seconds. This

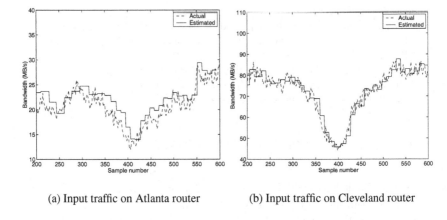

(a) Input traffic on Atlanta router   (b) Input traffic on Cleveland router

Fig. 4. Estimation performance

provides flexibility to the network manager to obtain more precise and instantaneous measurement results. The network manager should decide the optimal measurement period based on the traffic characteristics, the required granularity for the measured values and the appropriate time-scale of the application utilizing the measured values.

Next, we describe the methodology in running the experiments. The algorithm for available bandwidth estimation on a link does not make any assumption about the traffic models. The available bandwidth estimation algorithm works based on the measurements obtained from the network link. Thus, we do not need a network simulator. Instead, we can apply the algorithm to traffic traces obtained from real networks. In the following, we present the traffic profile predicted by available bandwidth estimation algorithm together with the actual profile. We do not present the predicted available bandwidth profile because that can be calculated by taking the difference of the link capacity and the utilization and thus it does not present significant information, when compared to the predicted utilization.

The choice of the thresholds Th_1, Th_2, etc. and h_{min}, p_{max} used for updating the values of p and h in Section 4 has to be made by the network operator depending on the conservativeness requirements of the network operation. We have obtained the following results by choosing $Th_1 = 0.9$, $Th_2 = 0.7$, $Th_3 = 0.5$, $Th_4 = 0.3$ and $h_{min} = 10$, $p_{max} = 50$. All the traffic traces used in the following results have been obtained from Abilene, the advanced backbone network of the Internet2 community, on March 13, 2002. In Fig. 4(a), the available bandwidth estimation algorithm is applied to the input traffic on the Atlanta router of the Atlanta-Washington D.C. link. In Fig. 4(b), same is done for the input traffic on the Cleveland router from the Cleveland-NYC link. As we can see, in both cases, the utilization estimation obtained by taking the peak prediction provides a conservative estimate. Also, when h_{min} is increased, the estimation becomes worse (see Fig. 5), in the sense that it does not follow the sequence closely but is still very conservative. We propose the use of overestimation as a metric to quantitatively measure the performance of the scheme for available bandwidth estimation. For

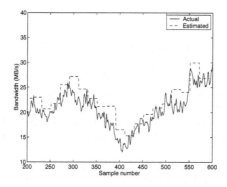

Fig. 5. Input traffic on Atlanta router ($h_{min} = 20$)

the case in Fig. 4(a), the mean overestimation is 1.31 MB/s for the estimation. Similar values are obtained for the case depicted in Fig. 4(b).

When compared with MRTG, we provide the available bandwidth estimates less frequently without a large compromise in the reliability of the estimate. In other words, the utilization profile obtained as a result of MRTG coincides with the actual traffic profile in Figs. 4(a) and (b), but our algorithm provides an estimate of the link utilization which is nearly accurate. Overall, our path selection algorithm has a higher computational complexity than a simple scheme based on nominal link utilizations. However, in that case, the links will be underutilized as the user agreements are based on the maximum traffic profile.

The performance of the proposed path selection algorithm is better than the shortest widest path routing algorithm because the available bandwidth estimation algorithm provides more accurate estimates of the link available bandwidth than just taking an instantaneous sample. In the worst case, when the link traffic is highly variable, the prediction interval for the available bandwidth estimation algorithm is small. In effect, it is similar to taking instantaneous samples. Thus, the worst case performance of the path selection algorithm is similar to the shortest widest path routing algorithm.

6 Conclusions

We have presented an algorithm for path selection in MPLS networks. The algorithm is based on estimation of the available bandwidths on links of MPLS networks. The available bandwidth estimation algorithm predicts the available bandwidth and tells the duration for which the estimate is valid with a high degree of confidence. The available bandwidth estimation algorithm dynamically changes the number of past samples that are used for prediction and also the the duration for which the prediction holds. The path selection algorithm utilizes these estimates as weights in the shortest widest path routing algorithm. In general, the performance of the path selection algorithm is better than the shortest widest path routing algorithm. In worst case, the path selection algorithm gives a performance similar to shortest widest path routing algorithm.

References

[1] T. Anjali, C. Scoglio, J. de Oliveira, I. Akyildiz, and G. Uhl, "Optimal Policy for LSP Setup in MPLS Networks," *Computer Networks*, vol. 39, no. 2, pp. 165–183, June 2002.

[2] Z. Wang and J. Crowcroft, "Quality-of-Service Routing for Supporting Multimedia Applications," *IEEE Journal on Selected Areas in Communications*, vol. 14, no. 7, pp. 1288–1234, September 1996.

[3] R. Guerin, A. Orda, and D. Williams, "QoS Routing Mechanisms and OSPF Extensions," in *Proceedings of 2nd Global Internet Miniconference (joint with GLOBECOM'97)*, Phoenix, USA, November 1997.

[4] J. de Oliveira, C. Scoglio, I. Akyildiz, and G. Uhl, "A New Preemption Policy for DiffServ-Aware Traffic Engineering to Minimize Rerouting,," in *Proceedings of IEEE INFOCOM'02*, New York, USA, June 2002.

[5] V. Jacobson, *Pathchar*, ftp://ftp.ee.lbl.gov/pathchar/, 1997.

[6] S. Keshav, "A Control-Theoretic Approach to Flow Control," in *Proceedings of ACM SIGCOM'91*, Zurich, Switzerland, 1991.

[7] K. Lai and M. Baker, "Nettimer: A Tool for Measuring Bottleneck Link Bandwidth," in *Proceedings of 3rd USENIX Symposium on Internet Technologies and Systems*, San Francisco, USA, March 2001.

[8] M. Jain and C. Dovrolis, "Pathload: A Measurement Tool for End-to-End Available Bandwidth," in *A workshop on Passive and Active Measurements*, Fort Collins, USA, March 2002.

[9] Iperf Tool, *http://dast.nlanr.net/Projects/Iperf/*.

[10] MRTG Website, *http://people.ee.ethz.ch/ oetiker/webtools/mrtg/*.

[11] I. Matta and A. Bestavros, "A Load Profiling Approach to Routing Guaranteed Bandwidth Flows," in *Proceedings of IEEE INFOCOM'98*, San Francisco, USA, April 1998.

[12] K. Kar, M. Kodialam, and T. V. Lakshman, "Minimum Interference Routing of Bandwidth Guaranteed Tunnels with MPLS Traffic Engineering Application," *IEEE Journal on Selected Areas in Communications*, vol. 18, no. 12, pp. 2566–2579, December 2000.

[13] P. Barham, S. Crosby, T. Granger, N. Stratford, M. Huggard, and F. Toomey, "Measurement Based Resource Allocation for Multimedia Applications," in *Proceedings of ACM/SPIE MMCN'98*, San Jose, USA, January 1998.

[14] J. T. Lewis, R. Russell, F. Toomey, B. McGuirk, S. Crosby, and I. Leslie, "Practical connection Admission Control for ATM Networks based on On-line Measurements," *Computer Communication*, vol. 21, no. 1, pp. 1585–1596, 1998.

[15] M. Grossglauser and D. Tse, "A Framework for Robust Measurement-Based Admission Control," *IEEE/ACM Transactions on Networking*, vol. 7, no. 3, pp. 293–309, 1999.

[16] N.S.V. Rao, S. Radhkrishnan, and B. Y. Cheol, "NetLets: Measurement-Based Routing for End-to-End Performance Over the Internet," in *Proceedings of International Conferenceon Networking'2001*, 2001.

[17] R. Guerin, "QoS Routing in Networks with Inaccurate Information: Theory and Algorithms," *IEEE/ACM Transactions on Networking*, vol. 7, no. 3, June 1999.

Providing QoS in MPLS-ATM Integrated Environment[1]

Sergio Sánchez-López, Xavier Masip-Bruin, Josep Solé-Pareta,
Jordi Domingo-Pascual

Departament d'Arquitectura de Computadors, Universitat Politècnica de Catalunya, Barcelona
(Spain)
{sergio, xmasip, pareta, jordid} @ac.upc.es

Abstract. Multiprotocol Label Switching (MPLS) is an advanced forwarding scheme, which allows the network to achieve the benefits provided by Traffic Engineering (TE) techniques. The establishment of an end-to-end LSP between two IP/MPLS networks interconnected through an ATM backbone is still an open issue. This paper focuses in an MPLS-ATM environment, and addresses the problem of providing a fast LSP establishment, with certain QoS (Bandwidth guarantee), between two MPLS subnetworks interconnected through an ATM backbone. The Private Network to network Interface (PNNI) is used in ATM backbone as a routing and signaling protocol. In order to achieve the paper objectives, new PNNI elements are defined and evaluated.

1. Introduction

The current Internet popularization and growth imposes the necessity of optimizing the use of the network resources. Moreover, a redefinition of the Internet capabilities is being required due to the request of services with a certain Quality of Service (QoS) together with the progressive increase of traffic.

Best effort services, provided by conventional Internet, are not enough to absorb the QoS requested by new emerging applications. In this way, mechanisms such as the Resource Reservation Protocol (RSVP), differentiated services model, Multiprotocol Label Switching (MPLS) [1], Traffic Engineering (TE) and Constraint-based Routing (CR) have been defined by the Internet Engineering Task Force (IETF) in order to allow applications such as high-quality videoconference, real-time audio, etc. to guarantee their performance features.

Traffic Engineering pursues to accommodate the offered traffic in a network in order to optimize the resources utilization while avoiding the congestion. To achieve this, some tools such as Constraint-based Routing and MPLS have been defined. The main CR goal is to compute the optimum route between two nodes using QoS parameters. MPLS is an advanced forwarding scheme based on the allocation of a short fixed length label at the IP packets. MPLS sets up a Label Switched Path (LSP). A label Distribution Protocol (LDP) is used to distribute labels among the nodes in the path. Nevertheless, data transport in Internet is currently supported by a heterogeneous set of network technologies, which have to coexist. One of these, the Asynchronous

---

[1] This work was supported by the CICYT (Spanish Ministry of Education) under contract TIC0572-c02-02 and by the Catalan Research Council (CIRIT) 2001-SGR00226

B. Stiller et al. (Eds.): QofIS/ICQT 2002, LNCS 2511, pp. 215–224, 2002.

Transfer Mode (ATM) technology is widely deployed in the backbone and in the Internet Service Provider environments. ATM can provide Traffic Engineering, QoS and fast data forwarding. The PNNI (Private Network to Network Interface) is a standard routing protocol over ATM defined by the ATM Forum [2]. Moreover, a PNNI Augmented Routing (PAR) and Proxy PAR was defined in order to allow non-ATM information to be distributed through an ATM backbone.

This paper analyses the case of two IP/MPLS networks interconnected through a backbone network, assuming MPLS as the mechanism to provide TE in the IP networks, and ATM technology in the backbone. In such a scenario, the interoperability between both MPLS and ATM to achieve MPLS connectivity through the ATM backbone is still an open issue. More specifically, the problem is how to set up an end-to-end LSP with QoS guaranties between two LSRs located in different MPLS domains. In this paper we address this problem. The remainder of this paper is organized as follows. In Subsections 1.1 and 1.2 a new mechanism to distribute MPLS label through an ATM network based on integrating MPLS and ATM is suggested. Moreover, these Subsections define the necessary mechanisms to set up an end-to-end LSP between two LSRs belonging to different domains. In Section 2 some proposals to provide QoS in these LSPs are suggested. In Section 3 a performance evaluation is carried out. Finally, Section 4 concludes the paper.

1.1.   MPLS and ATM Integration

The existing solutions described in [3,4] require either having an IP/MPLS Router on top of the ATM switches, or establishing a tunnel across the ATM backbone. A more appropriate and general solution would be that integrating both MPLS and ATM technologies. In this Section we deal with such a solution based on the PNNI Augmented Routing (PAR)

As it was indicated in [5], PAR is a common work performed by ATM Forum and the IETF. PAR is based on the PNNI [2], which is a routing and signaling protocol used in ATM networks and defined by the Forum ATM. PAR allows non ATM information to be transported through the ATM network. This information is transparent to non PAR-capable ATM switches. Our objective is to use the PAR mechanisms in order to transport the MPLS information between MPLS subnetworks connected through an ATM backbone.

PAR uses specific PNNI Topology State Elements (PTSEs) in order to transport the non-ATM information. PTSEs are encapsulated within PNNI Topology State Packet (PTSP), which is flooded throughout the ATM backbone. PAR uses the PNNI flooding mechanisms. So far, PAR defines specifics PAR Information Groups (PAR IG) in order to describe IPv4 and VPN information. We consider that the ATM Border Switches (BSs) are PAR capable and the external devices (IP Routers, LSRs, etc.) are directly connected to BS. An External device has to register its information in BS in order to be distributed to other BSs. Moreover, each external device has to obtain the information from the BS, to which is attached. The Proxy PAR [5] is the protocol defined in order to allow external devices to register and obtain information to and from the BSs. A Proxy PAR works as client and server mode. While the client mode is a simple procedure installed on the external device, the server functions are performed on the BS.

In order to use the PAR mechanisms to transport MPLS information, a new PAR IG has to be defined and the architecture of the Border Router (BR) has to be modified. According to this, we suggest defining the so-called PAR MPLS services Information Group. This new PAR IG allows MPLS labels to be distributed through ATM backbone. Concerning the BR architecture, we suggest to use a Proxy PAR Capable Label Switching Router (PPAR-LSR) consisting of the following elements: A LSR with routing and specific MPLS functions. A Proxy PAR client, added to the LSR in order to register and obtain information of the Proxy PAR server. An ATM Switch as a Proxy PAR server. And a forwarding table in order to establish a relation between MPLS labels an ATM outgoing interfaces.

A more detailed description of this solution can be found in [6].

1.2.   LSP Establishment in MPLS/ATM Environments

Let us consider the scenario depicted in Fig. 1. Once we have a solution to transport MPLS information from a MPLS subnetwork (ND1, Network Domain 1) to other MPLS subnetwork (ND3, Network Domain 3) through an ATM backbone (ND2, Network Domain 2), the next step is to set up an end-to-end LSP between two LSRs, each one in a different MPLS subnetwork. The solution suggested in Subsection 2.1 has the advantage that allows a path to be set up between an LSR of the ND1 and an LSR directly connected to a BR of ND2. The problem appears when there are more than one LSR until to reach the destination LSR, i.e. the destination LSR belongs to a MPLS subnetwork (e.g. LSRy in ND3). In this situation is needed to use a LDP in ND3 in order to set up the LSP until the destination LSR. In order to solve this problem, we suggest proceeding as follow [7]:

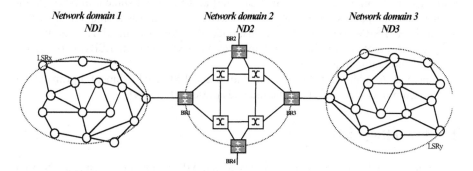

Fig. 1. Network Topology scenario

The Proxy PAR client on each BR registers the MPLS protocol along with labels, and all the address prefixes, which can be reached. Every server bundles its state information in PAR PTSEs, which are encapsulated within a PTSP, which is sent to a neighboring peer.

Using the received PAR MPLS devices Definition IG, every server side generates an MPLS topology database. Each client side will use the query protocol to obtain information about registered services by other clients.

LSRx decides sets up a connection with LSRy. Therefore, an LSP has to be created between both nodes. An RSVP-TE is used as LDP to distribute labels along the path.

We assume that an Ordered LSP Control performs the establishment of the LSP. Because of the implemented functions in the BRs both MPLS networks acts as if the BRs were an end node. As a consequence, after a Path message with a label request object has been sent along the path, the ingress BR1 returns a Resv message to the source MPLS node as if it were the end node along the path. Moreover, the egress BR triggers a Path message to the destination MPLS node, as if it were the source node.

When the RSVP Path message reach ingress BR1 in ND2, RSVP Resv message is returned to the source LSRx with the requested label. Simultaneously, PNNI signaling protocol sets up a VC from the ingress BR1 to the egress BR3. A UNI 4.0 signaling subset is used by the PNNI as signaling mechanism. Therefore, a call set up is triggered in the ingress BR1 when the RSVP path message reaches the ingress BR1. A SETUP message is sent to the egress BR3 in order to establish the connection. Just as was mentioned in Subsection 2.1, the SETUP message can contain some Generic Identifier Elements (GITs). They are transparently transported through the ATM backbone. Therefore, they can be used to carry non-ATM information such as LDP information. In this way, we suggested in [7] using a new GIT in order to start a LDP in ND3. We proposed adding a new GIT in the SETUP message with the following characteristics. An Identifier related standard/application field with a value (0x60), which corresponding to MPLS. An Identifier Type with a value (0x02) assigned to Resource. In this case, the identifier corresponds to the MPLS VCID [3]. However, in our case the VCID is not necessary because we have used the PAR mechanisms, explained in Subsection 2.1, in order to distribute MPLS labels. Thus, we proposed replacing the VCID field with the destination LSR IP address. In this way, when the SETUP message reaches the egress BR3, it has enough information about the protocol used in the connection (MPLS), type identifier (Resource) and destination IP address. With this information, the egress BR3 can start an LSP establishment in ND3. At the moment that the path is set up, the ND2 behavior is as if it was a unique LSR, called Virtual LSR (LSRV). This node is the last node along the path in ND1, while in ND3 it is responsible of setting up the path towards a destination node LSRy, carrying out the source node functions

This method, in addition to provide an end to end LSP establishment, optimizes the time of setup in comparison with the methods presented in [3,4].

2.   Providing QoS

The conventional way to compute routes in the MPLS domain is using a dynamic routing protocol, such as Shortest Path First (SPF). Since SPF does not take in consideration any QoS attributes (e.g. available bandwidth), LSPs computed by this protocol could lead the network to congestion situations. Although through longer paths (more hops counts), using TE based routing algorithms congestion could be avoided and network resources optimized. One tool to achieve this is the Constrain-based Routing (CR). The main goals of CR are both selecting routes with certain QoS requirements and increasing the network utilization. This Section is devoted to apply the CR concept to our proposal in order to reduce the congestion effects in the different MPLS domains composing our scenario.

Let us to consider the scenario shown in Fig. 1, and assume that the ATM backbone is able to set up a VC between the ingress and the egress BR with the QoS values requested by the LDP of ND1. For simplicity a single QoS parameter, the requested Bandwidth (Bw) is used and the topology and available resource information are considered perfectly updated in all the LSRs.

According to this, the mechanism to be used for setting up an LSP with Bw guarantees works as follows: 1) LSRx requests an LSP in order to set up a connection from LSRx to LSRy. 2) The required amount of Bw is requested in this LSP. Therefore, in order to compute the optimal route, the LSRx (step 3) computes the optimal route applying the Widest-Shortest Path (WSP) routing algorithm. The WSP calculates the path with the minimum hop count (shortest) among all the possible paths, and in the case that there is more than one with the same minimum hop count, it selects the path with maximum Bw available. 4) Once the WSP algorithm has selected the path, the LDP triggers a mechanism in order to distribute labels. Thus, an RSVP Path message is sent to LSRy. 5) When RSVP Path message reaches the ATM ingress BR1, it finds a label associated to the destination IP address (in this case, a subnetwork IP address). Then, an RSVP Resv message is returned to LSRx with the label and a relation <label in, label out> is set up in each LSR of the path in order to establish the LSP. 6) Simultaneously, the ingress BR1 triggers a SETUP message with the new GIT proposed in Subsection 2.2 with a Generic Identifier Element containing the destination IP address. Therefore, when the SETUP message reaches the egress BR3, the LSR side of the BR3 can use the destination address in order to compute the LSP. Now, assuming that within the ATM network a VC linking BR1 and BR3 with the amount of Bw requested by LSRx could be set up, the routing algorithm existing in the ND3 (step 7) calculates a LSP in this domain to reach the final destination. 8) Once this path is computed, a RSVP Path is sent to LSRy. The setup LSP will be finished when the RSVP Resv will return from LSRy to the egress BR3. 9) Simultaneously, the egress BR returns a CONNECT message to the ingress BR1 in order to establish the corresponding ATM VC.

At the point of the process, where the SETUP message reaches the egress BR3 (step 6) the routing algorithm used to calculate the path within the remote domain (ND3) cannot be a CR algorithm because the SETUP message do not contain any QoS parameters. So then routing algorithm, such as the SPF, should be used there. As a consequence, the LSP set up in ND3 could not offer the Bw requested by LSRx and in case of congestion in that LSP the loss of information would be unavoidable (no QoS is providing).

Obviously the ideal case is using the same constrain-based routing algorithm (for example WSP) in ND1 than in ND3. Therefore, the egress BR has to know, in addition to the destination IP address, the Bw requested by the LSRx. For fixing this issue, we suggest adding a new Identifier Type in the Generic Identifier Element proposed in Subsection 2.2. This can be easily done because an Identifier related standard/application may have multiples Identifier Type and the number of identifiers is limited by the GIT maximum length, which is 133 octets (see Subsection 4.1 of [4]).). We suggest to define the following identifier: Identifier Type = Resource, Identifier Length = 3 octets and Identifier Value = requested BW in Mbps. The format of the new GIT is shown in Fig. 2.

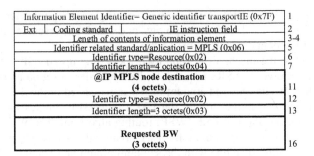

Information Element Identifier– Generic identifier transportIE (0x7F)	1
Ext	Coding standard	IE instruction field	2
Length of contents of information element	3-4
Identifier related standard/aplication = MPLS (0x06)	5
Identifier type=Resource(0x02)	6
Identifier length=4 octets(0x04)	7
@IP MPLS node destination **(4 octets)**	11
Identifier type=Resource(0x02)	12
Identifier length=3 octets(0x03)	13
Requested BW **(3 octets)**	16

Fig. 2. Generic Identifier with BW

In this way, when the RSVP Path message reaches the ingress BR1, the destination IP address and the requested Bw by LSRx are both transferred to the ingress BR1. These values are carried in the GIT and they are transported by the ATM SETUP message. Now, the egress BR3 has a value for a QoS parameter, which allows it to compute the path using a CR (as for example WSP) algorithm. Now we can come out with to different situations, namely finding a route with the requested Bw, or not finding it. As a consequence the last step of the algorithm described above, (step 9) has to be reformulated as follows: 9) If within ND3 a LDP with the requested Bw could be set up, the egress BR3 would return a CONNECT message to the Ingress BR1 in order to establish the corresponding ATM VC. If that LSP could not be found, the egress BR3 would return a RELEASE message to the ingress BR1, the VC would not be created, and a RSVP ResvTear message would be sent from the ingress BR1 to the LSRx in order to reject the connection request.

2.1. An Optimized Solution

The solution proposed so far still has a drawback, which is the possibility of wasting a lot of time. This happens when, after establishing the first segment of the LSP (the one of the subnetwork 1) and the VC through the ATM backbone, the end-to-end path cannot be completed due the impossibility of finding a LSP with the required Bw in the second subnetwork. In order to overcome this drawback, we suggest an optimized solution consisting of define a mechanism able to distribute information on the whole network topology and available resource between all the MPLS subnetworks interconnected through the ATM backbone. In this way, the subnetwork 1 would know in advance the possibility to establish an end-to-end path. Nevertheless, since Network Operators commonly do not like to share such an amount of information with others Network Operators, this solution have no chance to be applied when the different MPLS subnetwork belong to different administrative domains.

Let us consider again the scenario shown in Fig. 1 and assume that the egress BR3 has a Traffic Engineering Database (TED) where the ND3 topology and available resource information is kept. 15 nodes and 27 links compose the MPLS subnetwork topology. We only consider the link available bandwidth as available resource. Therefore, the TED could be 27x2 matrix where a metric (available Bw) is assigned to a link (couple of nodes). A node is represented as a natural number between 0 and 14. Moreover, the considered range is between 0 and 2.5Gbps where the metric is indicated in Mbps. Thus, the couple link-metric could be represented by 4 bytes and, consequently, the TED size is 108 bytes. We assume that the relation between the natural number assigned to a node and the node IP address is known by the BRs. Therefore, the node IP address is not included in the TED.

In order to transport the TED through the ATM backbone, we suggest a new PAR PTSE named PAR MPLS TED Service Definition IG. The element format is shown in Fig. 3.

C	IG Name	Nested in
768	PAR Service IG	PTSE (64)
776	PAR VPN ID IG	PAR Service IG (768)
784	PAR IPv4 Service Definition IG	PAR VPN ID IG (776) / PAR Service IG (768)
792	**PAR MPLS Services Definition IG**	**PAR Services IG (768)**
800	PAR IPv4 OSPF Service Definition IG	PAR IPv4 Service Definition IG (784)
801	PAR IPv4 MOSPF Service Definition IG	PAR IPv4 Service Definition IG (784)
802	PAR IPv4 BGP4 Service Definition IG	PAR IPv4 Service Definition IG (784)
803	PAR IPv4 DNS Service Definition IG	PAR IPv4 Service Definition IG (784)
804	PAR IPv4 PIM-SM Service Definition IG	PAR IPv4 Service Definition IG (784)
805	**PAR MPLS TED Services definition IG**	**PAR MPLS Services Definition IG (792)**

Fig. 3. PAR MPLS TED Services definition IG

Now the process to establish the LSP end to end is as follows. Each BR uses the Proxy PAR client to register the MPLS information and the TED. Then Proxy PAR server floods the PTSEs throughout the ATM backbone. Once the flooded information reaches all the BRs, each one uses the Proxy PAR client to obtain that information. Considering the Fig. 1 as scenario, ND1 distributes the ND3 TED using its own flooding mechanisms.

From here on, LSRx computes the end to end path using both ND1 and ND3 TED. The CR algorithm uses ND1 TED to compute the path from LSRx to ingress BR1 and it uses the ND3 TED to compute the path from Egress BR3 to LSRy. If there are both routes then a RSVP Path messages is sent to ND3. RSVP path message contains the Explicit Routing Object (ERO) in order to contain the routes. The ND3 route is distinguished from ND1 route by the ND3 subnetwork IP address.

Next, the ingress BR1 triggers a SETUP message in order to set up a VC. We can observe that now it is not necessary to include the required bandwidth in the SETUP message because the ND3 path has been computed using the required BW. Therefore, the GIT has only to include the computed path. A new GIT format is suggested and it is shown in Fig. 4.

Finally, when SETUP message reaches the egress BR, the route included in the messages is used by the RSVP Path message in order to set up the LSP.

We have considered that the TEDs are always updated. Therefore, the path has always the Bw specified in the TEDs. Nevertheless, this is not always the case. It may be possible that the TED would not be updated, for example due to fast changes in the link states. The impact of possible network information inaccuracy in the mechanisms proposed in this paper will be a subject for further studies. Preliminary we have suggested in [8] a new QoS routing algorithm in order to reduce LSP blocking probability and the routing inaccuracy without increasing the routing control information overhead and avoiding rerouting.

Information Element Identifier= Generic identifier transportIE (0x7F)	1
Ext	Coding	IE instruction field	2
Length of contents of information element	3-4
Identifier related standard/aplication = MPLS (0x06)	5
Identifier type=Resource(0x02)	6
Identifier length=4 octets(0x04)	7
@IP MPLS node destination **(4 octets)**	11
Identifier type=Resource(0x02)	12
Identifier length=5 octets(0x05)	13
PATH	
	18

Fig. 4. Generic Identifier with PATH

3. Performance Evaluation

The topology used to evaluate the routing algorithms behavior is shown in Fig.1, where two different link capacities are used. Links represented by a light line are set to 12Mbps, and links represented by a dark line are set to 48Mbps. The incoming requests arrive following a Poisson distribution and the requested bandwidth is uniformly distributed between two ranges, (16-64kbps) to model audio traffic and (1-5Mbps) to model video traffic. The holding time is randomly distributed with a mean of 120 sec. Finally, the existent Topology and available resources database in each border router is assumed perfectly updated.

Results shown in Fig. 5 have been obtained using an extension of the ns2 simulator named MPLS_ns2 developed by the authors of this paper. In this way, in Fig. 5 the LSP Blocking ratio as a function of the traffic load is analyzed. This value is defined according to expression (1):

$$LSP\_Blocking\_Ratio = \frac{\sum_{i \in rej\_LSP} LSP_i}{\sum_{i \in tot\_LSP} LSP_i} \tag{1}$$

where rej_LSP are the set of blocked demands and tot_LSP are the set of total requested LSPs.

Two different routing algorithms are applied in order to demonstrate the QoS parameters influence in the number of rejected LSPs. Thus, the SPF and the WSP behavior are shown for audio and video traffic.

The WSP selects the route in accordance with the residual bandwidth. The SPF does not take into consideration any QoS parameters to select the path. Thus, SPF algorithm does not block any LSP and all the traffic runs by the same shortest path. This situation produces congestion when the Bw of the traffic is higher that the Bw of the path. Therefore, all the LSPs requested from this moment will lose information. In our simulations we consider that the LSPs computed by the SPF, which lose information due to the congestion, are as blocked LSPs. Fig. 1 exhibits the goodness of the WSP in front of the SPF algorithm. We can see that the blocking is lower when a QoS parameter is included in the path selection process, that is, when the WSP is used.

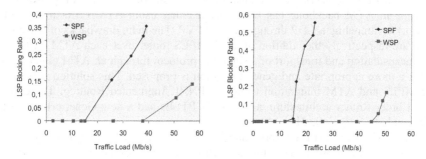

Fig. 5. LSP Blocking Ratio for audio and video traffic load

Once the solution proposed in this paper is implemented in the topology shown in Fig.1, we evaluate the following situation. The first segment of the LSP (the one of the ND 1) and the VC through the ATM backbone are established. Then, the end-to-end path cannot be completed due the impossibility of finding a LSP with the required Bw in the network domain 3, ND3. The results of several simulations are presented in Fig. 6. Three simulation groups have been performed. Each group corresponds to the average of five simulations. Each simulation differs from other in which is the destination node.

In this evaluation, rej_LSP are the set of blocked demand due the impossibility of finding a LSP with the required Bw in the network domain 3, ND3 and tot_LSP are the set of total requested LSPs.

The optimized solution presented in Subsection 3.1 avoids this situation because the source node in ND1 has a TED with the topology and available Bw of the ND1 and the ND3. In this way, the route computed by the routing algorithm will be always established avoiding the LSP blocking.

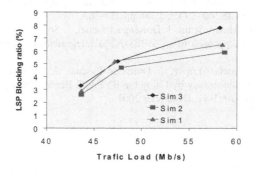

Fig. 6. LSP Blocking Ratio due the impossibility of finding a LSP with the required Bw in the ND3

4.   Conclusions

Different approaches have been considered in order to achieve the interoperability between MPLS and ATM technologies. Firstly, the distribution of MPLS information through an ATM backbone has been solved either using ATM-LSRs in the ATM network or tunneling a LDP through an ATM VP. The main drawbacks of these solutions are respectively the addition of an IP/MPLS router over each ATM switch and the encapsulation and transport of a signaling protocol through an ATM cloud. In this paper a more appropriate and general solution is proposed. This solution is based on the MPLS and ATM integration using the PNNI Augmented Routing. Therefore, a new Border Router architecture, a new PAR PTSEs and a new Generic Information Element have been defined.

Secondly, once the MPLS information has been distributed throughout the BRs, a new mechanism in order to set up an end to end LSP between two different MPLS domains, which are connected via an ATM backbone is suggested.

Finally, a method in order to apply the Constraint-based Routing concept to this scenario is proposed. As a consequence, we have achieved to set up an end to end LSP with the required QoS.

References

1. E.C. Rosen. A. Viswanathan and R. Callon, Multiprotocol Label Switching Architecture, IETF RFC 3031, July 2000
2. ATM Forum, Private Network-Network Interface Specification Version 1.0, af-pnni-0055.000, March 1996.
3. K. Nagami, Y.Katsube, N. Demizu, H. Esaki and P. Doolan, VCID Notification over ATM link for LDP, IETF RFC 3038, Jan.2001.
4. M. Suzuki, The Assignment of the Information Field and Protocol Identifier in the Q.2941 Generic Identifier and Q.2957 User-to-user Signaling for the Internet Protocol, IETF RFC 3033, Jan.2001.
5. ATM Forum, PNNI Augmented routing (PAR), Version 1.0, af-ra-0104.000, January 1999.
6. S. Sánchez-López, X. Masip-Bruin, J. Domingo-Pascual, J. Solé-Pareta, A Solution for Integrating MPLS over ATM, in Proc 15th International Symposium on Computer and Information Sciences (ISCIS2000), Oct. 2000, pp.255-303.
7. S. Sánchez-López, X. Masip-Bruin, J. Domingo-Pascual, J. Solé-Pareta, J. López-Mellado, A Path Establishment Approach in an MPLS-ATM Integrated Environment, IEEE GlobeCom, November 2001
8. X. Masip-Bruin, S. Sánchez-López, J. Domingo-Pascual, J. Solé-Pareta, Reducing LSP Blocking and Routing Inaccuracy by Using the BYPASS Based Routing Mechanism, Internal Report, UPC-DAC-2001-41, December 2001.

Business Modeling Framework for Personalisation in Mobile Business Services

Louis-François Pau[1,2]* and Jeroen Dits[2]

[1] Ericsson Core Networks AB, Sweden
louis-francois.pau@uab.ericsson.se, lpau@fbk.eur.nl
[2] Rotterdam School of management, The Netherlands

Abstract. This paper gives a formal six-level framework for personalization features in current and next generation mobile services, which can be used to drive the business modeling of M-business services from as service provider or system supplier points of view. I also analyzes the economic, sociological, information and psychic drivers for personalization in mobile services and why they sometimes differ from Web based services .A numerical case is provided from an operator introducing a location based personalized service.

Keywords: Personalization, Mobile services, M-business, Sociology, Location services

Introduction

Personalization was initially defined as the combined use of technology and customer information to tailor e-Commerce interactions and each individual customer (sic "Personalization Consortium" ,2001) . This definition however is rather superficial and does not identify:

- at service provider level, the production, authoring and storage ,nor the implications or not on individual tariffs (1)
- at personalized service level, the selection , the transformations, the structuring or the information exchange layout
- at user level, the usage , the feedback ,and the re-use
- at provider level, which tools and analytical applications to use to enable personalization

Furthermore, almost all reported work has been around Web site personalization (2, 3), and is characterized by a heavy reliance on simplistic user profile storage, usage ratings, and business intelligence/data mining tools, but not about personalization specifics when services are deployed via mobile terminals.

Several authors have mapped out the factors influencing the above process, but no one has defined and measured the features or characteristics of personalization,

---

* Contact: L-F Pau, Prof. Mobile business, Rotterdam School of management, Erasmus University, F1-28, Postbus 1738, NL 3000 DR Rotterdam, Netherlands; or General manager, Ericsson CNCP, Box 1505, S 12525 Älvsjö, Sweden

B. Stiller et al. (Eds.): QofIS/ICQT 2002, LNCS 2511, pp. 227-238, 2002.

beyond storing some user or group profiles with preferences and log files of selections (declared interest, usage history, interests and usage by affinity with other users).

After having discussed in Section 1, the sociological and economic drivers of personalization in mobile services, this paper gives in Section 2 a formal six-level framework for personalization features in current and next generation mobile services, which can be used to drive the business modeling of M-business services from as service provider or system supplier points of view. Section 3 discusses the estimation and valuation of personalization features, and the Case in Section 4 illustrates all above.

1. Sociological and Economic Aspects of Personalization in Mobile Services

1.1 Personalization Drivers: Economics

Personalization operates from the economic theory point of view as a re-balancing mechanism in market asymmetries, by giving buyers more advantages than those usually enjoyed by buyers due to information asymmetries (covered in the 2001 Nobel Prize awarded to George Akerlof, Michael Spence, and Joseph Stiglitz). More precisely, by opening up for personalized mobile services while observing a good or enjoying yet another service, mobile service users will be able to tell the goods/service advantages/disadvantages and to communicate about them, and leave the goods/service content on the market or take them out of the market, instead of relying on third party information and true worth formation

1.2 Personalization Drivers: Sociology

There is enormous, but ambiguous behavioral and emotional "power" in personalization, and personalized mobile services will clearly belong to the scope of consumer sociology, to which alternative marketing methods (4) apply most of which rely on the principle of "tribalism" or the principle of "authentic marketing":

- In line with the sociologist's Michael Maffesoli's work, "flash" relations can be established between individuals who a-priori have nothing in common. This is called the "linking value", and mobile communications is, alike some parties or other activities, a mechanism therefore.
- in "authentic marketing", from a simple pool of authentic concepts /products, traditions , with a detail enhancing rarity or scarcity ,it is possible to transform the user's desire to be "rooted " again into an advantage for a brand or a service .The interesting point is here the additional remark that research proves that some mobile services have or will use this principle as well as relays of the authentic concept/product/tradition or location, by extending its reach

There is one more principle, called "marketing by proxy", where a group of users hijack a brand or service intended for one target group, or target circumstances, for use in another context chosen by that group. We have not collected any evidence of this process, so far, in the current or planned personalization of mobile services.

1. 3 Effects of Personalization on Motivation

Mobile services personalization also opens one route to revisit the 1954 Maslow pyramid, whereby the needs of individuals (or employees) where aggregated hierarchically, and wherein it was believed that by acting on the needs one could influence the behaviors. But labor sociology has long ago shown that no one can decide the individual motivations, and that there is no direct relation between individual motivations and performances. Current sociological thinking (5) is rather that employees motivate themselves, as employees alone know which actions can meet their needs. Thus, the idea is to influence work /employment contexts and situations (6), and not to act on the Maslow needs and individual availabilities to have them fulfilled. Mobile services in a Mobile VPN context are thus essential means of deployment of this idea. This is called the concept of "stakes" which affects both the execution of the work tasks (interest, difficulties), its organization, relations (integration, isolation, networks, and tensions), the work context (company culture, perspectives), and finally salaries. The motivation, which is an emotional investment, leads to mix whom you are with what you do, while companies need contributors rather than pure motivated employees. Contributing workers work in teams, place relations before the personal image building, like responsibilities rather than individual performances. Mobile services in an enterprise environment not only enforce this "stakes" concept, but offer also the neutrality and trust needed for its deployment.

1.4 Mobile Services and Electronic Rumors or Disinformation

While some research, services or sites are tracking and tracing rumor formation on the Web, almost no research has neither looked into the diffusion patterns for rumors (and thus the control issues) using mobile services (voice or data) (7). The relative low user friendliness of mobile e-mail has so far been a limiting factor, but this is bound to change, and to be severely correlated with the extent of personalization in the mobile services. In this paper, it is enough to highlight that the proposed features and methodology allow to value for the user as well as for the supplier, the whole rumor emission, transmission and receipt, while it also at Levels 1, 4 and 5 offer ways to value denial of rumors, of course to the extent quantified values are sufficient to handle these issues.

1.5 Mobile Services as They Influence Psyches

Whereas some individual psychiatrists have shown the role of the Web as a means to confront or amalgamate reality and virtual fiction, and have focused the types of illusions in this context, mobile services offer paradoxically the possibility not to mix images or illusions with reality , by being "on the spot", or "zapping" between sites and locations; consequently mobile services in general maintains with the user the mastering of the images and illusions, instead of just letting this user be subject to their power as often on the Web or in virtual realities (8). Psychiatrists also start to point at the stabilizing role of mobile services, as tools to be re-assured, and thus obviously mandate a personalized dimension as commoditized services will not have the same role.

2. Framework for Business Modeling of Personalization in Mobile Services

2.1 Model

2.1.1 Personalization Feature
A personalization feature is here defined as any service characteristic, often embodied in a service specification (UML, SDL), offered to individual mobile service subscribers outside a standard service bundle subject to regulatory oversight (typically: mobile voice, SMS, voice mailbox, access to subscriber support) .All such features are categorized and organized into personalization levels; there may be several personalization features at a given level offered by the same service provider.

2.1.2 Personalization Levels
The proposed, and already widely tested personalization feature model below, organizes the personalization features into the following six levels of a personalized M-commerce/business service:
- Level 1: authentication level, at information security level but also within the context of cybercultures (9)
- Level 2: service structure level (location, neighbors, priority, codec range,...)
- Level 3: content level (search based, unfiltered push and filtered push)
- Level 4:personality profile of user
- Level 5: affinity level and affinity groups
- Level 6:M/e-Consulting level

There is not room to explain here how the personalization features are grouped into these levels, nor how a bundle of personalization features at a given level allows for a tarifing of personalized services per personalization level.

2.1.3 Individual User Value
The business value for a service provider of a personalized M-business service depends on the "Individual user value", or value to the user from the personally perceived angle .For this individual user value, six factors are identified, which include price (dependent on the tariff fixed by the service provider for personalized services), response time to access the feature (determined essentially by network provisioning and management characteristics levels, possibly selected by the user

from guaranteed QoS levels) and four other components with are : frequency of use , popularity , effectiveness, and an ergonomic attribute. A calibration is carried out, and reference values have been collected for different user groups and some communities.

The last four Individual user value factors are defined below:

1. "Frequency of use" is the first factor. If a person uses a service only once, it might have high value individual value to that user at that time ("immediate/occasional use"), but it has not lasting value to the person. If a person uses a service every day, it has much higher personal or user value.
2. Effectiveness: A second criterion that determines the user value of a feature is how efficiently and effectively a personalization service feature is able to meet the needs of the user. This can be called the "individual fit" .For example, the calculation of user value of SMS/MMS falls directly into the assessment of the individual fit quantification.
3. Popularity: The third criterion that determines the user value is connected to the "popularity of the service". SMS for example is not very efficient or effective but it is still very popular. A popular service has a value that is not directly connected to the effectiveness and efficiency. The reverse may also apply, especially in enterprise/Mobile VPN services, i.e. that the enterprise wants to motivate its employees by offering to them a mobilising mobile service (see Section 1.3). Detailed modelling exits to justify an individual user's decision to join or not electronic groups based on his assessment of the information quality and costs (10).
4. User friendliness: A fourth factor is the "friendliness of the user interface", which greatly influences the experience and therefore the value of the personalization feature to the user. This subject is widely researched, e.g. in (11).

2.1.4 Personalized Service Characteristics of a Mobile Service

At service provider level, a personalized service is defined as having some or all of the quantified characteristics defined in Table 1 below, applicable to any supported personalized feature as defined in Section 2.1.1 and categorized in Section 2.1.2.

Table 1. Quantified characteristics of a Personalised service of a mobile service.

Characteristic	Definition and Units
Price of feature	The revenue to the operator of the personalisation feature per session in Euro.
Frequency of use	The amount of times the service is used relative to all sessions of one user (in percentage)
Number of users	Number of users of the feature relative to the total number of people that have access to the feature (in percentage)
Business Value of feature	Total revenue to the operator per subscriber (in Euro)
Density of mobile terminals	Number of mobile subscribers of operator per km2
Total number of sessions	The total number of sessions per year per user
Annual revenue of personalised service	Annual revenue to the operator per km2 (in Euro)

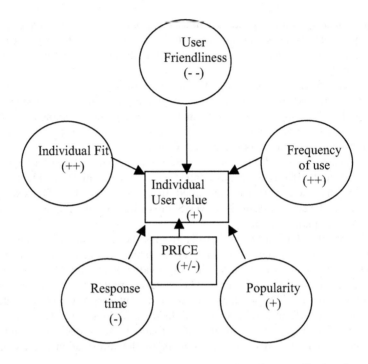

Fig. 1. Factors Factors influencing the Individual user value of personalisation features; the +/- labels are explained in Section 4 and pertain to the Case

2.1.5 Business Value to Service Provider

The business value for a personalized service provider is an indicator, but not the business revenue, of that service or of one feature of that service.

The business value is the product of the revenue to the service provider per session (according to the tariff structure for personalized features as communicated and accepted by users), multiplied by the frequency of use, by the number of users of the service, by the individual user value. All these constituents can be evaluated either for a personalized service altogether, or for a specific personalization feature/ characteristic as defined in next section.

```
Business value (of the personalised feature)=

(Tariff based Feature revenue to operator / session) *
(Frequency of sessions using feature ) * ( Number of
users of the feature ) * (Quantitative weight derived
from individual user value of that feature )
```

Please go to Section 3.2 for explanations as to the last factor.

2.2 Methodology and Tools

The individual user value is estimated from a causal graph with attributes, and from the personalized tariffs with associated access/response time classes.

It is very important to stress that by decoupling individual user value components from business value characteristics, it is possible to evaluate a personalized feature without first evaluating the number of subscribers; in other words, this allows for a bottom up user driven analysis.

The personalized service characteristics are derived from subscriber data.

The business value for the service provider is thereafter calculated, and allows the operator to tune its M-business service for optimal revenue, given intervals for the business value characteristics. It also allows to compare the business values of alternative personalized services, to generate a roll-out plan with best assured revenue, by timing right the sequence of the deployment of the personalized features.

3. Values of Personalization Features

It should be noted that this paper does not address the social value of mobile services or their personalization.

3.1 Business Value Vs. Individual User Value

It is important to make a distinction between business value to the service provider, and individual user value. The business value is probably the easiest to calculate, but it greatly depends on the Individual user value components (Figure 1) if the business value estimate is to be robust. For example, the business value of SMS services was very small when they were introduced, and in the beginning, so was the user value. But somehow the user value started to rise, and with that the business value also.

The Business value is expressed by simple characteristics. Of course these characteristics have to be determined per personalization feature, but they rely on the personalised service as a whole, with a heavy dependence on the dominant personalization level most of its features belong to.

User value is different. This is about the individual value per person. It is not important to know how many people use the service, as from a user's point of view this is not the dominant factor, only from a business point of view. The only personalisation factor where these two aspects contradict each other, and on purpose, is the Popularity characteristic. Therefore the individual user value can be estimated more easily, but at the expense of being more abstract as it is a compound indicator.

3.2 Estimation of Individual User, and Business Values Resp.

It is necessary to outline the process whereby the individual user feature values, and business values are been estimated, when a real business plan is calculated.

For all quantitative data, as used in this framework, mostly for the business values, data are usually collected by operator market research groups using internal as well as external information sources .They can be supplemented by interval statistics and probability distributions.

For all features based on a qualitative assessment, and resulting in a label (see e.g. Figure 1):

- Definition of features : this is done usually by the operator or value added service supplier as part of their technical and product decisions
- Definition of label values and ranges : this is done by the modelling group
- Expert estimation of individual user values: an individual or a group of experts give for each feature the qualitative label values representing the individual user value; if this is done by a group, consensus labels are chosen
- For target group individual feature value estimations, either one can rely on survey data in that group via a custom survey ,or one can use the segmentation done by marketing profiling companies or institutes ;this step is necessary to give the statistical distribution which expert estimations may not give

The calibration of the individual user values across features is usually done by estimating, for the same label values and ranges, the means across features, and a correction via the bias for each specific feature to ensure evenly spread fractiles.

3.3 Link between Business Value and Individual User Value

As stated in Section 2.2, the individual user value is calculated for each feature once this feature has obtained a label in terms of each of the factors in Section 2.1.3 (or Figure 1).

It belongs to the service provider to define:
- the range of qualitative labels
- the composition rule between labels
- a table which translates compound labels into a weight to be given to a feature

in that these three elements must be the same for all features. The Case in Section 4 provides one such example, but many other have been applied. Needless to say these calculation principles are open for criticism, but common to any multi-attribute decision making with qualitative elements.

4. Case: Personalized Location Based Services

The paper includes data from a Case study for a Mobile operator, which was about to deploy, and has since deployed, M-business location based services which include user location as a personalization characteristic made available to the user for these services

As lifestyle benefits, and limitations such as suspicions among users about revealing their location, are key elements of personalised location based services (12), it was relevant to work on the business elements thereof for this European operator. Below are given example values and calculations of the individual user features, and of the business value, from that Case. They serve only as illustrations of the methodology above in Section 2, and no conclusions are given here as to the service itself.

It should be noted that, in this precise case, the analysis is also of value to suppliers of equipment and systems of personalised wireless multi-channel location services proposing one-to-one marketing solutions.

The individual user value labels are given in Figure 1, for the personalization feature "Location information", ranging from a major negative effect "- -", to no effect on user value "+/-", to major positive effect "++". The overall individual user value is in this case the majority of those labels, i.e. 6 times "+" and 4 times "-", leading, by a trade-off table chosen by operator, a compound label "+" with corresponding weight 1 assigned to that feature in its Business value.

The principal personalised service features were the following, and the corresponding estimated Business values per feature those in Table 2,and the ranking per feature that of Table 3:

- subscriber authentication, and authentication for the personalised location service
- presentation of location/space data and distribution to list of value added providers selected by user
- tariff/price per managed location information
- information search based on user location, and mapping information on server, with criteria selected by user
- information pushed down to user based on user location, without user selected categories of information he/she is willing to receive ,but based on agreements between operator and information providers
- information pushed down to user based on user location, from user selected categories of information he/she is willing to receive
- personality features (proprietary)
- affinity selection by user of peers based on mobile instant messaging protocols
- M-consulting, that is within one general area of expertise (medical, navigation/weather, lifestyle tastes, etc...), ability of user to get on-line or short-delay advice based on instantaneous user location

It will be noted that the individual feature values are causing reordering amongst business values of personalization features, for that service .A calculation is also given of the actual annual revenue linked to terminal density.

5. Conclusion

Next generation mobile networks require next generation services, and these next generation services in turn depend crucially on radical changes in the definition, provisioning and tarifing of these mobile services. The present research puts forward personalization, via its economic, sociological and psychic drivers, as mobile specific enhancements, and offers a framework to evaluate the corresponding business models. In particular, it is proposed that a personalized service is built up from one or several personalization features, eventually bundled in personalization levels, which can be evaluated separately as to their individual usage value as well as their business value. The theoretical difficulties are linked to multi-attribute qualitative decision making, but casework with several operators has shown that the relative flexibility in the corresponding algorithmic parts allows them to differentiate in a traceable way. Likewise, the individual user values/estimates can be assessed and calibrated differently in different regions or cultures wherein these operators provide such personalized services.

Table 2. Case: Business value per feature, and individual feature values

Level	-Personalisation feature -Individual feature value	Tariff of feature (Euro)	Frequ ency of use (%)	Number of users of feature (%)	Business value of feature (Euro) (without applying individual feature value weighing)
1	Authentication (+ +)	0.08	5	10	0.0004
2/ Structure	Location information (+)	1.00	10	50	0.05
2/ Structure	Price of location content acquisition (++)	0.20	10	20	0.004
3/ Content	Information requested by search (+)	0.02	25	90	0.0045
3/ Content	Information pushed without filters (-)	0.00	10	100	0
3/ Content	Information pushed with user filters (content based selection) (+)	0.05	30	40	0.006
4	Personality profiles (++)	0.1	50	25	0.0125
5	Affinity – community (+)	1	5	5	0.0025
6	M-Consulting (+/-)	30	1	1	0.003

Table 3. Ranking of business values of personalisation features

	Personalization Feature	Business value with weighing by individual feature value)	Business value without weighing
1.	Location information	0.25	0.25
2.	Personality	0.0625	0.0125
3.	Information pushed with user filters	0.06	0.006
4.	Information requested by search	0.045	0.045
5.	Price of location content	0.2	0.040
6.	M-consulting	0.0015	0.003
7.	Affinity	0.0025	0.0025
8.	Authentication	0.002	0.0004
9.	Information pushed without filters	0	0

Table 4. Top three features personalization features by annual revenue by km2

	Personalization feature	Business value with weighting (Euro)	Density /km2 (high)	Total number of sessions per year per user	Annual revenue per km2 (Euro)
1	Location information	0.25	5000	3000	3 75 0 000
2	Personality	0.0625	5000	3000	937500
3	Information pushed with user filters	0.006	5000	3000	90 000

References

1. L-F Pau, The communications and information economy: issues, tariffs and economics research issues, Journal of economic dynamics and control, Vol. 26, nos. 9-10, August 2002, pp. 1651-1676
2. Special issue on "Personalization and privacy", IEEE Internet computing, Vol. 5, no. 6 , November/December 2001,pp 29-62
3. C.Hall, The personalization equation, The Software magazine, January 2001, pp. 26-31
4. V.Cova, B. Cova , Alternatives marketing, Dunod, Paris, 2002
5. P.Morin,E. Delavallee, Le manager a l'ecoute du sociologue, Editions d'organisation, Paris , 2000
6. P.Raghavan, Social networks: from the Web to the enterprise, IEEE Internet computing, Vol. 6, no. 1, Jan/February 2002, pp. 91-94
7. Web sites on rumors and disinformation:
   www.hoaxbuster.com , www.urbanlegends.about.com,hoaxbusters.ciac.org ,
   www.netsquirrel.com/combatkit,vmyths.com
8. S.Tisseron, Petites mythologies d'aujourd'hui, Aubier, Paris, 2000
9. D.Bell, Introduction to cyberculture,Routledge, Cambridge, 2001

10. M.Bacharach,O. Board, The quality of information in electronic groups, Netnomics , Vol. 4, no. 1, May 2002, pp. 73-97
11. Special issue on "Usability and the Web", IEEE Internet computing, Vol. 6, no. 2, March/April 2002, pp. 44-85
12. N. Wells, J. Wolfers , The business of personalization: finance with a personalised touch. Communications of the ACM, Vol. 43,no. 8, Aug. 2000, pp. 30-34
13. I.Cingil, A. Dogac, A. Azgan , A broader approach to personalization, Communications of the ACM, Vol. 43 ,no 8,Aug.2000, pp 136-141
14. J. Kramer, S. Noronha, J. Vergo, The human element: a user-centred design approach to personalization. Communications of the ACM, Vol. 43, no. 8,Aug 2000, pp. 44-48

Charging Control and Transaction Accounting Mechanisms Using IRTL (Information Resource Transaction Layer) Middleware for P2P Services

Junseok Hwang, Praveen Aravamudham,
Elizabeth Liddy, Jeffrey Stanton, and Ian MacInnes

CST 4-291, School of Information Studies, Syracuse University,
Syracuse, NY, 13224, U.S.A.
{jshwang, paravamu, liddy, jmstanto, IMacInne}@syr.edu
http://web.syr.edu/~jshwang/IRTL/

Abstract. One of the many challenges associated with p2p computing is providing mechanisms for control over accounting transactions among peers. With the increase in the number and variety of p2p applications, there is a need to distinguish between services that can be charged and services that can operate without charging mechanisms. In this paper we present an Information Resource Transaction Layer (IRTL) middleware architecture that addresses some of the technical challenges associated with heterogeneous resource transactions in the p2p-computing environment. We propose to handle charge control, transaction accounting, reputation management and several other p2p parameters through the IRTL.

1 Introduction

P2P computing has effectively made the Internet into a giant copy machine, allowing users and consumers to communicate and swap resources far more freely than businesses that produce content ever expected. One result is abundantly clear: With the increase in the number and variety of p2p applications, there is a need to distinguish between services that can be charged and services that do not require charging mechanisms. While some p2p services such as Napster allowed free and unrestricted copying of files, such services failed to differentiate between resources that are copyright protected and those that are not. The challenge of pure P2P computing - that is, sharing content without central control, organization or intermediaries - is that no one seems to control it so no one seems to be able to make money from it. This raises a number of interesting questions about p2p services: Could a p2p service successfully implement a payment mechanism similar to those adopted by business-to-business or business-to-consumer companies? But traditionally payment mechanisms for accessing resources through the Internet have not found many ways to generate revenue or, more importantly, to distribute revenue to content creators. So what are some of the workable ways to control charge and maintain accounting for resources accessed by peers? This is the set of questions raised by planners of services such as "Mojo Nation" and this is the set of questions we hope to explore by providing a research platform for experimentation with p2p control and charging management.

B. Stiller et al. (Eds.): QofIS/ICQT 2002, LNCS 2511, pp. 239–249, 2002.
© Springer-Verlag Berlin Heidelberg 2002

In this paper we present an IRTL (Information Resource Transaction Layer) middleware architecture that addresses some of the technical challenges associated with heterogeneous resource transactions in the p2p-computing environment. This IRTL facilitates discovery, valuation, negotiation, coordination, charging and exchange of resources among peer users as a middleware platform. We present the IRTL design and architecture and describe the charging and transaction accounting mechanisms using the IRTL. Finally, we close the paper with a short discussion on future works and implications.

2   Background and Related Works

Over recent years, the application and service nature of the Internet has gradually begun to change from a predominantly client-server orientation to an environment that also allows for peer-to-peer (p2p) services. The continuing evolution and growth of peer-to-peer services has provided distinctive challenges in the optimal use of computing resources, the management of quality-of-service, and the coordination of transactions between peers [1]. Various peer-to-peer service architectures and service mechanisms have been developed in response to these challenges.

Many peer-to-peer systems are based on coordination and allocation systems, including virtual machine and agent environment, that limit or balance access of the system to each peer's computing power, data and other resources. Sun Microsystems's JXTA project [2] provides a set of protocols to discover other peers and peer groups as well as to establish message and information channels between and among peers. As an enabling layer, JXTA aims to provide an architecture that can integrate and thus provide authentication, access control, audit, encryption, secure communication, and non-repudiation [2].

Grid computation tries to integrate and utilize dispersed resources to perform one task. Necessary services for grids include authentication, authorization, security and grid-wide name spaces for resources and resource registration/discovery. These services further facilitate resource accounting, job scheduling, and job monitoring. One notable example in the area of grid computing is the Globus project (http://www.globus.org/), which provides fundamental architecture and services for computational grids. A main focus of Globus is the open source middleware development platform, the Globus Toolkit. This toolkit provides infrastructure software to build grids and to provide solutions for many aspects of grids.

The Open Grid Services Architecture (OGSA) [3] builds on concepts and technologies from the Grid and Web Services [4] communities, the architecture defines a uniform exposed service Semantics (the Grid service); defines standard mechanisms for creating, naming, and discovering Transient Grid service instances; provides location transparency and multiple protocol bindings for service instances; and supports integration with underlying native platform facilities. The Open Grid Services Architecture also defines, in terms of Web Services Description Language (WSDL) [5] interfaces and associated conventions, mechanisms required for creating and composing sophisticated distributed systems, including lifetime management, change management, and notification. Service bindings can support reliable invocation, authentication, authorization, and delegation, if required.

Early peer-to-peer systems typically relied on the generosity of users to open their systems to others. A peer-to-peer network without transactions and accounting is, in socio-economic terms, a common pool resource. Such common resources can experience a "tragedy of the commons" unless appropriate institutions exist to counteract destructive incentives for overuse [6]. Accounting and transaction systems can resolve the free rider problem by providing incentives for users to provide access to their resources [7]. Monetary pricing or other valuation schemes can ration access in a similar fashion to traditional markets. For such a system to be economical for a large volume of electronic transactions, however, a protocol for electronic micro-payments must exist. Proposals for services (e.g., Mojo Nation) that include such systems of micro-payments, attempt to solve the tragedy of the commons problem by placing "financial fences" around scarce resources. Such systems can involve one or more separate currencies developed and maintained for this specific purpose. The protocol would include provisions for assigning units to currencies, for determining whether and how denominations are expressed (e.g., granularity), and for associating user-generated names with denominations.

Although some peer-to-peer systems depend on common or freely shareable resources, sustainable future systems will probably have to center on the exchange of valuable resources whose owners have a stake in receiving compensation for their creation (e.g., [8]). Intellectual property owners will want to ensure that network users pay for information that they access, borrow, rent, or lease. A peer-to-peer information resource transaction system must include mechanisms to facilitate tracking and transfer of ownership rights.

Few middleware platforms supporting charging and accounting control for p2p computing and networking services. Adar et al. [9] conducted an extensive user traffic analysis on Gnutella [10] to show the significant problem of free riding. The authors argued that the free riding problem eventually degrades the system performance and increases the system vulnerability. To over come the free riding problem, many have suggested that a market based architecture that imposes structures for charging and transaction accounting control is essential. To date, however, no such architecture has established itself as a predominant model. Although a number of projects (some of them open source) have attempted to develop a workable model, we believe that considerably more research and experimentation must occur to resolve the major challenges involved. Our proposed Information Resource Transaction Layer provides a test-bed for such experimentation.

3   IRTL (Information Resource Transaction Layer) Middleware

The middleware platform creates a "glue" layer of various service managers that will integrate the services to fulfill different functional requirements [11]. This glue layer, termed the Information Resource Transaction Layer (IRTL), will integrate services from existing resource management platforms. Existing platforms such as Sun's JXTA can be used for developing a p2p infrastructure with the IRTL. It defines a set of protocols that can be used to build any p2p application with flexibility to create p2p environments for application specific requirements. [12] The IRTL can utilize those protocols to support the middleware for information resource management in the p2p environment.

The proposed IRTL architecture comprises four major overlaid sections called the Service Identification Layer, Manager Layer, XML Generator Layer and the Metadata Information Base Layer. These layers lie on top of an Application Layer and a typical Internet Protocol network as illustrated in Figure 1.

The Application Layer comprises of a number of p2p applications that make a connection to the Middleware layer. Typically, SOAP [13] [Simple Object Access Protocol] is the mechanism used to embed requests for resources.

The Service Identification Layer contains WSDL documents, which offer a description on the type of services managed by the middleware layer. Characteristics such as port types, service types and message types are contained inside a WSDL document.

The Manager Layer facilitates decisions about the use of resources and controls usage of resources on behalf of a peer. The Manager elements allocate and manage resource sharing among different peer systems and networks with different resource definitions and identifiers through metadata information. Resource characteristics embedded inside metadata are subsequently passed on between managers. A Manager also provides a mechanism for managing resources within a resource domain where it has direct control and provides a mechanism to manage resources with other peer domains. A Manager supports resource allocation and provisioning at the system boundary among multiple peers.

The Metadata Information Base layer is a distributed database that manages metadata information for the resources and profiles of the peers. Metadata provides an extensible markup mechanism for documenting the structure and content of heterogeneous information resources: Our architecture facilitates the use of automated application of markup and thus simplifies the related processes of resource discovery and resource inventory maintenance. Information resources for reputation management, trust assessment, charging and accounting control, privacy, and security will be maintained in the Metadata Information Base.

The managers will relate to one another through the Metadata Information Base for dynamic management of metadata services among transacting peers. Instantiating a "manager" that controls metadata information, or even transaction accounting as to each user's role in sharing data simplifies communication inside the middleware layer. Further, this manager can establish connections to other components inside the middleware layer through WSDL methods, and XML can be used as the primary markup language to embed data content. Instantiating several managers to perform specific tasks pertaining to the p2p application helps in controlling middleware services in general, and also in determining application specific information to help analyze if applications are being subjected to change or even non-existent within the p2p domain and hence knowing the state of the application. In the p2p example, we call the "Peer-to-Peer Manager", which primarily contains metadata information and transaction accounting information. There can also be a "Resource Manager", which controls user information, and user(s) role in resource sharing. These modules can act interdependently with each other sharing private and public data. Details about various managers can be found from the project website shown in the title of this article.

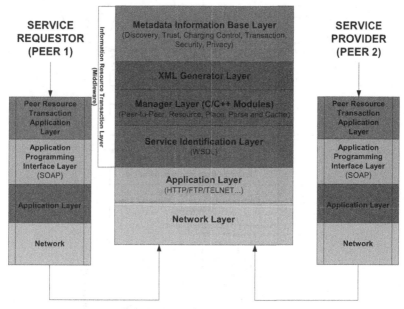

Fig. 1. IRTL Middleware Architecture

4 Charging Control and Transaction Accounting through IRTL

In this section, we present how the IRTL handles charge control and transaction accounting mechanisms with a simple example.

4.1 Charging Control

Figure 2 shows the breakdown of the different sub-layers inside the IRTL layer, with emphasis on the Peer-to-Peer Manager and the Resource Manager for charge control and Transaction Accounting. In Figure 2, the IRTL receives requests from Peer 1 and Peer 2 to retrieve a research paper. The internal logic schema for charge control is represented in Figure 3. Figure 3 also outlines the algorithms used in charging control with emphasis given to peer relationships that are used in request servicing.

The request traverses through the Service Type Identifier, detecting "Web Services" type of service application. The request is embedded inside a SOAP document, which is sent to the Service Type Identifier (STI) via HTTP. The STI sends a request to the Universal Description, Discovery and Integration (UDDI) [14] to invoke WS-Inspection Documents, which identify Peers that have published research papers similar to the request. Peer 3 is a possible Service Provider that a Peer identified through the UDDI layer. The assumptions made are as follows:

- o   Peer 2 is a "buddy" of Peer 3
- o   Peer 1 is an anonymous peer logged in for a one-time transaction.
- o   The Network is not BUSY
- o   There are no copyright or patenting rights on the information requested.

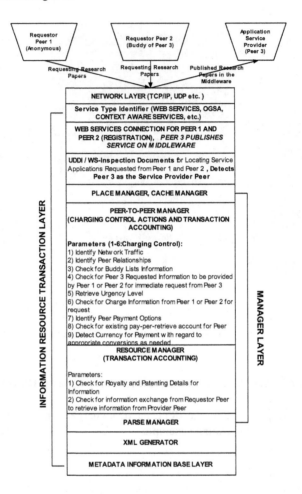

Fig. 2. Charging Control and Transaction Accounting with IRTL

The UDDI layer invokes the Place Manager, which detects the specific server as to where the Information is located. (We assume the Metadata Information Base Layer is comprised of many servers that are used to distribute the vast plethora of information types and their associated structures). The Peer-to-Peer Manager is called next which establishes a connection with the MIB layer. The p2p Manager requests "Buddy List" information on the requestor Peers. "Buddy Lists" is one of many schemas we propose for handling charge control. The mode of operation is similar to the ones present in Messengers (Yahoo! Messenger, MSN, AOL AIM Messenger etc.). A peer can add another peer to his buddy list upon consent from the peer. The Metadata Information Base Layer contains information on "Buddy Lists" on a simple database that adds in entries made from a peer each time a buddy is added or deleted. In our example, Peer 2 is a buddy of Peer 3, and Peer 1 is an anonymous peer. Also, Peer 1 does not offer payment or exchangeable information for access. Hence, Peer 2 is offered service, and then Peer 1 again based on other requestor Peers.

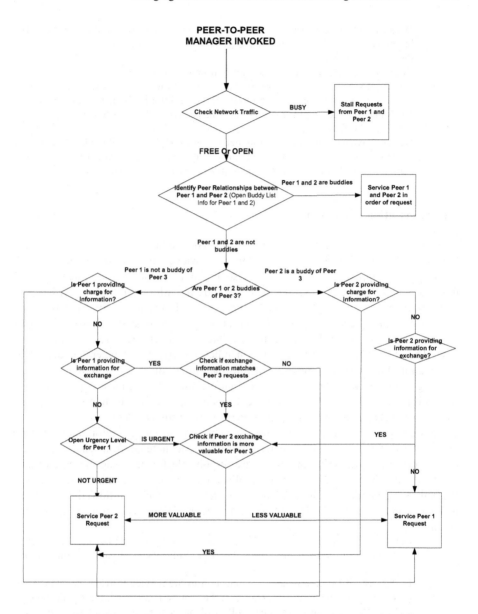

Fig. 3. Peer-to-Peer Manager Logic Representation for Charging Control

The different parameters through which the Peer-to-Peer Manager emphasizes charge control are:

- o Detecting Network Traffic (If Busy or not available, Stall Both requests from the Peers).
- o Detect Peer relationships between the Requestor Peers. If both peers are buddies, then handle request based on a First-in-First-out basis.

- o   Detect Buddy List information between Requestor Peers and the Provider Peer. If a buddy is matched, more preference is given to the buddy peer, and less preference to the anonymous peer.
- o   Retrieve Urgency Level of Request made from the Requestor Peers.
- o   Detect information that could be exchanged for information. For example, if the Provider Peer has submitted a request to the IRTL for information that is not present, the IRTL makes a record of the request, and matches them for possible exchangeable information for future requests.
- o   Detect "Payment" information from peers. For example, if the Provider Peer submits a charge for information that needs to be accessed, the Information becomes "pay-per-view"; the Requestor Peers need to pay in order for information access.

4.2   Transaction Accounting

Transaction Accounting is primarily managed inside the Peer-to-Peer Manager and the Resource Manager. The Peer-to-Peer Manager retrieves information about Peer payment options setup. The IRTL offers a system similar to the PayPal system to Peers through which Peers can setup payment methods for information access. Other payment mechanisms like "Split-Payment" are also made available. Split-Payment is a schema by which Peers can pay based on an installment or a split basis. Discounts, and Deal Information on Peers are also available for Peers for bulk information access. The Resource Manager primarily handles copyright and patenting details on information. Patented and Copyrighted Information are disclosed to Requestor Peers, with algorithms in place for detection and enforcement. Since, most p2p models rely on "Information Exchange", we propose a model capable of detecting specific information for exchange from the Requestor Peers to the Provider Peers.

Figure 4 primarily deals with cases encountered during p2p information transfer noting the different types of transfer algorithms encountered from charging to exchange principles. Charging principles are responsible for detecting the type of payment ensued, while exchange principles follow a more complex routine in determining the nature and validity of information. The different parameters followed by the Peer-to-Peer Manager and Resource Manager for Transaction Accounting is as follows (Figure 4).

- o   Detection of Payment Type for Transaction by the Resource Manager. (There are 3 payment types considered: Micro Payment, Split Payment and Information Exchange).
- o   In the case of Micro Payment, a connection to the Service Provider is made to disclose payment amount for acceptance. If the request is accepted, the Application Requestor Peer is serviced. The Peer-to-Peer Manager sends a request back to the Service Type Identifier Layer through XML'ised data embedded inside a SOAP response generated by the STI. If the Service Provider denies the payment amount put forward by the Requestor Peer, the service is terminated for the Requestor Peer.
- o   In the case of Split Payment also, the Service Provider is notified for a response. Note: the Requestor upon connection sends the Split Payment details via SOAP, which traverses through to the Peer-to-Peer Manager. If accepted, the Application Requestor Peer is serviced. If the Provider denies the transaction, a request is formulated through the Peer-to-Peer Manager to be sent to the Requestor seeking other Payment options. If the Requestor Peer changes Payment type to Micro Payment or

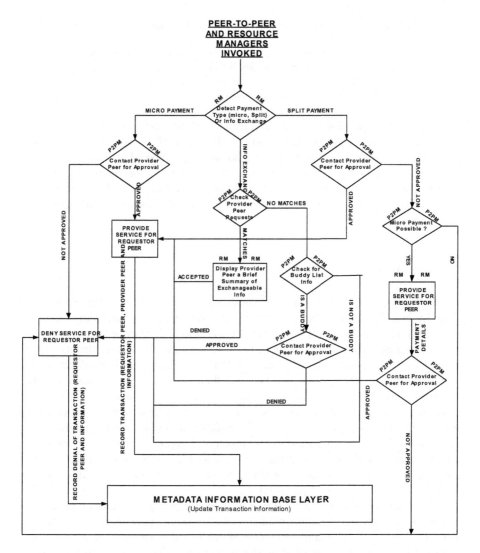

Fig. 4. Transaction Accounting Logic Representation

Information Exchange, then once again the Provider is notified of the change seeking a response. If approved, the Requestor Peer is serviced, and if denied, the service is terminated. We do consider Micro Payment for approval once Split Payment is denied, since we aim at offering Requestor Peer several opportunities for service.

o Information Exchange is more complicated compared to other options, since the extent of information, which can be termed as "valuable", can never be appropriately determined. The approach we follow is to collect Service Requests made by the Service Provider at an earlier stage to determine "valuable information". Since records of information requests are continuously updated in the Metadata Information Base Layer, a simple searching scheme to match the information exchangeable by the Requestor Peer to data already present in the Metadata Information Base Layer yields "valuable" information. If a match does occur, the requestor peer is serviced, and if not, service is denied.

o "Buddy List Information" can also be used for transaction accounting. Suppose the Requestor Peer is found in the buddy list of the Provider Peer, a response is made to the Provider Peer seeking approval for transaction. If approved, the Requestor Peer is serviced, and if denied, the service is terminated.

The Metadata Information Base Layer records every transaction occurrence for information request. This helps in maintaining a "History" of Peer requests for identifying possible exchangeable information and also for Reputation Management. Additionally, copyrighted and patented information need to be considered for access. Since security for information can vary from peer to peer, different levels for security are enforced in the Resource Manager identifying types of information that can be accessed corresponding to peer's.

5 Conclusion

In future p2p computing, the heterogeneous resources available for sale or barter will naturally include the basic building blocks of communications bandwidth, data storage, and computational power but eventually will also include more abstract resources such as databases, directories, catalogs, resource maps, security keys, identities, membership lists, price lists and many other information resources. For information resource transaction services in networked, distributed computing environments, a middleware approach would provide consistent, standardized solutions to problems of transaction security, directory services, quality of service, discovery of resources, and accounting. In this paper, we presented a middleware platform architecture and design, the IRTL (Information Resource Transaction Layer), for information transaction services that provide access and integration across heterogeneous p2p service platforms. In particular, we demonstrated one possible use of IRTL for charging control and transaction accounting of p2p services.

One of the many challenges offered by p2p computing is effective handling of reputation management. A usable reputation management system must include mechanisms that allow for automatic correction of publicized reputations. In a peer-to-peer system there is a further complication that public reputation information must exist in distributed form without a central administrator or repository. In the IRTL, this function can be utilized by instantiating a "Reputation Index" (RI) for Peers. A reputation index is a count of activities made by a peer when accessing the IRTL. We are currently investigating this reputation management issue further to identify generic factors affecting the RI in relation to the transaction accounting of peer-to-peer service environments.

References

1. Oram, A. (Ed.) Peer-to-Peer: Harnessing the Power of Disruptive Technologies. O'Reilly and Associates. (2001)

2. Gong, L. Project JXTA: A Technology Overview. Sun Microsystems, Inc. Palo Alto, CA, USA, (2001) http://www.jxta.org/

3. Foster I., Kesselman, C., Nick, J. and Tuecke, S. The Physiology of the Grid: An Open Grid Services Architecture for Distributed Systems Integration, the Fourth Global Grid Forum, Toronto, Canada, (2002) http://www.globus.org/research/papers/ogsa.pdf

4. Web Services: W3C Architecture Domain http://www.w3.org/2002/ws/

5. Christensen, E., Curbera, F., Meredith, G. and Weerawarana., S. Web Services Description Language (WSDL) 1.1. W3C, Note 15, (2001) www.w3.org/TR/wdsl.

6. Ostrom, E. Governing the Commons: The Evolution of Institutions for Collective Action. Cambridge: Cambridge University Press. (1990)

7. Gupta, A., Stahl, D., Whinston, A. The Internet: A Future Tragedy of the Commons? Paper Presented at the Conference on Interoperability and the Economics of Information Infrastructure (1995) http://cism.bus.utexas.edu/alok/wash_pap/wash_pap.html.

8. Korba, L. Towards an Agent Middleware Framework for E-commerce. Netnomics, 2 (2), (2000) 171-189

9. Eytan Adar, and Bernardo A. Huberman, "Free Riding on Gnutella," FirstMonday, (2000)

10. The Gnutella Homepage, http://gnutella.wego.com/

11. Internet2: Middleware Initiatives http://middleware.internet2.edu/

12. JXTA: The JXTA Solution to p2p technology October 2001
http://www.javaworld.com/javaworld/jw-10-2001/jw-1019-jxta.html

13. SOAP: W3C Simple Object Access Protocol Version 1.1, May 8, (2000) http://www.w3.org/TR/SOAP/

14. UDDI: Universal Description, Discovery and Integration http://www.uddi.org/

Design and Implementation of a Charging and Accounting Architecture for QoS-differentiated VPN Services to Mobile Users[*]

Thanasis G. Papaioannou and George D. Stamoulis

Department of Informatics, Athens University of Economics and Business (AUEB),
76 Patision Str., 10434 Athens, Greece.
telephone: +30-10-8203549, telefax: +30-10-8203686
{pathan, gstamoul}@aueb.gr

Abstract. In the emerging context of mobile Internet, the importance of VPN services is rapidly increasing. Provision of this service is among the subjects of IST project INTERNODE. Besides the necessary technical means, charging and accounting also are key related issues, and constitute the subject of this paper. Only by dealing successfully with charging and accounting, VPN providers can recover their provision costs, increase their profits, and provide the right incentives to their users, thus leading to efficient operation of their network. In this paper, we first study the chargeable characteristics of QoS-differentiated VPN services offered to mobile users w.r.t. transport, security and mobility. Then, we define a complete charging scheme that is fair for the users and provides them with the incentives to use only the resources they really need. This scheme is based on the time-volume charging approach by Kelly; the adoption of this approach is justified in detail in the paper. We then show how the contributing providers can share the total charge earned by each VPN service instance in a fair way, with each provider collecting the portion of charge that corresponds to the consumption of his own resources for the service. This is also a very important issue for the commercial viability of VPN services to mobile users, given that its provision spans multiple domains. Finally, we specify an appropriate charging and accounting architecture pertaining to the specified charging scheme for VPN, to the mechanism for revenue sharing, and to the technical implementation of the VPN services studied. This architecture is compliant to the relevant standards and can serve as a basis for applying other charging schemes as well.

1 Introduction

The globalisation of commerce as well as the ease in human transportation has increased the mobility of professionals and tourists. Also, in recent years the use of mobile phones has grown tremendously. The increased terminal capabilities as well as

---

[*] The present work has been carried out as a part of the IST project INTERNODE (IST-1999-20117) funded by the European Union, through a subcontract with INTRACOM S.A. © Copyright the INTERNODE consortium: BYTEL, EI-NETC, GMD, BALTIMORE, CR2, INTRACOM, UPC.

B. Stiller et al. (Eds.): QofIS/ICQT 2002, LNCS 2511, pp. 250-262, 2002.
© Springer-Verlag Berlin Heidelberg 2002

the development of a great number of SMS and WAP applications have brought the notion of mobile Internet into reality. In this new networking environment, mobile users need to retain seamless and secure connectivity while in a visited domain, as if being at home. Also, mobile users should be able to form private working groups independently from their respective point of attachment to the Internet. These requirements are fulfilled by the provision of VPN services to mobile users. Furthermore, VPN services should be customized in order to satisfy certain user preferences regarding levels of security and quality of service (QoS). We use the term QoS-differentiated in order to imply that there are different possible QoS levels for traffic transport together with the possibility of Best Effort. On the other hand, VPN providers should account and charge for their services, in order to recover their provision costs, increase their profits (while being competitive), and provide the right incentives to their users, thus leading to an efficient operating mode of their network. The users should be provided with the right incentives to use as many network resources as they really need, while they should be charged in a fair way; i.e., pay exactly for the network resources they actually use. Moreover, the charge assigned to each instance of VPN service should be shared among the providers involved in an efficient and fair way. In this paper, we discuss and fully specify a complete, yet lightweight, charging and accounting architecture for QoS-differentiated VPN services offered to mobile users (referred to as M-VPN services). We analyze the chargeable characteristics of M-VPN services, and justify the adoption of a proper charging scheme that satisfies the requirements discussed above, and simplifies the fair sharing of revenue resulted by a M-VPN service instance to the various contributing providers. As already mentioned, the charging scheme adopted provides users with the right incentives. This helps providers to: i) set competitive tariffs that match user needs, and ii) thus maintain their position in a competitive market such as that emerging in reality. Finally, we compare our work to other related articles and clarify our contribution. This work is part of the IST project INTERNODE [1], which studies the provision of QoS-differentiated VPN services to mobile users.

2   The INTERNODE Approach for VPN Provision

In this section, we describe the context of IST project INTERNODE for providing QoS-differentiated VPN services to mobile users (M-VPN). INTERNODE designs, specifies and implements a platform, which will be used to create multi-domain VPN services for mobile users (e.g. e-business, mobile Internet and intranet access, personalized services, etc.). Specifically, VPN connectivity is handled by a VPN provider that owns a VPN Service Provisioning and Support Platform (SPS) capable, among others, of automatically configuring and managing the VPNs on behalf of the VPN subscribers. A customer subscribes to the VPN SPS and registers a number of mobile users to use the M-VPN services in a way specified according to a contract. Specifically, the customer establishes a VPN contract with the VPN provider where the security, QoS (flow-level SLAs), and other characteristics of the M-VPN services for each registered user are defined. Through the SPS platform, the VPN provider performs VPN access control, charging and accounting, and management of the VPN termination points, so as to provide the security and the QoS levels declared in the

VPN contract. Security is provided by means of the IPsec protocol [2] on a gateway-to-gateway basis, while QoS is provided using the Differentianted Services (DiffServ) architecture [3]. Implicitly assumed is the existence of inter-domain DiffServ SLAs among CPs that are established through Bandwidth Brokers [4] and maintained by certain mechanisms as those in [5]. Note that QoS differentiation inside the IPsec tunnel can be accomplished if the DS field of each internal packet is copied or mapped to the DS field of the external packet, as in the case of IP-within-IP tunnelling for Mobile IP [9].

The VPN provider fulfils the above tasks through federated APIs with the potentially visited and home (edge) CPs as depicted in Figure 1. Note that all edge CPs have to apply the security and mobility support technologies adopted by the VPN provider. Therefore, we assume that according to such federation agreements, the VPN provider leases the following equipment to each of the aforementioned CPs: i) an enhanced router that supports IPsec, which is referred to as Security Gateway (SG), and ii) an enhanced router supporting mobility, which is referred to as Home Agent (HA) [resp. Foreign Agent (FA)] if the CP serves as home (resp. visited) network. SG and HA/FA are used as mediation devices by the federated CPs, in order to associate resource usage to particular users (and thus to customers), support the mobility and security features of the M-VPN services, and monitor the conformance with the VPN contracts that the VPN provider has established with its customers. The traffic generated by a mobile user is forwarded to his home domain and from there to its final destination, as specified by reverse tunnelling for Mobile IP [10]. Also, it has to be noted that for a secure duplex end-to-end communication, the creation of two IPsec tunnels is necessary, i.e. one for each direction of the traffic between the sender and the receiver. Note that there also exist other implementation alternatives of providing VPN services to mobile users [6], both when Mobile IPv4 [7] and when Mobile IPv6 [8] are in use. In these cases only the position of end-points of VPN tunnels would change, as opposed to the management procedures applied by the VPN provider. The technical problems stated in [6] fall off the scope of the present work.

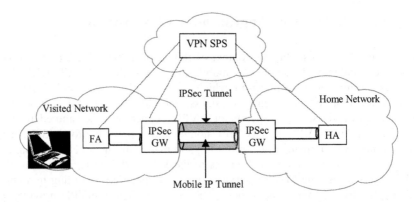

Figure 1. VPN provision to a mobile user from a visited domain to his home domain

3   Analysis of Business Roles

In this section, we define the business roles related to providing QoS-differentiated VPN services to mobile users (M-VPN) offered in the way described in Section 2. We present this part prior to the charging scheme, since the structure of the latter (i.e. which role accounts and charges for what) is influenced by the business roles involved and by the interactions among them. As discussed in Section 2, the VPN provider owns a VPN Service Provisioning and Support Platform (SPS) capable (among others) of automatically configuring and managing the VPNs on behalf of its customers. The VPN tunnel may traverse through any number of intermediate CPs between the home and the visited networks, by which it is treated as specified by the corresponding inter-domain DiffServ SLAs. Moreover, each edge (home or visited) CP has the obligation, due to federation agreements with the VPN provider, to support the M-VPN services provision in terms of QoS, mobility, security and accounting/charging. Thus, the federated CPs serve as 3rd-party providers, each offering a certain part of the M-VPN service instance. On the other hand, recall from Section 2 that the VPN provider is responsible for management and "packing" of the M-VPN services. Each federated CP has to send the information on the charge that arises from its participation in the service provision to the charging and accounting subsystem of the VPN SPS. Specifically, the charging and accounting subsystem of each 3rd-party CP, in accordance with his federated agreement with the VPN SPS, has to: i) measure the usage for the subscribed users to the M-VPN services and associate it appropriately with its users, in order to allocate individual charges, and ii) send this information to the charging and accounting subsystem of the VPN SPS. It is a responsibility of the federated CPs to charge (or set the charging tariffs) for the resource usage in their network domain, according to the VPN contract. On the other hand, the VPN SPS calculates the total charge for providing the M-VPN services.

4   Specification of the Charging Scheme

In this section, we present our approach for charging QoS-differentiated VPN services provided to mobile users (M-VPN) as described in Section 2. In particular, we adopt *additive* charging over all flows arising within a VPN; this is appropriate for charging VPN services, as explained in [14]. The charging scheme for individual flows should reflect the actual usage of network resources. The main issues concerning charging of individual flows inside an IPsec tunnel of a VPN are analyzed below.

Charging for the transport of traffic: As already mentioned, the DiffServ architecture is used for QoS provision to the individual IP flows and/or the aggregates of traffic traversing the IPsec tunnels. Two charging schemes are applicable in this context: i) the scheme proposed by Courcoubetis and Siris [11] for charging DiffServ SLAs, and ii) the *time-volume* approach (also referred to as "*abc* approach") proposed by Kelly [12]. The latter applies to services with quality guarantees, including ATM VBR and DiffServ; the charge equals $aT + bV + c$, where T is the time duration of a flow, V is the corresponding volume of traffic, and a, b, c are the tariff parameters. Both of the charging schemes above are applicable to paths comprising a single-link only, as well as to longer paths; they can be applied for such paths by focusing (for charging purposes) on the bottleneck link and ignoring all others [11]. Both charging schemes use the on-off bound as a proxy for resource usage, which is the effective bandwidth of an on-off source with certain mean and peak rate, and serves as an upper bound to the effective bandwidth of every source with the same mean and peak rate. We have adopted the time-volume approach for charging IPsec flows served by DiffServ QoS classes, because this approach benefits considerably from a priori information of the traffic properties. Indeed, such information can be made available for an IPsec flow through the specific type of user that generates that flow. In particular, we assume that the user identity is part of the VPN contracts and it is indicative of the expected network usage. Specifically, we assume that all users corresponding to the same identity or to a certain group of identities are of the same type, e.g. administrative employees or technical employees etc. A certain type of users corresponds to a certain type of traffic source; for example, administrative employees usually use videoconference applications, whose traffic volume and time duration can be monitored, thus leading to statistical information that can be used in order to optimize the selection of tariff parameters (see below). As the time-volume approach is used for the computation of the charge for the traffic of an IPsec flow, it is necessary to specify the way that the proper a, b, c tariff is selected for charging this flow. As already mentioned, the identity of the user sending/receiving traffic over an IPsec flow determines the type of the user, and consequently the traffic source (i.e. the group of applications that the user may use together with statistics of the associated traffic). We assume that for each different type of user and for each different application there are predefined leaky bucket parameters and an estimate m of the mean rate value for each QoS class that can serve this application. Note that a M-VPN service instance is bi-directional, i.e., it involves the creation of two IPsec flows. We take as mean rate of an application the mean rate of the IPsec flow that conveys the content of the application (e.g. the flow delivering video-frames), rather than of the flow conveying control signals. Note that if both sending and receiving IPsec flows of an application convey useful content for a user type (e.g. the two flows involved in video-conference), then an estimate of the mean rate is kept for each flow. On the other hand, the VPN contract determines the DiffServ QoS class that serves the flow of a particular application for a specific type of user. Thus, the pair [user identity, VPN contract] determines the eligible tariffs for the computation of the charge of each IPsec flow. As explained in [12], the set of tariff parameters a, b, c that *minimizes the expected charge* is selected on the basis of the available estimate of the mean rate m of the application traversing the IPsec flow. (The various connectivity providers (CPs) have the incentive to try to minimize the charge for their users in order to be competitive.) Recall that, initially, the charging module has an estimation of the mean rate for each application and for each type of users. The mean rate m for each application for a certain type of users from each particular point of attachment is constantly monitored and its "future" value is predicted, e.g. as a weighted average of past measurements. (The weights may be larger for more recent measurements.) Note that estimating the mean rate per application for each user type induces a storage and monitoring overhead. Another alternative with less overhead would be to monitor the aggregate mean rate per user type. This however may result in considerable inaccuracy in the prediction of the actual mean rate of an IPsec flow, while all such flows will have to be charged with the same a, b, c parameters. According to [12], this will result in a higher total expected charge for the entire M-VPN service instance. Thus, there is a *trade-off* between accuracy of the prediction of the mean rate and the monitoring overhead. Finally, if an IPsec flow is served Best Effort, then the time-volume formula is still applicable for computing the charge using either the total volume or some other volume measure expressing the *burstiness* of the flow [13].

Charging for security: Each federated provider that its network domain serves as a source or sink of the M-VPN services provision is provided by the VPN Service Provisioning and Support (SPS) platform with a Security Gateway (SG). There is a computational overhead for SGs in the establishment of a new security association (i.e., IPsec tunnel). Clearly, aimless or malicious creation of IPsec tunnels can be prevented by adding a fixed charge per IPsec tunnel creation or alternatively a (smaller) fixed charge per individual IPsec flow insertion to the IPsec tunnel. Next, we discuss charging of the consumption of *computational resources* in the security procedure. There is a constraint in the service rate of the SG that encapsulates/ decapsulates packets according to IPsec, which is similar to the capacity constraint of a telecom link. This constraint should be automatically satisfied only when the traffic of the IPsec tunnels is served by DiffServ QoS classes, since in this case the traffic inserted is policed (i.e. already constrained) according to the VPN contracts of the customers. Thus, consumption of computational resources for such traffic should *not* be charged. However, in case of Best Effort traffic, an extra volume- or burstiness-based charge should be introduced so as to enforce the aforementioned constraint. Additionally, there should be a constant charge per time unit for each IPsec flow, because the identity [Security Parameter Index (SPI), IPsec protocol (AH or ESP) and IP destination address] of an IPsec tunnel is a "scarce" resource, since there can only be finitely many IPsec tunnels between two security gateways. Thus, aimless maintenance of IPsec tunnels is prevented through charging. Note that in order for the right incentives to users to be maintained, different fixed charges are assigned to different security levels that are offered by the M-VPN services. Finally, note that IPsec tunnelling results in increased volume and time (due to the induced computational overhead) for a M-VPN service instance, which are included in the computation of the transport charge.

Charging for mobility: We first consider the charging issues arising in the provision of M-VPN services in the context of terminal mobility. As explained in Section 2, the VPN SPS provides each federated CP involved with a Mobility Agent (MA, i.e., Home or Foreign Agent). As reverse tunnelling for Mobile IP is used, a Mobile IP tunnel is created for each direction between the FA and the HA. There is a constraint in the service rate of the MA that encapsulates/decapsulates the packets according to Mobile IP, which is again similar to the capacity constraint of a telecom link. This constraint is automatically satisfied in case of statistically guaranteed services and no charge is required, according to the discussion above for computing the charge for security. Again, in the case of elastic services, an extra volume- or burstiness-based charge should be introduced. Furthermore, each network domain has a finite set of IP addresses that a potential visited user can use interchangeably. The visitors should have the incentive to de-allocate their care-of IP addresses when not really needed. Charging for this "scarce" resource is accomplished by a fixed price per time unit for the usage time of a care-of IP address. (Note that this argument applies only to the case of Mobile IPv4, for which a care-of address is needed.) In case of personal mobility, the user leases a terminal for accessing the M-VPN services. This terminal can also be a "scarce" resource, as the number of the available terminals can be limited. Thus, leasing of this terminal should be charged by a fixed price per time unit throughout the leasing period. Note that Mobile IP tunnelling also results in increased volume and time (due to the computational overhead) for a M-VPN service instance, which are included in the computation of the transport charge.

Having discussed all key issues on charging M-VPN services, next we discuss the computation of the total charge and revenue sharing among the players involved.

Computation of the total charge: As already explained, the total charge is the sum of all contributions defined above. This summation spans all users of a particular instance of the M-VPN services that belong to the same customer. In particular, a user of a certain user type j using an application i that is served by a Diffserv QoS class q during a M-VPN service instance should be charged according to the formula below:

$$\left(p + a_{ijq}(m_{ijq})\right)T + b_{ijq}(m_{ijq})\ V + c_s \tag{1}$$

where T is the duration of the application and V is the corresponding volume transferred. p is the sum of prices for mobility and security support per time unit. $a_{ijq}(m_{ijq})$, $b_{ijq}(m_{ijq})$ are the transport tariffs for the user type j with estimated mean rate m_{ijq} for the application i served by the QoS class q, which are derived from the VPN contract. Last, c_s represents the sum of fixed costs associated to a new M-VPN service instance, and only depends on the security level s. In case the above application i of a user with user type j were served Best Effort, formula (1) should be modified according to the relevant charging schemes used for traffic transport, security and mobility; see above. A discount can be included in certain of the terms in formula (1) prior to the computation of the total charge on the basis of the identity of the customer and/or QoS/contract violations (see [16]).

Sharing of Revenue: As already explained, the charge is the result of the addition of various contributions, each of which corresponds to the consumption of resources owned by a player with a certain business role. A straightforward way to share revenue among these players is for each of them to collect the revenues corresponding to the consumption of his own resources; e.g. an edge CP contributing to a M-VPN service instance would collect the revenues resulting from the transport, mobility and security support offered by his own network resources. Business agreements however can enforce different methods of revenue sharing, since different players may have different market power.

5   The Charging and Accounting Architecture

5.1   Accounting Issues

In order for the charging scheme previously described to be properly applied, accounting has to measure the traffic of each IPsec flow and associate it with a specific type of users belonging to a customer. (Thus, we employ accounting per user, but charging per customer.) In order to accomplish these, accounting has to:

1. Discover the identity of the user and thus specify the type of the user as well as the identity of the customer that the user belongs to. The user identity is determined

by his home IP address. If the home IP address is private [17], then, according to [10], the user is uniquely identified by the pair (home IP address, Home Agent IP address). Some link layer information is also required for user identification, in case the mobile user is registered to multiple HAs.

2. Measure the traffic (volume and/or burstiness measure) inserted in an IPsec tunnel by each user. This accounting procedure is performed at the ingress of the tunnel. The Security Gateway (SG) performs this task, using header information of packets at both the input and output interfaces, i.e. before and after IPsec encapsulation.

3. Be able to provide to the customer feedback on his current charge; this is referred to as live accounting. This capability may include support for debit payments and/or pre-paid services, or warnings concerning the charge accumulated.

4. Support the capability of providing the customer with one total bill for the service (i.e. opaque billing) from the VPN provider or with separate bills from each contributing provider to the service provision (i.e. transparent billing).

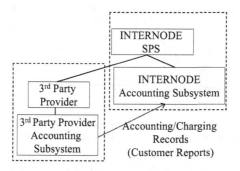

Figure 2. The overall Accounting Information Model between a CP and the VPN provider

The overall model of communication between the charging and accounting subsystem of the VPN provider and that of a 3rd-party provider is based on the principles of the AAA architecture [15]. Thus, all charging and accounting records are generated by the charging and accounting subsystems of the 3rd-party providers and forwarded to that of the VPN provider also in the form of Session Data Records (SDRs); see Figure 2. Recall that, according to AAA [15], a SDR contains a summary of the resources consumed by a user over an entire session. In the INTERNODE context, a SDR conveys the resource consumption of a registered user during a VPN session according to the VPN contract. Finally, both the way the charging and accounting records are generated and the information contained therein depend on the contractual agreements between the federated CP and the VPN SPS, according to the TeleManagement Forum [16].

5.2   The Building Blocks of the Architecture

In this section, we present the charging and accounting architecture, which is depicted in Figure 3 in a TINA hierarchy, for clarity. The architecture consists of the charging and accounting subsystems of the VPN provider and those of the contributing 3rd-party CPs. Below, we describe the functionality of the building blocks of a charging and accounting subsystem; see Figure 3.

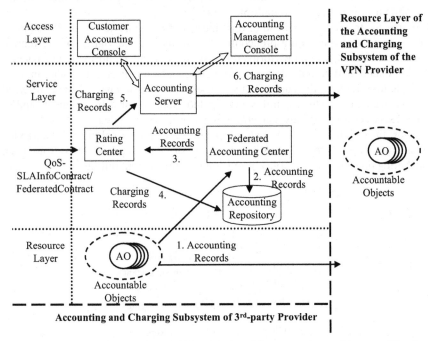

Figure 3. A high-level view of the building blocks of the accounting architecture. The charging and accounting subsystems of the VPN provider and the 3rd-party Connectivity Providers (CPs) have the same structure

Accountable Objects (AOs): For each creation of a new instance of QoS-differentiated VPN services to mobile users (M-VPN), AOs associated with the corresponding Mobility Agents (MAs) and the Security Gateways (SGs) for this bi-directional VPN tunnel are activated, and an AO associated with the M-VPN service instance itself is created. The AOs associated with MAs and SGs belong to the charging and accounting subsystems of the 3rd-party providers contributing to the M-VPN service instance. They collect the relevant accounting information and forward it both to the AO associated with the M-VPN service instance (belonging to the VPN provider) and to the Federated Accounting Centers of their respective charging and accounting subsystems, as depicted by arrow 1 in Figure 3. Thus, the AOs form a ladder (i.e. a hierarchy in the information flow, according to TINA accounting information model [18]) of accounting information for each M-VPN service instance, which is used for charging and accounting *auditing* by the VPN provider. This accounting ladder simplifies management of AOs and the aggregation of accounting events associated to a particular service instance. The AO associated to the entire M-VPN service instance forwards the accounting events to the Federated Accounting Center of the VPN provider.

Federated Accounting Center: The Federated Accounting Center receives all accounting information collected by the AOs associated with the M-VPN services provided. The received accounting information is stored to a database referred to as *Accounting Repository*, as depicted by arrow 2 in Figure 3. If the Federated Accounting Center belongs to the VPN provider, it also receives Charging Records (i.e. Accounting Records and charging information) by the charging and accounting subsystem of a 3rd-party CP. Subsequently, it categorizes usage and charging information on a per customer basis, produces Accounting Records (or Charging Records in case it received charging information), and forwards them to the Rating Center (see arrow 3 in Figure 3).

Rating Center: The Rating Center takes accounting information from the Federated Accounting Center and produces Charging Records taking also into account both the contract/QoS violations calculated according to [16] and the charging scheme (see Section 4) for the offered VPN contract. In case the Rating Center receives Charging Records (by the charging and accounting subsystem of a 3rd-party CP), the Rating Center performs auditing of the received information, calculates the final charge according to the charging scheme, and produces the corresponding Charging Records. The charging information is stored to the Accounting Repository, as depicted by arrow 4 in Figure 3.

Accounting Server: The Accounting Server is the reference point of the charging and accounting subsystem to other software systems. All charging information produced by the Rating Center is forwarded to the Accounting Server (as depicted by arrow 5 in Figure 3). The Accounting Server: i) either forwards the charging information to the Federated Accounting Center of the VPN provider according to the federation agreements (in case it belongs to the federated CP), as depicted by arrow 6 in Figure 3, or ii) it produces a bill and forwards it to the customer of the VPN provider (in case it belongs to the VPN provider) according to the customer preferences (e.g. continuously/periodically).

Finally, in the Access Layer of Figure 3, depicted are the applications enabling interaction of an administrator and a user (or customer) with the charging and accounting architecture; namely, the Accounting Management Console and Customer Accounting Console, respectively. The information communicated within the charging and accounting subsystem of the federated CP is different than the corresponding information in that subsystem of the VPN provider. Nevertheless, the structure of their charging and accounting subsystems is the *same*. Our architecture is a lightweight one, as it involves only the edge CPs contributing to the M-VPN services and the network domain of the VPN provider. Finally, note that in order for the proposed architecture to function securely over an open network (such as the Internet), a standardized inter-domain AAA protocol should be used for the communication of the charging and accounting subsystems.

6   Comparison with Related Work

So far, we have presented a charging scheme and a charging and accounting architecture appropriate for QoS-differentiated VPN services to mobile users (M-VPN). There is significant related work in the literature, which we discuss below.

The specification of an Authorization, Authentication Accounting and Charging (AAAC) architecture for QoS-enabled services offered to mobile users is presented in [19]. This architecture applies for an IPv6-based mobility-enabled end-to-end QoS environment for point-to-point communication (rather than VPN), and is based on the current IETF's QoS models, Mobile-IPv6, and AAA framework. According to [19], the accounting and charging procedures in a network domain are encapsulated in separate service equipment (i.e., software modules) and managed by an AAAC server via Application Specific Modules (ASMs). Certain AAAC servers of the network domains are traversed by the service traffic; these servers exchange authorization, authentication, accounting and charging information via a standardized inter-domain AAA protocol. Our architecture can be combined with the approach of [19], if we view our charging and accounting subsystem as a service equipment of the AAAC server.

Furthermore, [20] presents an accounting and charging architecture and a charging scheme for QoS-enabled VPN services without mobility support. This architecture is based on currently available protocols of the Internet protocol suite and focuses on secured, reservation-based approaches. The key idea of this approach is the establishment of QoS-enabled VPN SLAs through the separate negotiation of the QoS and the VPN parts of the service by respective brokers functioning in each network domain along the path that the traffic traverses. This negotiation results in a SLA establishment between the adjacent network providers along the traffic path. The accounting information is exchanged among the contracted network providers through a signalling mechanism. Our proposed charging and accounting subsystem could be used for charging and accounting inside each negotiating domain. A charging scheme is also specified in [20] in an abstract way; this scheme involves a fixed part, a time-based part and a volume-based part. Our analysis of the chargeable characteristics of the M-VPN services and the charging scheme we propose, which is of the same form with the one proposed in [20], pave the way for a detailed specification of a suitable charging scheme for the services considered therein.

7   Conclusions – Future Work

In this paper, we have studied and analyzed the charging and accounting issues involved in the multiparty provision of QoS-differentiated VPN services to mobile users (M-VPN). We have developed an appropriate charging scheme and an architecture for charging and accounting for such services that conforms to the existing standards for accounting. Charging is based on the time-volume approach by Kelly [12]. Using this charging scheme, each provider collects exactly the revenue arising due to the consumption of his resources in the service provision. On the other hand, the users are charged according to the resources they actually use. Thus, the charging scheme is fair for both the providers and the customers. It also provides users with incentives to use the M-VPN services according to their real needs, and indirectly (i.e. by means of tariff selection by users) give providers their predictions on future traffic. There are several interesting directions for further research. For example, additional functionality may be provided to customers by means of intelligence. This may include support for budget management in the customer side, and efficient VPN contract selection or re-negotiation by means of a customer agent.

References

1. INTERNODE (IST-1999-20117). "Interworking Service Architecture and Application Service Definition". Work Package 2: Deliverable 3, December 2000. URL: http://www.internode.org
2. S. Kent and R. Atkinson. "Security Architecture for the Internet Protocol". IETF RFC: 1825, November 1998.
3. S. Blake, D. Black, M. Carlson, E. Davies, Z. Wang and W. Weiss. "An Architecture for Differentiated Services". IETF RFC: 2475, December 1998.
4. K. Nichols, V. Jacobson and L. Zhang. "A Two-bit Differentiated Services Architecture for the Internet". IETF RFC: 2638, July 1999.
5. G. Dermler, M. Günter, T. Braun and B. Stiller. "Towards a Scalable System for per-flow Charging in the Internet". In Proceedings of *Applied Telecommunication Symposium*, Washington D.C., U.S.A., April 17-19, 2000.
6. F. Adrangi, K. Leung, Q. Zhang and J. Lau. "Problem Statement for Mobile IPv4 Traversal Across VPN Gateways". IETF Internet Draft <mobileip-vpn-problem-statement-00>, March 2002.
7. C. Perkins. "Mobility Support for IPv4". IETF RFC: 3220, January 2002.
8. D. B. Johnson and C. Perkins. "Mobility Support in IPv6". IETF Internet Draft <mobileip-ipv6-17>, March 2002.
9. C. Perkins. "IP Encapsulation within IP". IETF RFC: 2003, October 1996.
10. G. Montenegro. "Reverse Tunneling for Mobile IP, revised". IETF RFC: 3024, January 2001.
11. C. Courcoubetis and V. Siris. "Managing and Pricing Service Level Agreements for Differentiated Services". In Proceedings of *IEEE/IFIP IWQoS'99*, London, United Kingdom, May 31 – June 4, 1999.
12. F.P. Kelly. "Tariffs and effective bandwidths in multiservice networks". In: J. Labetoulle and J.W. Roberts (eds), The Fundamental Role of Teletraffic in the Evolution of Telecommunications Networks, *14th International Teletraffic Congress, ITC 94*, volume 1a, pages 401-410. Elsevier Science B.V., June 1994.
13. P. Reichl, S. Leinen and B. Stiller. "A Practical Review of Pricing and Cost Recovery for Internet Services". In Proceedings of *IEW'99:2nd Internet Economics Workshop*, Berlin, Germany, May 28-29, 1999.
14. C. Retsas. "DSL Technology, DLS Service Models, and Charging DSL Services". M.Sc. Thesis (in Greek), Computer Science Department, University of Crete, Heraklion, Greece, November 2000.
15. B. Aboba et al. "Introduction to Accounting Management". IETF RFC: 2975, October 2000.
16. TeleManagement Forum. TOM Application Note: "Mobile Services: Performance Management and Mobile Network Fraud and Roaming Agreement Management". Document: GB910B, version 1.1, September 2000.
17. Y. Rekhter, B. Moskovitz, D. Karrenberg, G. J. de Groot and E. Lear. "Address Allocation for Private Internets, BCP 5". IETF RFC: 1918, February 1996.
18. TINA-C. "Network Resource Architecture Version 3.0: Accounting Management". Document No. NRA_v3.0_97_02_10, pages 137-180, February 1997.

19. Hasan, J. Jähnert, S. Zander and B. Stiller. "Authentication, Authorization, Accounting, and Charging for the Mobile Internet". *Mobile Summit 2001*, Barcelona, Spain, September 2001.
20. B. Stiller, T. Braun, M. Günter and B. Plattner. "The CATI Project: Charging and Accounting Technology for the Internet". *5th European Conference on Multimedia Applications, Services, and Techniques (ECMAST'99)*, Madrid, Spain, May 26-28, 1999.

MIRA: A Distributed and Scalable WAN/LAN Real-Time Measurement Platform*

Ricardo Romeral[1], Alberto García-Martínez[1], Ana B. García[2],
Arturo Azcorra[1], and Manuel Álvarez-Campana[2]

[1] Department of Telematic Engineering
Carlos III University of Madrid (UC3M)
28911 Leganés, SPAIN
{rromeral, alberto, azcorra}@it.uc3m.es
[2] Department of Telematic Systems Engineering
Technical University of Madrid (UPM)
28040 Madrid, SPAIN
{abgarcia, mac}@dit.upm.es

Abstract In this paper we describe MIRA, a distributed and scalable architecture for flow-based monitoring and traffic analysis. MIRA relies on data inspection to provide advanced monitoring services such as Acceptable User Policy auditing. It is based on a low cost hardware platform that can be deployed in both LAN and WAN environments. The distributed architecture of the measurement platform is designed to provide real-time global results from a complex network. Processing rate can be seamlessly increased by the addition of new hardware elements. The MIRA architecture has been thoroughly tested in a field trial on RedIRIS, the Spanish National Research Network.

1 Introduction

Network services are playing a capital role in our society, and this trend is steadily gaining momentum. As a consequence, demand for network resource monitoring is increasing to asses a given Acceptable Usage Policy (AUP) in the network, to prevent, for example, the transport of inappropriate contents or the abuse of network bandwidth by certain users.

In this article we present the architecture of MIRA, a distributed and scalable measurement platform based on a novel approach for IP traffic analysis. While most analysis tools rely on address and port inspection, MIRA incorporates real-time pattern search to inspect the transferred data applying user definable heuristics to obtain more elaborated information, for example, to classify the traffic into acceptable or non-acceptable.

The modular and distributed architecture allows traffic inspection to be performed over complex networks comprised of physically remote links. Although

---

* This research was supported by the MIRA (Methods for IP traffic Analysis) project, funded by the Spanish National R&D Programme under contract CICYT 2fd-97-2234-c03-01.

B. Stiller et al. (Eds.): QofIS/ICQT 2002, LNCS 2511, pp. 263–272, 2002.

the system was initially designed for ATM STM-1 WAN environments, MIRA has extended its functionality to Ethernet LAN inspection, enabling combined backbone and campus network analysis.

Data processing is performed in a low cost hardware platform, based on conventional PC boxes equipped with a free Unix operating system. Scalability, in terms of analysed traffic, is achieved by replication of the capture and processing elements.

The platform has been adapted to monitor IPv6 packets, since this protocol is expected to spread in the near future. The MIRA platform allows simultaneous monitoring of networks carrying a mix of IPv4, IPv6, and tunnelled traffic such as IPv4 over IPv4 and IPv6 over IPv4 (expected in the initial transition stages to IPv6).

The development of this platform has continued the work initiated by previous Spanish R&D projects CASTBA [1] and MEHARI [2]. The presented architecture is complemented by the analysis tools and GUIs support developed by the CCABA group at Universitat Politècnica de Catalunya.

The remainder of the paper is structured as follows: in section 2 we summarize related work. In section 3 we present the functional architecture of the MIRA platform, describing in detail each functional module. Section 4 is devoted to the study of distribution and scalability, with some examples of possible configurations. Finally, conclusions and future work are presented in section 5.

2   Related Work

Several tools have been developed for traffic classification, starting with the well-known tcpdump. However, few of them are based on content pattern inspection, and less allow measurement distribution and scaling. We will review some of the related measurement approaches.

Snort [3] is a traffic analysis and packet-logging tool that performs protocol analysis as well as content search to perform security hazard detection. It uses a flexible rule language to describe the traffic that it should collect. However, it cannot be neither distributed nor easily scaled, so it is limited to LAN environments. Bro [4] and the Network Flight Recorder [5] are similar tools presenting the same drawbacks.

CoralReef is a software suite developed by CAIDA that provides a programming library to collect traffic using PC boxes. It supports IP traffic capture over several link layers, including ATM (with the monitor components OC3MON and OC12MON [6], that have been used for monitoring the vBNS network [7]). CoralReef includes software for header-based analysis and web report generation, but no content inspection is performed and no special support is provided for measurement distribution.

Commercial software is available for traffic measurement, mainly based on address/port flow analysis. An example is NetFlow [8], from Cisco Systems, aimed to collect resource utilisation on a per flow basis. However, NetFlow does not perform content inspection. Further processing can be performed by applications

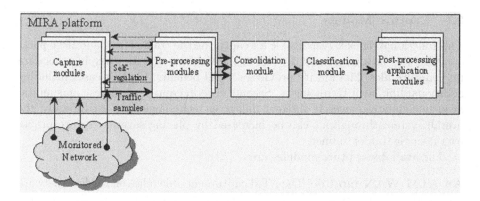

Fig. 1. Basic MIRA architecture

running on PCs or Workstations, either with Cisco proprietary software or with free-software tools like cflowd, but data capture scalability is not considered.

RTFM's NeTraMet [9] proposes a distributed measurement architecture comprised of *meters, meter readers, applications* and *managers*, to allow several network traffic processing applications to access to the same data. However, meters only provide traffic information based on addresses and ports, and data capturing scalability is not tackled.

RMON [10] defines an SNMP MIB interface for accessing the data measured on a network probe, including address and port inspection, and for programming trace recording. However, the architecture is not intended for intensive data inspection, since access to captured data is performed using SNMP.

The NIMI (National Internet Measurement Infrastructure) project [11] has defined a distributed performance measurement architecture. Internet performance is characterised by means of tests that are performed among the monitoring platform nodes, to obtain end-to-end figures. This approach is completely different to ours, since its aim is network performance characterisation.

3   Functional Architecture of the MIRA Platform

The basic architecture of the MIRA system is shown in figure 1. The capture and analysis functions are split into several modules, which are arranged in cascade. Each module is implemented by one or several processes, which can be hosted in different physical devices. Communication between processes is performed by means of files that are periodically written and read. When the modules are to be hosted in different physical hardware platforms, NFS is used to allow file sharing.

In the rest of the section we will describe the functionality of each module.

3.1 Capture Module

This functional block is responsible for capturing the traffic (IPv4 and IPv6 packets) that will be analysed by the rest of the modules and for dumping this data into capture files. Modularity provides several benefits. First, it eases including additional link layer technologies. Second, several capture modules can simultaneously collaborate even if different link layer technologies are used.Finally, the overall system throughput can be increased by placing several capture probes over a single link or subnet.

The available capture modules are:

An ATM WAN module The ATM capture module relies on PC Fore network interface cards connected to passive splitters that probe each direction of a send/receive fibre pair. Splitters are used to asure that the measurement process does not affect negatively network performance. A FreeBSD driver has been adapted for the capture of complete IP packets. The capture can be performed either in promiscuous mode or applying VPI/VCI filtering. Cells are reassembled into AAL5 frames, that carry the whole IP packet. The capture driver adds some information, such as the VPI/VCI pair used for the transmission and the capture timestamp.

A LAN module This module for Ethernet and Fast-Ethernet networks is based on the services provided by the libpcap library.

A capture module can generate one or more sequences of constant-size files, each one containing a specific subset of the captured data. The captured file sequence determines the unit of data that can be separately handled in the following steps of the process, so it is an important parameter for scalability. Different criteria can be used for the generation of these sequences: in ATM, VPI/VCI filtering, outgoing and incoming traffic (depending on the fibre of the pair used); for Ethernet, grouping can be based on IP/MAC origin/destination addresses.

A MIRA *flow* is defined as a group of IP packets that have the same pair of IP address and application port both at source and destination. Inside the file, the data captured for a given flow is marked with a label to distinguish among different aggregated flows. The aggregated flow is in most cases the unit of data for which final results are going to be generated. Different flows can be assigned to the same aggregate. Typically, it will be defined taking into account geographical or organisational criteria. For ATM, VPI/VCI pairs can be used for defining aggregated flows. For Ethernet, IP/MAC origin/destination address filters can be used, allowing for example the definition of incoming, outgoing, transit or internal traffic with regard to a given network.

3.2 Pre-processing Module

The capture module generates large amounts of sampled information. It is necessary to process this information as soon as possible in order to reduce the storage demand discarding unnecessary data. Data privacy also justifies fast data removal. Fortunately, early data deletion is possible if the final goal is traffic classification or accounting, although selective packet logging can still be programmed if required. For each captured file, the pre-processing module extracts the relevant parameters from the samples and deletes the files resulting from the capture process when they are no longer necessary, generating a much smaller file. This module carries out several tasks:

Fast classification For some purposes, IP addresses and TCP/UDP ports can provide the required information. For example, if one of the IP address corresponds to a known server of inappropriate content, no additional information is required for classification, so pattern scanning is not performed.

Symptom detection This is the core activity of the MIRA system, and also the one that consumes most resources. Its aim is to identify *symptoms*. A *symptom* is a group of patterns, represented by a label, that refer to a similar usage profile.

A pattern can be a character string or a binary sequence. An example of a possible symptom could be the "MAIL" symptom, defined to mark all the flows that are expected to belong to SMTP email transference. Some patterns associated to the symptom "MAIL" could be the presence of the string "RCPT To:", or the presence of the four letters "EHLO" at the beginning of a line. The association among patterns and symptoms is defined in a configurable database. As we can see, in general, several patterns can be associated to the same symptom. Whenever a pattern is found on a packet, the corresponding symptom is added to the characterisation of the corresponding flow, and a record of the number of occurrences of each symptom is maintained (note that the patterns themselves or its occurrences are not stored). For the specification of the patterns that define a symptom, heuristics, including natural language strings, application protocol commands or binary data format (e.g. delimiters of an MP3 audio file), can be used.

Flow aggregation Aggregation, i.e. accumulation of statistics and symptoms related with a flow stored in different captured files, is performed periodically, with a period that can be adjusted by the administrator. For tunnelled packets, flow aggregation is not based on the outer packet header data, but on the tunnelled addresses and ports. For each flow, the pre-processing module computes the number of captured packets, the number of captured octets, and a list of the symptoms found and the number of occurrences.

Pre-processing is intensive in computing resources. To prevent the fast producer capture module from flooding the slow consumer pre-processing module, a self-regulation mechanism is introduced that stops the capture when the number of files that have not been pre-processed exceeds a given threshold.

3.3   Consolidation Module

Although MIRA is a real-time distributed traffic inspection platform that allows several monitoring systems to be placed in different network locations, the results obtained on each meter element, analysing different flows, are gathered together into a single report. This process will be performed in one consolidation module per MIRA platform. The fact that each meter can work at a different capture rate has to be taken into account by this process to guarantee that the consolidated results are correct.

Before generating a single report, bi-directional flow correlation is performed (obtaining the so called *biflows*), to account for IP communication involving both IP directions. This process increases classification accuracy, since in many cases relevant symptoms only arise on one direction. Flow correlation allows, for example, linking the symptoms derived from natural language text patterns found in HTTP responses with HTTP requests in which no relevant content for the considered symptoms is transmitted.

Note that information gathering should not be performed at a prior stage: if the data were consolidated before pre-processing, large amounts of data, holding the whole captured data, should be delivered from possible many different places to the central host in which consolidation is performed.

3.4   Classification Module

The aim of this module is to classify each biflow according to usage categories, if it has not been classified before by the pre-processing module (see fast classification, section 3.2). These categories can be later used to audit network usage or to enforce Acceptable Usage Policies. The MIRA administrator can define the categories to process, and the rules to apply to classify a flow into one category, that are based on relations among the number of symptom occurrences (majority of symptoms, comparison of the number of symptoms, etc.). For example, we can define a Leisure class for flows in which there is a majority of symptoms such as MP3, GAMES, VIDEO, SPORTS, etc.

Note that several combined strategies could be required for the detection of the stealthy usage of particular applications, such as MP3 audio distribution in environments where port restriction applies. Different ports and even different applications could be used for data transference, so port identification alone may be useless. In this case, the detection of the MP3 symptom (defined to identify Leisure traffic) could be performed by a search for the string ".mp3" combined with the pattern-based detection of the usage of ftp (that would not necessarily be linked to the Leisure class determination) or other file distribution applications, along with the binary inspection of the transferred data to detect a sequence of eleven consecutive bits with the value of 1 [12].

3.5   Post-processing Application Modules

The MIRA architecture allows the development of application modules that extend the functionality of MIRA. The available data at this stage include not only the classification results, but the addresses, ports and symptoms detected.

An example of a post-processing application module is the Network Usage Statistics Module. This application gathers classification information based on aggregated-flow identifiers (VPI/VCI for ATM networks, or administrator-defined address aggregations for Ethernet), that usually refers to topological locations, for example, a city. The module accumulates data for each aggregated-flow, and extrapolates the results obtained by sampling to allow even comparison among different aggregated-flows.

Other applications organise the resulting data taking into account the source and destination Autonomous Systems, adapt Snort [3] detection rules to be used by MIRA to perform security intrusion detection, or provide billing and charging information [13].

4   MIRA Distributed and Scalable Measurement Architecture

In this section some guidelines for the deployment of MIRA in a distributed network are suggested. Scalability is also addressed, since it is a main concern when high performance networks are inspected.

4.1   Distributed Measurement

One capital problem that arises when distributed measurement is performed is avoiding the measurement of the same traffic in different physical locations. We can illustrate the problem with the example of an ATM WAN star-based network, with a LAN segment attached to it whose internal traffic is required to be inspected (figure 2). If all the traffic were captured on each link, data traversing from one inspected network to another one (e.g., from A to B) would be accounted twice. We would like to assure that each packet can be captured, and that it will not be captured more than once. A way of achieving this would be to capture on each link just the traffic going in one direction, either from the central node to the considered network or vice versa. The same policy should be forced in all the WAN probes. As a consequence, traffic going from network A to network B will only be captured once. Additionally, the whole traffic sent to and received from the central node should be captured, to assure that all the traffic is inspected once.

Note that, in the configuration described above it is useless to try to correlate the flows gathered on the same link, since different directions of the same traffic are probed on different links. However the flow correlation process performed in the consolidation phase will solve the problem.

To prevent the Ethernet MIRA probe placed in the shared Ethernet link from capturing duplicated data, this probe should be configured to capture and analyse only internal traffic. Another option to consider is the removal of the ATM probe placed on link C, delegating the capture of the traffic to the Ethernet probe (provided that traffic destined to or sourced at the C exit router and communicating with networks A or B is not relevant).

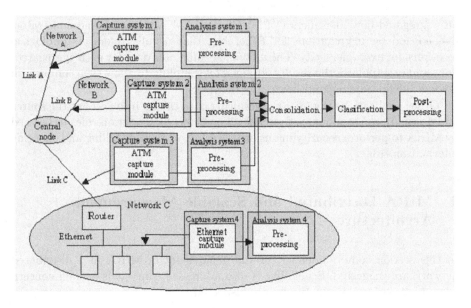

Fig. 2. Distributed Measurement Example

4.2   Guidelines for Deploying a Scalable System

One important concern about traffic inspection is system scalability, as large amounts of data have to be scanned when dealing with high capacity networks.

Although the whole MIRA platform can be hosted in a single device, there are several ways of increasing performance by the addition of new hardware. First of all, we have to stress that different modules can run in different machines, provided that file sharing is allowed. Since real-time consolidation, classification and post-processing are not highly demanding tasks, focus must be set on distributing capture and pre-processing. In the field trial performed in RedIRIS, six PCs (Pentium II, 650 MHz, 256 MB of memory) were arranged into three pairs of capture and analysis devices to inspect three pairs of STM-1 ATM fibres. With this configuration, the percentage of analysed traffic is close to 5%, despite of the CPU-intensive processing required. The computed statistical relevance of the data obtained allows assuring a 5% confidence interval for the mean daily results over a given month with a 95% confidence level. A deeper study on the relevance of the obtained data can be found in [14].

A step further to increase performance would be to split symptom detection into two devices (see figure 3). In this example, outgoing and incoming ATM captured traffic is separated to be pre-processed in two different machines, so two sample files are generated. The capture element can be configured to generate a different number of files that group traffic with the same criteria used for the definition of the aggregated flow. This possibility allows the optimisation of hardware usage; for example, in the test field with the equipment described above, we have measured that pre-processing takes around 2,8 times longer that

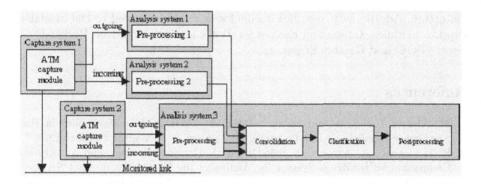

Fig. 3. Example of performance increase by new hardware addition

capturing; therefore, it makes sense to have from 2 to 3 machines performing pre-processing for each capturing element.

If performance has to be further increased, new capture elements (and associated pre-processor devices per probe) can be added. In this case, it must be assured that each probe samples different traffic, programming the probe to capture a different set of aggregated-flows.

5   Conclusions and Future Work

Acceptable User Policy monitoring, to allow subsequent enforcement, is a requirement for many networks. MIRA is a measurement platform especially aimed to fulfil this goal, relying on configurable pattern search to perform traffic classification, in addition to classic address and port analysis. The content search strategy implemented in MIRA broadens traffic inspection possibilities, allowing for example the detection of content formats transported over non-expected transport ports or application protocols (like MP3 over new peer-to-peer communication protocols).

To the best of our knowledge, this is the first distributed and scalable pattern-based analysis tool. MIRA's modular architecture allows achieving probe and analysis distribution, and processing scalability. Measurement distribution and scaling have been addressed in the paper, and some configuration guidelines have been issued. The first two stages of the MIRA architecture, capture and pre-processing, can be distributed at administrator convenience. This allows the increment of analysed traffic, since the analysis process can be speeded via hardware replication, and also enables measurement on complex network topologies. WAN and LAN combined analysis, based on the availability of ATM and Ethernet/Fast Ethernet capture modules, has also been discussed. The distributed measurement system has been field tested in RedIRIS, the Spanish National Research Network.

As further work, we should highlight that additional experience in heterogeneous networks is required to fully validate the generality of the MIRA architecture. Additionally, new link technologies could be added to the available capture modules. Interesting choices for WAN analysis would be Packet Over Sonet (POS) and Gigabit Ethernet.

References

1. M. Álvarez Campana and et al. CASTBA: Medidas de tráfico sobre la Red Académica Española de Banda Ancha. In *Telecom I+D*, Madrid, October 1998.
2. P. J. Lizcano, A. Azcorra, J. Solé-Pareta, J. Domingo-Pascual, and M. Álvarez Campana. MEHARI: A System for Analysing the Use of the Internet Services. *Computer Networks and ISDN Systems (ISSN 0169-7552)*, 31(10), November 1999.
3. M. Roesch. Snort - Lightweight Intrusion Detection for Networks. In *USENIX Systems Administration Conference (LISA)*, November 1999.
4. V. Paxson. Bro: A System for Detecting Intruders in Real-Time. In *Seventh USENIX Security Symposium*, pages 31–35, San Antonio, Texas, January 1998.
5. M. J. Ranum and et al. Implementing a generalized tool for network monitoring. In *USENIX Systems Administration Conference (LISA)*, San Diego, CA, Octover 1997.
6. Claffy and et al. OC3MON: Flexible, Affordable, High-Performance Statistics Collection. In *USENIX*, September 1996.
7. J. Jamison and R. Wilder. vBNS: The Internet Fast Lane for Research and Education. *IEEE Communications Magazine*, pages 60–63, January 1997.
8. Cisco Systems. NetFlow Services and Applications. White paper, June 2000.
9. N. Brownlee, C. Mills, and G. Ruth. Traffic Flow Measurement: Architecture. RFC 2063, January 1997.
10. S. Waldbusser. Remote Network Monitoring Management Information Base. RFC 1757, February 1995.
11. V. Paxson, J. Mahdavi, A. Adams, and M. Mathis. An Architecture for Large-Scale Internet Measurement. *IEEE Communications*, 36(8):48–54, August 1998.
12. ISO. Coding of Moving Pictures and Associated Audio for Digital Storage Media at up to about 1.5 MBit/s. International Standard IS-11172, October 1992.
13. C. Veciana and et al. Server Location and Verification Tool for Backbone Access Points. In *13th ITC Specialist Seminar IP Traffic Measurement, Modelling and Management*, Monterey, September 2000.
14. C. Veciana, A. Cabellos-Aparicio, J. Domingo-Pascual, and J. Solé-Pareta. Verifying IP Meters from Sampled Measurements. In *IFIP 14th International Conference on Testing Communicating Systems*, Berlin, March 2002.

Traceable Congestion Control

Michael Welzl

Leopold Franzens Universität Innsbruck
Institut für Informatik
Technikerstr. 25/7, A-6020 Innsbruck, Austria
michael.welzl@uibk.ac.at

Abstract. A new, easily traceable congestion control scheme which only requires bandwidth information from rare packets that query routers is presented. In contrast with TCP, the (very simple) rate calculation can be performed by any node that sees the occasional feedback message from the receiver; this facilitates load based charging as well as enforcing appropriate behaviour. The control law, which is based on logistic growth, quickly converges to a fixed and stable rate. In simulations, the mechanism showed greater throughput than TCP while maintaining a very small loss ratio and a smaller average queue length.

1 Introduction

We present a slowly-responsive congestion control mechanism which only uses rare bandwidth querying packets to calculate the rate and needs no other feedback; since it does not depend on the loss ratio or the round-trip time (RTT), the (very simple) rate calculation can be performed by any network node which sees acknowledgments from the receiver. Our scheme has a number of additional advantages: it converges to a stable state instead of a fluctuating equilibrium, has a smooth rate and showed greater throughput and less loss than TCP over a wide range of parameters in simulations.

Signaling is carried out with the *Performance Transparency Protocol (PTP)*, which resembles ATM ABR Explicit Rate Feedback, but is scalable and lightweight: the code to be executed in routers is reduced to the absolute minimum and does not involve any per-flow state — all calculations are done at end nodes. PTP packets carrying information requests are sent from the source to the destination and are updated by intermediate routers. In order to facilitate packet detection, the protocol is layered on top of IP and uses the "Router Alert" option [1]. Upon detection, a router adds a timestamp, the address and nominal bandwidth of the outgoing link[1] and a traffic counter to the packet; single intermediate missing routers are detected via the "TTL" field in the IP header and compensated for by taking the incoming link into account. The receiver builds a

---

[1] This value represents the maximum bandwidth available to flows from a particular incoming link to a particular outgoing link; it may actually be less than the link bandwidth, but it is fixed and predefined.

B. Stiller et al. (Eds.): QofIS/ICQT 2002, LNCS 2511, pp. 273–282, 2002.

table of available bandwidth information from two consecutive PTP packets and determines the available bandwidth at the bottleneck during the period between the two packets. The nominal bandwidth and traffic of the bottleneck dataset are fed back to the sender in a standardized manner, where they are used by the congestion control mechanism; feedback packets are just as easy to detect as any other PTP packet and can be used to calculate the rate similar to the sender somewhere else in the network. For an in-depth description of the protocol, the reader is referred to [2] and [3].

In the next section, we motivate the design of the endpoint control law and show that, assuming a fluid model and equal round-trip times, our mechanism converges to an asymptotically stable equilibrium point. We present simulation studies in section 3, usage scenarios for QoS support, differentiated pricing and incremental deployment in section 4 and conclude with an overview of related and future work.

2   Control Law

The endpoint congestion control scheme builds upon the *Congestion Avoidance with Proportional Control (CAPC)* ATM ABR switch mechanism, which solely relies on available bandwidth information — this is very uncommon for such schemes. Convergence to efficiency is achieved by increasing the rate proportional to the amount by which the traffic is less than the *target rate*, a rate that is slightly less than the bottleneck link bandwidth. If the total traffic is higher than the target rate, rates are decreased proportional to the amount by which the target rate was overshot. Additional scaling factors ensure that fluctuations diminish with each update step, while upper and lower limits for the multiplicative rate update factor increase robustness of the scheme [4].

Trivially, if all RTTs are equal, CAPC must also work if these calculations are done by end systems. The key to finding the necessary code changes for the mechanism to work in the asynchronous RTT case was a simulation based design method for studying two users with a fluid model and a single resource. A detailed description of this process, which is based on the vector diagram analysis of AIMD in [5], can be found in [6]. With this method, it became evident that in order to converge to fairness (equal sharing of the single resource), a user should not just adapt the rate proportional to the current load but to the relationship between the current rate and the available bandwidth. In other words, if a user whose load accounts for 90% of the current total load would increase the rate very slightly, and a user whose load accounts for 5% of the total load would increase the rate drastically, the system would converge to fairness. This relationship prevails even if both users experience a different feedback loop delay.

Applying these changes to CAPC led to a control law that is scalable because it does not depend on the round-trip time: the mechanism still works if users "artificially prolong" their RTTs by restricting the amount of generated measurement packets. If these packets do not exceed x% of the generated payload, PTP traffic does not exceed x% of the total traffic in the network; thus, it scales linearly with traffic. In conformance with its distributed nature, we call our modified version of CAPC *"Congestion Avoidance with Distributed Proportional Control" (CADPC)*. The control law can be described with the following expression:

$$x_i(t+1) = x_i(t)\left(2 - x_i(t) - \sum_{j=1}^{n} x_j(t)\right) \tag{1}$$

Here, $x_i(t)$ is the normalized rate of user i at time t (the time between two PTP measurement intervals – a time unit – and the target rate are 1) and there are n users involved. The sum of the rates of all users models the measured traffic which is obtained from PTP at time t. Since all sources obey the same control law, the total traffic seen during the last measurement interval is simply n times the user's own rate in the synchronous case. The control law can then be shown to stem from the equation for logistic growth:

$$\dot{x}(t) = x(t)a\left(1 - x(t)/c\right) \tag{2}$$

For $a > 0$ and $c > 0$, this equation has the unstable equilibrium point $\bar{x} = 0$[2] and the asymptotically stable equilibrium point $\bar{x} = c$ [7]. It was popularized in mathematical biology and applies to a wide range of biological and socio-technical systems; the equation can be used to describe the growth of populations ranging from bacteria colonies to animals and humans [8]. Logistic growth is characterized by its S-shape, which is the result of beginning exponentially and slowing the growth rate with the "negative feedback term" $(1 - x(t)/c)$. It seems to make sense that this equation, which generally models the growth of species populations with no predators, but limited food supply can be applied to congestion control in computer networks with limited bandwidth.

Replacing a with 1 and c with $1/(1 + n)$, we get the continuous-time form of eqn. 1. In discrete time, a must be less than 2 in order for the mechanism to remain stable. In particular, for values above 2.5699456..., the equation yields a chaotic time series [9]. The rate of a user converges to c while the total traffic converges to $nc = n/(1 + n)$, which rapidly converges to 1 as the number of users increases. [3] CADPC can also be expected to properly adapt to background traffic: if we assume a constant background load b, the measured traffic becomes $nx(t) + b$, which yields the equilibrium point $(1 - b)/(1 + n)$ for the rate of a user. The total traffic then converges to $1 - b$.

Even though the target rate should normally be less than the bottleneck link bandwidth to allow queues to drain, we can use the bottleneck link bandwidth

---

[2] This must be taken care of in implementations: the rate of a user can be arbitrarily small, but it must never reach 0.

[3] One might be tempted to replace eqn. 1 with the logistic growth equation $x_i(t+1) = x_i(t)\left(2 - \sum_{j=1}^{n} x_j(t)\right)$, which converges to $1/n$ in the synchronous RTT case; with this control law, the property of adapting to the relationship between the current rate and the available bandwidth is lost, leading to unfair rate allocations of a single resource in the asynchronous RTT case.

because the traffic will never reach the target rate. In spite of the low values for very few users, this behaviour should be significantly better than the behaviour of TCP: in real data networks, situations where only 2 or 3 users are actually able to use all of the available resources are very rare. We study the behaviour of CADPC under varying conditions by simulation.

3   CADPC Performance

CADPC was implemented for the *ns* network simulator; PTP code for ns and Linux can be obtained from [3], where we will make CADPC available as well. During early tests, it became evident that receiving and reacting to feedback more often than every 2 RTTs causes oscillations. This is easy to explain: as a fundamental principle of control, one should only use feedback which shows the influence of one's actions to control the system [10]. Since CADPC differs from TCP in that it reacts more drastically to precise traffic information, increasing the rate earlier can cause the traffic to reach a level where the control law tells the source that it should back off and vice versa. The best results were achieved by calculating the PTP period (when we would send a new PTP packet) similar to the TCP retransmit timeout, which is approximately 4 RTTs; in fact, the scheme performed almost equally when we used exactly 4 RTTs. The round-trip time was estimated via PTP packets only.

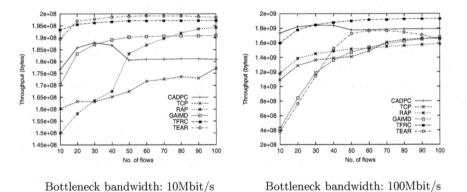

Bottleneck bandwidth: 10Mbit/s   Bottleneck bandwidth: 100Mbit/s

Fig. 1. Throughput of CADPC and TCP-friendly congestion control mechanisms

3.1   CADPC Vs. TCP(-friendly) Congestion Control

To evaluate the performance of CADPC, its behaviour was compared with TCP Reno and several TCP-friendly congestion control protocols in simulations — RAP [11], GAIMD [12] with $\alpha = 0.31$ and $\beta = 7/8$ (which was realized by

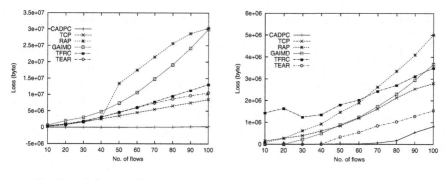

Bottleneck bandwidth: 10Mbit/s   Bottleneck bandwidth: 100Mbit/s

Fig. 2. Loss of CADPC and TCP-friendly congestion control mechanisms

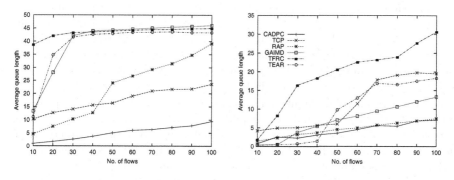

Bottleneck bandwidth: 10Mbit/s   Bottleneck bandwidth: 100Mbit/s

Fig. 3. Avg. queue length of CADPC and TCP-friendly congestion control mechanisms

tuning RAP parameters), and the slowly-responsive TFRC [13] and TEAR [14]. Flows of one particular kind at a time shared a single bottleneck link in a "dumbbell" topology. For all the results in figures 1, 2 and 3, the simulation time was 160 seconds and link delays were 50ms each, which, neglecting queuing delays, accounts for a RTT of 300ms — thus, a simulation took approximately 500 RTTs when the network was lightly loaded. The bottleneck bandwidth was 10 Mbit/s for the diagrams on the left-hand side and 100 Mbit/s for the diagrams on the right-hand side while the bandwidth of all other links was 1000 Mbit/s so as to avoid limiting the rate anywhere else in the network. The queuing discipline was drop-tail, payload packet sizes were 1000 bytes and PTP was used in a mode where packets have an initial size of 32 bytes and grow by 16 bytes at each router [2]. Only the payload flow is shown in the figures.

Note that in our simulation scenario, unresponsive senders which transmit at a high data rate would always achieve the greatest total throughput; therefore, high throughput can only be considered a good sign if the packet loss is low. For example, the high throughput of TFRC is not necessarily a good sign because it corresponds with high loss. Yet, TFRC appears to work better with many flows than RAP because RAP shows less throughput as well as more loss.

CADPC significantly outperformed TCP in our simulations: it achieved more throughput (and, in the case of the 100 Mbit link, more throughput than almost all other mechanisms) while maintaining the lowest loss. Also, while CADPC is the only mechanism that does not acknowledge every packet, it almost always showed the smallest average queue size — it should be no surprise that CADPC also worked very well in simulations with Active Queue Management, which we did not include due to space restraints.

3.2   Dynamic Behaviour of CADPC

Figure 4 shows that CADPC is also superior in terms of convergence speed and rate smoothness. The outlier after approximately 6 seconds is the result of an overreaction of the very first flow. Since all flows are started at the same time but are nevertheless queued consecutively, only flow 1 sees a completely "empty" link and thus reacts more drastically than all other flows. This behaviour, which we only encountered in the somewhat unrealistic special case of flows starting at the same time, is more severe at higher link speeds and with a larger number of flows — in such a case, it can significantly delay convergence. It should be possible to eliminate this effect: simulation experiments have shown that CADPC can be tuned to behave less aggressively by changing the constant parameter a of the control law to a smaller value > 0 without disturbing the basic behaviour of CADPC (i.e. the point of convergence); it also remains intact if the rate increase or decrease factor is limited by a maximum or minimum value. Tuning these parameters represents a trade-off between convergence speed and robustness against various link delays and the level of statistical multiplexing.

A somewhat more realistic scenario is shown in fig. 5: here, CADPC flows are started with a delay of 30 seconds. In the case of a 1 Mbit/s bottleneck link, CADPC shows minor fluctuations; these are caused by the fact that packet sizes limit the granularity of rate adaptations in the case of a non-fluid model. These fluctuations are diminished at higher link speeds.

4   Usage Scenarios

CADPC has two obvious disadvantages: i) it requires every other router to support the PTP protocol, and ii) any ISP interested in calculating the load should be in control of at least one router along the backward path — which is not always the case for highly asymmetric connections, where the uplink can be a modem and the downlink can be a satellite connection. CADPC will still work in such a scenario, but tracing its theoretical bandwidth becomes more difficult. In the following section, we briefly discuss how CADPC could be used within a controllable domain, where these two disadvantages are no issue. Then, we describe ideas which we think should be pursued in order to overcome the first problem.

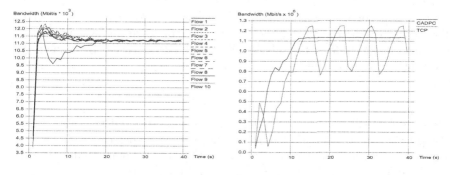

Bottleneck bandwidth: 1Mbit/s   Bottleneck bandwidth: 10Mbit/s

Fig. 4. CADPC startup behaviour: 10 single flows, 10 * CADPC vs. 10 * TCP

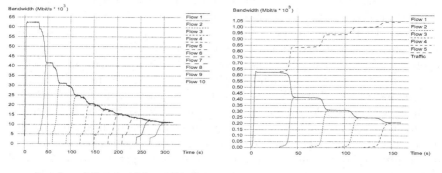

Bottleneck bandwidth: 1Mbit/s   Bottleneck bandwidth: 10Mbit/s

Fig. 5. Staggered start of sources

4.1   CADPC in a Controllable Environment

In the simplistic scenario of a packet network where PTP could be used instead of TCP and an ISP is in control of all routers, CADPC can be expected to show the advantages seen in simulations: it should be more efficient than TCP in several aspects, feasible for streaming media applications and remain scalable. Additionally, appropriate behaviour can be enforced through traffic shaping or policing, and it is easy to detect malicious senders, which facilitates the prevention of denial-of-service attacks.

Load-based charging requires a router to trace the bandwidth of a flow — with CADPC, this can be done by merely executing the control law for every feedback packet. It should be noted that this approach may not work under conditions of extreme load because even PTP packets may be lost in such cases. While these occurrences can be expected to be very rare due to the small loss and average queue length seen with CADPC, it may be necessary to design a special back off behaviour for the case of a lost PTP feedback packet.

As an additional aid, if the number of sources is known, the rate which a flow is supposed to achieve (unless flows enter or leave the system) can be determined instantly from the solution of equation 2, which has the same point of convergence [8] as our discrete control law:

$$x(t) = \frac{c}{1 + e^{-at}} \tag{3}$$

Since the calculated rate of CADPC is deterministic and traceable, bandwidth pricing can be differentiated by allowing a sender to behave like m senders; this method, which is similar in spirit to the "MulTCP" approach described in [15], merely involves multiplying the rate by m in our case.

4.2   Incremental Deployment Ideas

One way of deploying CADPC in an environment where its behaviour is traceable and well-defined (i.e. protected from the more aggressive behaviour of TCP and unresponsive UDP flows) would be the definition of a new DiffServ class. While DiffServ ensures scalability by state aggregation, the necessity for congestion control remains within a behaviour aggregate. Using heterogeneous or TCP-friendly congestion control mechanisms within such an aggregate leads to traffic bursts and packet loss; therefore, it is difficult to provide meaningful per-flow QoS with DiffServ. This problem, which could be called *"QoS in the small"*, might be solved by means of a scalable and enhanced congestion control mechanism which is supported by routers. The service could then include a rule such as "we offer QoS and provide router support, you use CADPC and get a good result (and we can calculate your rate, charge you accordingly and make sure that others behave well, too)". This kind of service would of course only be meaningful across domains which support it.

Another way of integrating CADPC with DiffServ would be signaling between edge routers in support of flow aggregates; a thorough description of such an approach can be found in [16]. In this scenario, CADPC is required to be advantageous even if flows are not greedy; this could be investigated by studying, for example, TCP with a rate limit from CADPC (TCP "over" CADPC). CADPC could then also be "translated" into TCP and vice versa by introducing an edge node which acts as both a TCP sender and CADPC receiver at the same time.

The mechanism in [17] is described to be deployable in a cloud-based approach similar to that proposed by Core Stateless Fair Queuing; due to the similarities between this scheme and ours, this method can be expected to apply to CADPC as well.

5   Related and Future Work

The advantage of moving a step further from binary to multilevel feedback is outlined in [19]. Actual ABR-like signaling based on RTCP is used with per-flow state in the "Adaptive Load Service" (ALS) [20] and with a window-based approach, no per-flow state but more strenuous router calculations where the header of every payload packet is updated in [17]. In [16], the focus is on the interaction between ABR-like signaling and DiffServ. Work based on binary feedback includes TCP-friendly approaches with a smoother throughput such as TFRC and TEAR [13] [14]; while relying on packet loss instead of explicit signaling ensures high scalability, all TCP-friendly mechanisms need to exceed the available bandwidth in order to receive a congestion notification and react properly.

CADPC can be expected to converge to max-min fairness because, in an initially empty network, it works like the "progressive filling algorithm": all sources which share a bottleneck see the same feedback and their rates converge towards the rate of this bottleneck. From the perspective of these flows, other flows which traverse the same link but have a different bottleneck are merely background traffic. These flows, in turn, increase their rates until their respective bottleneck is filled, leading to a max-min fair rate allocation (or, more precisely, a rate allocation which converges to max-min fairness with a growing number of flows). Future work includes testing this claim via simulations with multiple bottlenecks.

A max-min fair rate allocation is achieved by maximizing the linear utility function of all users. Although this kind of utility function may in fact be feasible for streaming media applications, services like file transfer and web surfing have a logarithmic utility function; as of yet, it remains an open question whether CADPC could be extended to support such utility functions and converge to proportional fairness [21].

So far, CADPC was only studied with a single dumbbell where all sources behaved in a similar manner. Future work includes testing the mechanism with different background traffic, different topologies and CADPC flows which are not greedy; possible extensions to multicast scenarios will also be examined. Most importantly, it is planned to pursue the ideas for gradual deployment which were outlined in the previous section.

6   Conclusion

In this paper, a novel approach to congestion control based on explicit and lightweight feedback was presented. The control law converges to a fixed and stable point, which accounts for higher throughput, less loss, less fluctuations and thus better QoS than TCP. Moreover, the rate calculation is simple and can be performed wherever feedback packets are seen; it is easy to enforce appropriate behaviour and implement load-based and differentiated pricing. While our mechanism has the obvious drawback of requiring router support, we believe that the significant performance improvement combined with the ideas for gradual deployment can lead to a future service that is fit for the Internet.

References

1. D. Katz: *IP Router Alert Option*, RFC 2113, February 1997.

2. Michael Welzl: *A Stateless QoS Signaling Protocol for the Internet*, Proceedings of IEEE ICPADS'00, July 2000.
3. The PTP website: *http://fullspeed.to/ptp*
4. A. W. Barnhart: *Explicit Rate Performance Evaluations*, ATM Forum Technical Committee, Contribution ATM Forum/94-0983 (October 1994).
5. D. Chiu and R. Jain: *Analysis of the Increase/Decrease Algorithms for Congestion Avoidance in Computer Networks*, Journal of Computer Networks and ISDN, Vol. 17, No. 1, June 1989, pp. 1-14.
6. Michael Welzl: *Vector Representations for the Analysis and Design of Distributed Controls*, MIC 2002, Innsbruck, Austria, 18-22 February 2002.
7. David G. Luenberger: *Introduction to Dynamic Systems - Theory, Models, and Applications*, John Wiley & Sons, New York 1979.
8. Perrin S. Meyer, Jason W. Yung and Jesse H. Ausubel: *A Primer on Logistic Growth and Substitution: The Mathematics of the Loglet Lab Software*, Technological Forecasting and Social Change 61(3), pp.247-271, Elsevier Science 1999.
9. Mark Kot: *Elements of Mathematical Ecology*, Cambridge University Press, Cambridge, UK, 2001.
10. Raj Jain and K. K. Ramakrishnan: *Congestion Avoidance in Computer Networks with a Connectionless Network Layer: Concepts, Goals and Methodology*, Computer Networking Symposium, Washington, D. C., April 11-13 1988, pp. 134-143.
11. Reza Rejaie, Mark Handley, and Deborah Estrin: *RAP: An End-to-end Rate-based Congestion Control Mechanism for Realtime Streams in the Internet*, Proceedings of IEEE Infocom 1999, New York City, New York, 21.-25. 3. 1999.
12. Y. Richard Yang and Simon S. Lam: *General AIMD Congestion Control*, Technical Report TR-2000-09, Dept. of Computer Sciences, Univ. of Texas, May 2000.
13. Sally Floyd, Mark Handley, Jitendra Padhye, and Jörg Widmer: *Equation-Based Congestion Control for Unicast Applications*, Proceedings of ACM SIGCOMM 2000.
14. Injong Rhee, Volkan Ozdemir, and Yung Yi: *TEAR: TCP emulation at receivers – flow control for multimedia streaming*, Technical Report, Department of Computer Science, North Carolina State University.
15. Philippe Oechslin and Jon Crowcroft: *Differentiated End-to-End Internet Services using a Weighted Proportional Fair Sharing TCP*, ACM CCR, 1998.
16. Na Li, Sangkyu Park, and Sanqi Li: *A Selective Attenuation Feedback Mechanism for Rate Oscillation Avoidance*, Computer Communications, Vol. 24., No. 1, pp. 19–34, Jan. 2001.
17. Dina Katabi, Mark Handley, and Charlie Rohrs: *Internet Congestion Control for Future High Bandwidth-Delay Product Environments*, ACM SIGCOMM 2002, Pittsburgh, PA, 19-23 August 2002.
18. Ion Stoica, Scott Shenker, and Hui Zhang: *Core-Stateless Fair Queueing: Achieving Approximately Fair Bandwidth Allocations in High Speed Networks*, ACM SIGCOMM 1998, Vancouver, British Columbia, September 1998.
19. Arjan Durresi, Mukundan Sridharan, Chunlei Liu, Mukul Goyal and Raj Jain: *Traffic Management using Multilevel Explicit Congestion Notification*, Proceedings of SCI 2001, Orlando Florida 2001.
20. Dorgham Sisalem and Henning Schulzrinne: *The adaptive load service: An ABR-like service for the Internet*, IEEE ISCC'2000, France, July 2000.
21. Frank Kelly: *Charging and rate control for elastic traffic*, European Transactions on Telecommunications, 8. pp. 33-37. An updated version is available at http://www.statslab.cam.ac.uk/frank/elastic.html

Applying the Generalized Vickrey Auction to Pricing Reliable Multicasts

Ashish Sureka and Peter R. Wurman

North Carolina State University, Raleigh, NC 27695, USA,
asureka@unity.ncsu.edu, wurman@ncsu.edu

Abstract. Reliable multicast networks scale back the speed at which content is delivered to accommodate the slowest receiver. By extracting information about the receiver's capabilities, the service provider can select which receivers to admit, and at which speed to operate the network. We examine the application of the Generalized Vickrey Auction to pricing reliable multicast and present a distributed method to compute the GVA prices.

1 Introduction

Multicast protocols allow a message to be sent to many receivers at the same time more efficiently than a unicast protocol and less wastefully than broadcast protocols. In a multicast network, the messages are sent only to identified subscribers. The rate at which any individual subscriber receives data depends on many factors, including the type of machine and the throughput of its connection to the internet. In standard multicast protocols, hosts or network elements that cannot handle data at the rate being sent by the server will simply drop packets.

While packet dropping is satisfactory for some applications, in some situations the sender wants to be sure that everyone receives all of the data in a timely manner. Potential application that require this reliability include some uses of video conferencing, stock quote feeds, auctions, online computer games, and other real-time applications that demand symmetric and complete information delivery. *Reliable multicast* protocols are designed to satisfy this need. In order to achieve reliability and source-order data delivery, the sender must send at a speed acceptable to the slowest receiver. Thus, the rate at which all subscribers receive packets is determined by the capabilities of the slowest subscriber.

We assume that the service provider has the ability to charge subscribers for the content, either to recover her costs (and make a profit) or to discourage network abuse. In addition, each subscriber has a value for the service that depends upon the speed at which it is offered. The service provider must determine both the speed and the price at which to offer her service. From the point of view of the provider, the marginal cost of adding subscribers is zero. Thus, we assume that the service provider will choose to admit anyone into the subscriber pool who has the capability to receive data at a rate at least as fast as the provider's chosen speed. However, the provider runs the risk of having a user overstate his reception rates in order to gain admittance to the subscriber pool. Once the user is in the pool, the reliable multicast protocol will automatically

B. Stiller et al. (Eds.): QofIS/ICQT 2002, LNCS 2511, pp. 283–292, 2002.

compensate for his slower speed and failure to acknowledge packets by decreasing the sending rate for everyone.

In this paper, we consider the task of admitting potential receivers into the subscriber pool and selecting a service speed that maximizes the social welfare (as opposed to maximizing the revenue). The task is simplified greatly if the service provider can extract the true capabilities and values from potential subscribers. The Generalized Vickrey Auction (GVA), a mechanism well-known in the economics literature, has the property that it is a participant's weakly dominant strategy[1] to truthfully reveal his value for, in this case, receiving data at various speeds and the limitations of his capabilities.

Although in general the GVA is NP-hard, we demonstrate that in the case where there are a linear number of potential operating speeds, it is computationally feasible for the agents to report their valuations for the different network speeds and for the service provider to determine the various payments. In addition, we show that the computation can be distributed across the network in a manner that reduces the overall number and size of the messages that must be communicated.

2   Model

We consider a simple model of a reliable multicast network. The network is a tree rooted at the service provider. The leaves of the tree represent clients who have values for receiving the multicast. The internal nodes of the tree represent the routers.

In the general model, the service provider, k, is capable of transmitting the the signal at a speed chosen from the set S. Let I represent the set of possible receivers. An individual receiver, i, has a value for receiving the transmission at speed s denoted $v_i(s)$. Let J be the set of routers, with an individual router denoted j.

Receiver i has a threshold, t_i, above which he cannot keep up with the transmission. If the service is provided at a speed greater than t_i, receiver i will start dropping packets and the reliable multicast protocol will automatically rein back its delivery speed. Thus, the service provider would like to know the value of t_i for each potential recipient. It is convenient to model the threshold as $v_i(s) = 0$ for all $s > t_i$.

The service provider has a cost for broadcasting at speed s denoted $\gamma(s)$. We consider the case where the service provider desires to offer the service at the speed that maximizes the social welfare. That is,

$$V^* = \max_s -\gamma(s) + \sum_{i \in I} V_i(s). \tag{1}$$

Let s^* be the speed resulting in V^*. Every receiver capable of receiving data at or above speed s^* is admitted.

---

[1] In a game a strategy is said to weakly dominate all other strategies in the player's strategy set if, regardless of what he expects his opponents to do, this strategy always yields a payoff that is at least as good as any other of his strategies.

2.1 Generalized Vickrey Auction

The Generalized Vickrey Auction (GVA) [9] is a well-known mechanism in economics which has the surprising and desirable property that each participant has a weakly dominant strategy to truthfully reveal his true values for all options. Informally, the GVA achieves this property, called *incentive compatibility*, by computing a payment for an individual participant that is independent of his stated values; the participant's statement determines whether she will receive anything, but not how much she will pay.

To compute the GVA payment, the service provider computes the allocation V^*_{-i} that maximizes the sum of utilities of the other participants when i is excluded from the allocation.

$$V^*_{-i} = \max_s -\gamma(s) + \sum_{h \in I, h \neq i} v_h(s).$$

Receiver i's payment when speed s^* is chosen is

$$p_i = V^*_{-i} + \gamma(s^*) - \sum_{h \in I, h \neq i} v_h(s^*). \tag{2}$$

The GVA computes a payment per individual, not a price per resource. Thus, two individuals can pay different amounts for the same resource, a fact that is particularly clear in the reliable multicast setting in which all admitted receivers get the same data at the same rate.

In many practical settings, the service provider would be more concerned with selecting the speed and the pricing method that will generate the greatest profit (revenue minus cost). However, there are arguments in the economics literature [2] that suggest that socially efficient mechanisms will generate as much revenue as those that try to maximize revenue because efficient mechanisms will attract greater participation and competition.

Economic arguments aside, in this paper we are concerned with removing the incentive to misrepresent operating capabilities that will later degrade the experience for the entire user population. By using an incentive compatible mechanism, we can extract truthful information from the participants regarding their values and capabilities and set the operating speed appropriately.

Example 1 The simple example shown in Figure 1 has a service provider, K, seven receivers (R1 to R7) and 6 routers (N1 to N6). For simplicity, assume the service provider can transmit the data at three different speeds (fast, medium, and slow). Table 1 shows the costs for the service provider and the values for each of the receivers for each of the three speeds. The network on the right branch of the tree is slower than the left, and the receivers on the right cannot receive at the fast rate. In place of the threshold, we have set the right receivers' utility for the fast speed to zero.

It is easy to see that fast is the optimal speed and produces a total utility of 39 (versus 37 for medium and 11 for slow). The GVA payments are computed according to equation (2). For receiver R1, we first compute the value of the optimal solution that is credited to other agents: $V^* - v_{R1}(s^*) = 39 - 9 = 30$. Next, we compute the utility the other agents would have if R1 were not present. Without R1, the optimal speed is

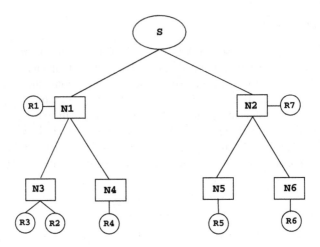

Fig. 1. Example network model.

	Fast	Medium	Slow		Fast	Medium	Slow
K	(3)	(2)	(1)	R4	11	3	1
R1	9	4	1	R5	0	8	1
R2	14	7	2	R6	0	7	2
R3	8	6	3	R7	0	4	2

Table 1. The costs for the service provider K, and the utilities of receivers R1 to R7 for three speeds.

medium and $V_{-i}^* = 33$. Thus, R1's GVA payment is 3. Similarly, we can compute that R2 pays 5, R4 pays 6, and all of the other agents pay zero.

3 A Distributed Protocol for Computing Payments

A straightforward implementation of a GVA mechanism in this environment is for each receiver to send his utility function to the "auctioneer", either the provider or a designated third party. The auctioneer then computes each receiver's payment and informs each receiver individually. This scheme has several drawbacks. First, as we shall see below, the amount of information that needs to be communicated is unnecessarily high. Second, the details of each user's utility function is passed all the way to the top of the tree. An observer near the top of the tree would see the utility function of every user on that branch. The alternative scheme described below aggregates information, thereby making it more difficult for an observer to discern a particular agent's utility.

For the reliable multicast network model, we can compute the GVA payments in a distributed fashion. The solution computes each receiver's payment in the router at which they are connected to the tree. For the router to make this computation, it must know the value of operating at all speeds, and the effect of adding this particular receiver on the choice of network speed. Let I_j be the set of receivers directly connected to router j, $C(j)$ be the routers directly connected to j as child nodes, and $p(j)$ be the parent node of router j. Note that all of the receivers not connected to j can be reached by a traversal through one of the nodes in the set $C(j) \cup p(j)$. Let $I_{p(j)}$ and $I_{c(j)}$ be the set of receivers reachable from j through the parent and a child, c, respectively. We can rewrite equation (1), from the perspective of receiver j, as

$$V^* = \max_s -\gamma(s) + \sum_{i \in I_j} V_i(s) + \sum_{i \in I_{p(j)}} V_i(s) + \sum_{c \in C(j)} \sum_{i \in I_{c(j)}} V_i(s). \tag{3}$$

We can rewrite equation (2) in a similar fashion. It is clear from the above decomposition that to compute the GVA payment of $i \in I_j$, router j needs to know only the aggregate utility at each speed for all of the nodes reachable through its parent and its children. To make the following description more concise, let $V_{p(j)}$ be the vector of aggregate demands for the nodes reachable through j's parent, $V_{c(j)}$ the aggregate demand for the nodes reachable through j's child c, and V_{I_j} the aggregate demand of receivers directly connected to j. Our distributed protocol computes the necessary aggregate utility messages and propagates them through the network.

The protocol works as follows:

Step 1: Every receiver sends its utility vector to the router connecting it to the multicast tree. Each router j now has v_i for all $i \in I_j$ and can compute V_{I_j}.

Step 2: Each router with no children sends V_{I_j} to its parent. When a router with children receives $V_{c(j)}$ for each of its children, it computes $V = V_{I_j} + \sum_{c \in C(j)} V_{c(j)}$ and sends it to its parent. This step propagates aggregate information up the tree.

Step 3: Each router propagates aggregate information about the rest of the tree to each of its children. The composition of router j's message to its child c depends upon router j's position in the tree, as follows:

$$V = \begin{cases} V_{I_j} - \gamma + \sum_{z \in C(j), z \neq c} V_{z(j)} & \text{if } j \text{ is the root,} \\ V_{I_j} + V_{p(j)} + \sum_{z \in C(j), z \neq c} V_{z(j)} & \text{otherwise.} \end{cases}$$

This step propagates information both sideways and down the tree. The service provider may choose to include the optimal operating speed as part of the message, though every router will have enough information to compute it directly.

Step 4: At this point, each router has enough information to compute the payment for each receiver attached to it. The reader can verify that router j now has all of the terms required in equation (3). Once the router has computed the payments, a message of the form $\{i, p_i\}$ must be sent to both the receiver and the service provider.

The GVA computation is decomposable in this setting because the number of solutions is the number of operating speeds, and the "resource" is *non-rival*, that is, the object can be assigned to more than one participant. In fact, the very purpose of this enhancement to the reliable multicast system is to determine a common speed at which to operate. In the general case of assigning combinations of rival objects, all possible sub-solutions would have to be communicated throughout the tree, making the approach unworkable.

3.1   Example of the Distributed Protocol

To illustrate the protocol, consider router N1 in the example problem defined in Figure 1 and Table 1. During Step 1, N1 learns that $v_{R1} = [9, 4, 1]$. In Step 2, N1 receives message $[22, 13, 5]$ from N3 with the aggregate values of receivers R2 and R3, and message $[11, 3, 1]$ from N4. N1 then composes its message to the service provider, K, by summing the vectors it has received, producing $[42, 20, 7]$. The service provider receives $[42, 20, 7]$ from N1 and, after the protocol on the right side of the tree completes, $[0, 19, 5]$ from N2.

In Step 3 of the protocol, K constructs the message $[-3, 17, 4]$ to send to N1 by subtracting the cost vector from the message from N2. N1 now has the information shown in Table 2, which is exactly what it needs to compute that R1's GVA payment is 3.

	Fast	Medium	Slow
R1	9	4	1
N3	22	13	5
N4	11	3	1
K	-3	17	4

Table 2. Information at router N1 after Step 3 of the protocol.

Router N1 was also responsible for passing information down the tree in Step 3. N1's message, $[17, 24, 6]$, to N3 is the sum of the vectors describing R1, N4, and K. Similarly, N1 sends $[28, 34, 10]$ to N4. These two messages give N3 and N4, respectively, enough information to compute the GVA payments for R2, R3, and R4.

3.2   Computational Considerations

Above, we claimed that the distributed protocol had some advantages over the straightforward implementation in which all utility vectors are passed up the tree to the service provider, who then computes payments and sends them back down the tree to the receivers. In this section we elaborate on those claims.

Consider a symmetric tree of depth d in which each router has b children and l directly connected receivers. We will distinguish the computation phase, which ends when some element of the network knows the payment for each receiver, and the notification phase, which ends when both the service provider and receiver know the payments.

The communication costs can be measured in both the amount of information communicated, and the amount of time taken to perform the communication. First, we note that the notification phase has the same cost in both the centralized and distributed protocols because a complete message path has to exist between the service provider and the receiver to carry the payment notification to both parties. It does not matter whether the message starts at the service provider or the router near the receiver.

Thus, we concentrate on the amount of information communicated in the computation phase of the two protocols. We can eliminate the communication between the receivers and their local routers, as this is the same in both protocols. However, with respect to information passed among the routers, the distributed protocol has significant savings over the centralized protocol. In the straightforward implementation, the number of vectors that a router in level n of the tree must send to its parent is $l(\frac{b^{d-n+1}-1}{b-1})$. In contrast, the distributed protocol sends two vectors (one each in Steps 2 and 3).

Although the distributed protocol requires far fewer vectors to be communicated, it does require more traversals of the tree. The centralized protocol requires two traversals, one up from the bottom with all of the utility functions, and then one down from the top with all the payments. The distributed protocol requires three traversals, one each in Steps 2, 3, and 4. It seems likely that the benefits of reducing the number of messages that need to be communicated will compensate for the increased number of tree traversals.

In the centralized protocol, no information needs to be stored in the routers; each message can be sent directly up the tree and forgotten. In contrast, in the distributed protocol, every router needs to store information about the reported utilities in order to later determine the payments for the receivers connected to it. However, the router needs to store a utility, or aggregate utility, vector from only those receivers and routers directly connected to it. In the symmetric case, the router must store $b + 1 + l$ vectors of size $|S|$.

3.3 Other Considerations

One of the desirable characteristics of a pricing mechanism is its robustness to changes in the subscriber population. If the payment of one receiver changes drastically because another receiver left or joined the network, then the pricing mechanism is not robust. With GVA pricing, when a new receiver joins or leaves the network, the service provider recalculates the optimal speed at which the network should operate and determines the payments for each of the receivers. The addition or subtraction of one receiver can change the payments of other receivers and even cause the service provider to select a new speed, abandoning some of the previous recipients. The former problem is not serious, as the new payments will never be greater than a receiver's value for the service. However, the latter issue is of some concern. While we have not yet studied this problem, we suspect that a utility model that properly considers both the value of admission and the value of continuing service over time is a promising avenue to explore.

Two well-known drawbacks of the GVA are its susceptibility to *collusion* and *false-name* bids. Bidder collusion is a serious problem in many auctions; in our proposed mechanism the bidders can collude by deciding beforehand on values to report to the auctioneer that improve the utility of at least one of the agents. Table 3 shows the valuations of three receivers for three speeds. If the receivers truthfully report this information, the respective payments will be 4, 0, and 0. No receiver can unilaterally change the vector he reports and make himself better off.

However, if Receivers 1 and 3 collude and report the values shown in Table 4, the respective payments will be 3, 0, and 0. Thus, by reducing his stated value for medium and slow, Receiver 3 is able to lower Receiver 1's payment. Presumable, Receiver 1 will be willing to offer Receiver 3 some portion of the savings as a reward, thus improving both agents' overall utility. In the same spirit, it is possible for a receiver to

	Fast	Medium	Slow
Receiver 1	9	1	1
Receiver 2	0	6	3
Receiver 3	4	2	2

Table 3. Information at the service provider.

	Fast	Medium	Slow
Receiver 1	9	1	1
Receiver 2	0	6	3
Receiver 3	4	1	1

Table 4. Information at the service provider.

submit a bid under a false name and lower her payment [10]. However, to succeed with this charade in the reliable multicast scenario would also involve running, or spoofing, a real multicast client or the false bid would be quickly detected.

One final potential drawback of calculating payments using the GVA is *price discrimination*. Two agents might end up paying different prices for the same speed even when they have the same value for receiving data at that speed. There are likely to be reliable multicast applications in which price discrimination will be regarded as unfair and unacceptable by the receivers. The perceived negative benefits of price discrimination will have to be weighed against the potential loss of social welfare due to the loss of incentive compatibility.

4   Related Work on Network Pricing Schemes

Several researchers have proposed new ways of controlling network usage using economic mechanisms. The approaches address the questions of how to allocate bandwidth in a network and how to do admission and congestion control.

Smart Markets were proposed by MacKie-Mason and Varian [5] and introduced the concept of congestion-sensitive pricing. Rather than forwarding packets based on a FIFO policy, packets are selected to be forwarded based on the priority (or bids) associated with each packet. The proposed Smart Market is a GVA, and achieves incentive compatibility by charging the successful senders the priority of the highest priority packet that is dropped. However, because it requires accounting at the level of packets, the Smart Market is impractical to implement in a real network [8].

A experimental evaluation was done by Breker [3] on both static and dynamic pricing strategy. Breker's simulations show that when strategies are static, as the price per packet increases, the network utilization decreases. In the dynamic pricing strategy simulations, the price varies over time as the result of user demand for bandwidth. Breker also found, predictably, that as the number of users increases, the price per unit of bandwidth also increases.

Murphy et al. [6] propose a feedback scheme to increase network efficiency by taking advantage of the fact that users can modify their traffic inputs in response to feedback signals from the network such as cell loss and/or delay. Each user computes a demand curve that specifies her willingness to pay for various levels of bandwidth. The service provider dynamically adjusts the prices based on the current network conditions and sets the prices so that the marginal benefit the user places on his resource allocation is equal to the marginal cost of handling the resulting traffic in the network.

Shenkar et al. [7] critiques the optimality paradigm (achieving optimal welfare requires charging marginal congestion costs for usage) and propose that the pricing in computer networks should focus more on structural and architectural issues rather than the optimality paradigm.

Feigenbaum et al. [4] investigate cost sharing protocols for multicast transmission [4] and point to two distinct mechanisms, *Marginal Cost* and *Shapley value*, as the two solutions appropriate in the context of sharing the cost of a multicast transmission. In their paper, Feiganbaum et al. prove that *Marginal Cost* has a natural protocol that uses only two messages per link of the multicast tree and the Shapley Value requires a quadratic total number of messages. The paper analyzes the communication complexity of the two mechanism and the apparent gap between the communication complexity of the Marginal Cost and Shapley Value.

Our protocol also bears some resemblance to the COTREE algorithm proposed by Andersson and Ygge [1] which propagates aggregate (continuous) demand messages through a self-imposed communication tree. The COTREE algorithm is designed to propagate the necessary aggregate demand information to the root of the tree so that an equilibrium price can be computed.

5   Conclusions

In this paper we applied the Generalized Vickrey Auction to the problem of allocating bandwidth in a multicast network, and argue that it will lead to an efficient use of the resources. The computation of GVA payments in this environment can be efficiently computed in a distributed fashion that greatly reduces the amount of information that must be propagated through the network.

In the future, we plan to examine pricing in a more detailed model of reliable multicasting, and to investigate robustness of the mechanism against collusion and false name bids.

Acknowledgements

This work was supported by the Defense Advanced Research Projects Agency under contract F30602-99-1-0540 and E-Commerce@NC State program. The views and conclusions contained herein are those of the authors. We would like to thank Doug Reeves for his valuable comments.

References

[1] Arne Andersson and Fredrik Ygge. Managing large scale computational markets. In *31st International Conference on System Sciences*, pages 4–13, 1998.

[2] Lawrence M. Ausubel and Peter Cramton. The optimality of being efficient. May 1998.

[3] Leanne P. Breker. A survey of network pricing schemes. Technical report, Department of Computer Science, University of Saskatchewan, 1996.

[4] Joan Feigenbaum, Christos H. Papadimitriou, and Scott Shenker. Sharing the cost of mu-liticast transmissions (preliminary version). In *ACM Symposium on Theory of Computing*, pages 218–227, 2000.

[5] J. MacKie-Mason and H. Varian. Pricing the internet. Technical report, In Public Access to the Internet (Kahin B. and Keller J., Eds.), Prentice Hall 1995., 1995.

[6] L. Murphy and J. Murphy. Pricing for atm network efficiency. Technical report, Proc. 3rd International Conference on Telecommunication Systems Modelling and Analysis, Nashville, TN, March, p. 349-356. Available from URL http://www.eeng.dcu.ie/murphyj/atm-price/atm-price.html, 1995.

[7] S. Shenker, D. Clark, D. Estrin, and S. Herzog. Pricing in computer networks: Reshaping the research agenda. Technical report, Proc. of TPRC 1995., 1995.

[8] R. Singh, M. Yuksel, S. Kalyanaraman, and T. Ravichandran. A comparative evaluation of internet pricing models: Smart markets and dynamic capacity contracting. Technical report, Proceedings of 10th Annual Workshop on Information Technologies and Systems (WITS), Australia, 2000.

[9] H. Varian and J. MacKie-Mason. Generalized vickrey auctions. Technical report, Technical report, University of Michigan., 1995.

[10] M. Yokoo, Y. Sakurai, and S. Matsubara. Robust combinatorial auction protocol against false-name bids. *Artificial Intelligence*, 130(2):167–181, 2001.

InterQoS – Strategy Enterprise Game for Price and QoS Negotiation on the Internet[1]

Lilian N. Nassif[1,3], Luiz Henrique A. Correia[1], Carlos Frederico M.C. Cavalcanti[1,2], José Marcos Nogueira[1], Antonio A. F. Loureiro[1], Geraldo Robson Mateus[1]

[1] Computer Science Department, Federal University of Minas Gerais,
Av. Presidente Carlos Luz, 6627, Belo Horizonte, Brazil
{lilian, lcorreia, cfmcc, jmarcos, loureiro, mateus}
@dcc.ufmg.br
[2] Computer Science Department, Federal University of Ouro Preto,
Campus Morro Cruzeiro w/n Ouro Preto, Brazil
cfmcc@iceb.ufop.br
[3] Prodabel, Av. Presidente Carlos Luz, 1275, Belo Horizonte, Brazil
lilian@pbh.gov.br

Abstract. This paper discusses strategies issues for negotiating price and QoS guarantees between telecommunication companies, ISP (Internet Service Provider) and Internet users. There is an increase in telecom infrastructure cost when QoS guarantees are offered and, consequently, a conflict between providers and users of the system arises. Another important question is about end-to-end pricing since QoS guarantees have to be allocated end-to-end. This usually involves money repass to telecom and providers companies that are participating in the end-to-end solution. We elaborated a strategy game called InterQoS. The main objective of this game is to identify and treat technology and pricing problems boosted by adding QoS in the Internet. The idea of the InterQoS game is to provide an environment where players can experiment strategies in order to evaluate their decisions and actions.

1 Introduction

The use of QoS demanding applications on the Internet is incipient although important progress has been done in the last few years. DiffServ and IntServ [1] are important milestones in the evolution of Internet in direction of providing QoS guarantees. Constrained routing, traffic shaping and policies are examples of QoS strategies formerly used in non-IP networks that are now extensively discussed in the IP world.

There is an increase in telecom infrastructure cost when QoS guarantees are offered. In this scenario, a new pricing model will be associated with QoS services. Although pricing is a simple subject at the current technological stage of the Internet, it is not true when QoS capabilities are added in. Economical models, resource ac-

---

[1] This project is supported by the Brazilian Council for Science and Technology (CNPq) through the Brazilian Research Network (RNP). The group is composed by researchers from Federal University of Minas Gerais (UFMG) and Federal University of Pernambuco (UFPE).

B. Stiller et al. (Eds.): QofIS/ICQT 2002, LNCS 2511, pp. 293-302, 2002.

counting by differentiated services, and models of money distribution among network providers are new components of this scenario.

Therefore it is necessary to define an interaction model of all relevant entities in the Internet expliciting the relationship among them. This paper discusses this problem and has the objective of introducing a game, named InterQoS, where Internet users, ISP and telecommunication operators, according to their roles, can experiment strategies in order to evaluate their decisions and actions. The game takes into account technology facilities, agreements among players and marketing strategies. A better understanding of those questions and their relationships can provide some outcomes such as regulations and decision-support information.

We propose to build InterQoS as a simulation game. This provides a great flexibility since different situations can be experimented and evaluated. Depending on the actions made by the players it is possible to have scenarios that show up problems that are not easily identified.

This paper is organized as follows. Section 2 presents some games related to enterprise simulations. Section 3 describes the characteristics of the InterQoS game. Section 4 discusses technical aspects of the game. Finally, Section 5 presents our concluding remarks.

2   Enterprise Games

We use the term "enterprise games" to express the class of games concerned with comprehension and study of the enterprise environment in real cases. This kind of game is largely used in business administration. In the following we briefly discuss two enterprise games.

Markstrat3 [2] is a simulator that has been designed for teaching strategic marketing concepts. It allows students or executives to practice their skills in a concentrated amount of time before trying them out in their real business environment. Students can go beyond strategy formulation and implement their strategic plan through decisions at the tactical level, a challenge not provided by traditional pedagogical methods. In particular, Markstrat3 claims the possibility of assessing the success or failure of marketing implementation over time as an important benefit offered.

"Boom and Bust Enterprises Management Flight Simulator" [3] is a game oriented to the life cycle of new products and illustrates fundamental principles of a corporate strategy including the learning curve, time delays in capacity expansion, competitive dynamics, and market saturation. The strategic environment captured in the game is particularly applicable to toys and games, fashions, fads, and other products that experience rapid growth and saturation.

3   InterQoS Characteristics

InterQoS game can be classified as a non cooperative game according to Game Theory [4]. InterQoS has three different categories of players: service providers called

ISPs, infrastructure providers called telecom companies and Internet users. The game is designed to be run with at least three players and becomes more exciting when there are various players from each category. The players have to open an individual session, preferably in a different computer, and start to compete with other players of the same category.

At the beginning of the game, it is defined a geographic area where the competition takes place. Each game lasts for a predefined period of time. Each category defines objectives to be achieved. At the end, the winner of each category is the player who had the best performance according to the objectives.

3.1 Players' Profile

Each category of player has a profile comprised by objectives and strategies. In the following is described the player's profile for each category.

ISP Player. The main objective of this category is to offer end-to-end services with QoS guarantees maximizing the number of clients and minimizing the operational cost. ISP clients are the Internet users. The ISP player strategy can consider questions such as: parameters to be used in the cost composition; costs dependent on geographic locations; subscription plans to be offered to the clients, possibly including differentiated services; number of connections to the Internet; and economical model to be used to money transfer.

Telecommunication Operator Player. The main objective of this category is to sell links with QoS guarantees to ISPs maximizing the use of the installed capacity and minimizing the operational cost. All telecom operator players share the same market and all of them compete among themselves in order to increase their profitability. The Telecom operator strategy can consider questions such as: parameters to be used in the cost composition; subscription plans to be offered to the clients considering fixed or variable bit rate, possibly including classes of services according to periods of time; and economical model to be used to money transfer.

Internet User. The main objective of this category is to maximize the ratio between information received with QoS over money spent. The Internet User strategy can consider questions such as: choose an ISP provider, possibly considering access rate, services offered, subscription plans, payment options, and QoS management; and content service providers.
Note that both the ISP and Telecom Operator must monitor and guarantee the service contracted by their clients.

3.2 How the Game Works

Before the game starts, the configuration of the environment must be defined:
1. Internet users;
2. ISPs with their classes, locations and connections to Telecom operators;

3. Telecom operators with their access points and connections to other Telecom operators. The types of links, with or without QoS guarantees, between access points of each Telecom operator have to be defined.

Each player starts playing the game with an amount of virtual money that can be spent to achieve the game objective. At the beginning, ISP players offer their plan of services that can be subscribed by the Internet users. Telecom players inform the price list for their service classes and it is available for the entire duration of the game.

After the game starts, prices can change and are valid for contracting new services. ISP players provide to Internet users a menu of services with QoS guarantees. For each kind of service, there is a price associated to it.

The services provided to Internet users by ISPs can be:

Video on Demand: The film price depends on the title and on the geographic location of the video server.

Voice over IP: The price depends on the geographic location of the destination and duration of conversation.

Videoconference: The price depends on the geographic location of the destination and duration of session. This service is based on a reservation of resources among the participating nodes. Internet users have to ask for this service in advance and ISPs reply with the price of the service.

Interactive games: The price depends on the kind of the game and duration of the match. This kind of service need QoS guarantees.

3.3 What-If Situations and Scenario Creations

The InterQoS game is a rich environment to test new situations and scenarios. To illustrate the usefulness of the game, two situations are shown as example: (1) who pays for a service and how the provider get paid for the offered service; (2) how a Telecom operator deals with demands to connect locations not yet connected. We investigate what it would happen when some strategies are used.

To complete the environment, a Clearinghouse (CH) is introduced. It is responsible for all money transfer between ISPs, Telecom operators, and Internet users.

Situation 1. Who pays for a service and how the provider get paid for the offered service.

To verify the options of movement in the game related to the situation of What-If, we associate with each player the following set of decisions:

Internet user (U): possible ways of payment are:
1. pre-paid card;
2. credit card;
3. monthly invoice concerning the use in the past month;
4. fixed charge encompassing all services included in the subscription plan.

ISP (I): ISPs get paid for the provided services and pay back other ISPs and Telecom Operators. The payment transactions can be carried out by:
1. A service order is created and sent to a Telecom operator, and then to the final ISP, and finally to the CH (this case will be further explained below);

2. The Telecom Operator identifies the ISPs that have used its services and charge them;
3. The ISP identifies the ISPs that have used its services and charge them;
4. The ISP charges a fixed price for each ISP connected to it as happens in the current model of the Internet.

Telecommunication Operator (TO): Money transfer depends on the payment option chosen by the ISP connected to it. The TO must be prepared for any choice made by ISPs and has the following options:
1. It accounts the resource spent by other players for future claim credit from the CH;
2. It identifies the ISPs that have generated traffic flow and send an invoice to them. It will be used when an ISP has chosen the I2 option;
3. It charges monthly only the ISPs that have contracts based on a fixed value. It will be used when an ISP has chosen the I3 and I4 options.

Some scenarios can be formed from the set of strategies presented. For example, scenario A is a combination of options U1, I1, and TO1. The combination of strategies among user, ISP and operator, can provide different outcomes.

Situation 2. How a Telecom operator deals with demands to connect locations not yet connected.

The TO has the following options: 1) rent links from other telecom operators; 2) install new links; and 3) establish partnerships with other telecom operators. These options only involve a relationship among telecom operators and are independent of any agreement made by users and ISPs. In the next section, we describe how the money transfers must be carried out in scenario B. It represents a scenario when option 3 above is chosen as strategy.

The game has to be executed several times to get interesting results. It is necessary because the number of times the game is executed is proportional to the number of combinations experimented. The idea is to identify strategies that lead to better results or should be avoided.

There are other situations that can be tested in the game, beyond the situation of a What-If presented above. For example, the impact of a new variable on the calculation of price and cost and others situations that involve a set of alternatives. Currently the InterQoS game is under development and it is being designed to allow the addition of any new situation.

3.4 Pricing Internet Services on InterQoS – Money Transfer

Each player category operates differently depending on which scenario it is playing. We will show some scenarios and how the money transfer among players is carried out.

Scenario A: The Internet user has a pre-paid card to be spent on the Internet. As introduced previously, the Clearinghouse is responsible for all money transfer among the ISP player and telecom operator players. Figure 1 shows the interaction model of players considering this scenario.

When the user player asks for a service, the pre-paid card code (ID) has to be given to the ISP player that the user is connected. The ISP player checks the credit and negotiates the end-to-end QoS guarantees in real time. The subscription fee covers the connection cost. Then, each Internet user pays a subscription fee to use a best-effort service to be connected on the Internet. Other QoS services are charged to the pre-paid Internet cards.

The ISP records the end-to-end telecommunication costs and sends the value and the card ID (record) to the Clearinghouse that subtracts that amount from the user's credit. This approach considers that the telecom operator accounts the utilization of its services based on the Service Order (SO) number and length of each period of utilization and sends a request of payment to the CH.

The Clearinghouse is the entity in charge to sell the pre-paid cards and effectively transfers all money spent by the Internet users to the other players. It has all control over all operations done in order to transfer the correct amount of money to the other players. The CH receives the record with all services spent by a user associated with a specific ID card from the telecom operators and authorizes the credit transfers for each telecom operator involved in the communication and for the content supplier.

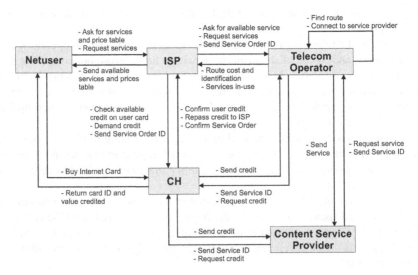

Fig. 1: InterQoS interaction model showing the money flow through InterQoS players

3.5 Content Distribution Network

The final price to the service depends on the cost associated with each link along the end-to-end path from the Internet User to the Content Provider. This price is called access cost and it can be decreased if the desired content (service) is offered by a closer content provider. This approach is always possible to be chosen using CDNs (Content Distribution Network) in order to decrease the access cost [5].

3.6 Starting and Finishing InterQoS

When the game starts, the administrator distributes randomly the service applications provided by the content providers over the nodes of the network, establishes the time duration of the game, the initial network topology and the initial credit given to each player. For each service application is associated a QoS requirement.

After the time duration of the game has elapsed, an evaluation is made assessing the participation of each player, the winners are found out for each player category and the game is finished. The assessment is based on the objectives of the game. The user winner is the one that got the biggest amount of information and paid the lowest price; the ISP winner got the biggest number of users and was the most profitable; the telecom operator winner got the biggest number of connections and was the most profitable.

4 Game Technical Approach

Besides representing an enterprise e-business logic, implementing providers and consumers interaction mechanisms, InterQoS game also implements new technologies that are widely argued in QoS forums. Some interesting problems emerge from this complex context, as described below.

4.1 Choice of Minimum Price

The inter-domain routing protocol used in current technological stage of the Internet is BGP-4 (Border Gateway Protocol) [6]. This protocol uses the number of hops as the only administrative metric to calculate the best route between source and destination nodes [7]. The number of metrics used in have an impact on the complexity of the routing algorithm.

A node running BGP-4 protocol sends to all its neighbours the number of hops to achieve the destination and also the path to reach it. Because each node has a table with distances to all other nodes, searching algorithms such as binary search and hash tables are used to implement BGP-4.

Our solution to find out the minimum price associated to an end-to-end path is based on searching the routing table associated to each BGP-4 node, and finding the shortest path. The access cost of an Internet user is the sum of the costs of all links belonging to the chosen path. Our proposal does not add new metrics to the algorithm. Instead, it calculates the prices based on the current BGP-4 metrics, i.e.: number of hops. It is necessary to use policy based routing to drive out the services to QoS guaranteed links [8].

4.2 Network Architecture

Structural relations between the entities of the game, which are the ISPs, Telecom operators and Internet users, are defined by a layered network architecture. The players that compete among themselves stay in the same layer. In the vertical sense, well done articulations between the players can lead to more profitable possibilities. In the architecture, each layer provides services to the upper layer. The network architecture considered in this game has the following elements as despicted in Figure 2:

1. A set of Internet users who interact with ISPs to gain access to the Internet. An user contracts a local telecommunication operator in order to have an access link to the chosen ISP.
2. A set of ISPs that provide services to Internet users. Each ISP is an Internet Autonomous System (AS) connected to one or more network provider. An Autonomous System is composed of a set of network elements such as routers, switches, and servers that are under the authority of a single administrative domain. An ISP can connect to network providers through one or more telecommunication links.
3. A set of network providers that offer Internet connectivity to ISPs. To do that, they contract links from telecom providers.
4. A set of telecom providers that own transport systems and offer links to network providers.

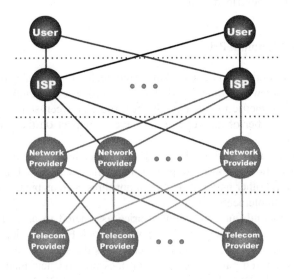

Fig. 2: Layered Network Architecture.

The architecture has the following additional characteristics:

☐ Each relationship between different Autonomous Systems is defined by a specification of the service to be provided (Service Level Specification [9]).
☐ End-to-end quality of service guarantees are established based on the guarantees of the links that belong to corresponding path.

☐ An Autonomous System assures quality of service in terms of delay, delay variation (jitter), packet loss rate, and bandwidth.

5 Conclusions

The Internet infrastructure needs considerable enhancements in order to support QoS needed in demanding applications, such as multimedia. A new Internet pricing scheme for these services can be a way to make effective the use of these applications.

Despite the advances of methods and techniques to provide QoS, the use of multimedia applications in the Internet is not widely available. A new pricing model can lead to a profitable approach to make e-business on the Internet. It brings security in taking decisions of new infrastructure investments.

The InterQoS game intends to represent this complex context when Internet users, ISPs and Telecom Operators have strong relationships among them. Such game allows the simulation of strategies, investments and interactions between the participants simulating problems that would be faced only in final implementation of services with QoS guarantees.

The InterQoS game can be classified as a non cooperative game. It is a complex and dynamic game, where the players do not have complete information about other players strategies. Although there are different types of players, they only compete with players of the same type. Because partnerships between them can be established, the game can be cooperative between partners.

In order to make investments in the Internet more attractive and profitable, it is necessary to coordinate the new QoS supporting architectures and new pricing models in an orchestrated way.

The dreamed vision of Internet as a multiservice network with quality of service appears to be approaching. The InterQoS game here proposed can contribute to this vision comes true.

References

1. Blake, S., Black, D., Carlson, M., Davies, E., Wang, Z., Weiss, W.: RFC 2475: An Architecture for Differentiated Services. IETF 12 (1988)
2. Larréché, J.C., Gatgnon, H.: Markstrat 3: the Strategic Marketing Simulation. International Thomson Publishing, (1997)
3. Paich, M., Sterman, J. D.: Boom, Bust, and Failures to Learn in Experimental Markets. Management Science 39(12), (1993) 1439-1458
4. Owen, G.: Game Theory.3rd ed.. Academic Press, Inc., London, (1995)
5. Qiu, L., Padmanabhan, V., Voelker, G.: On the Placement of Web Server Replicas. In Proc. 20th IEEE INFOCOM, (2001)
6. Rekhter, Y., Li, T.: A Border Gateway Protocol 4 (BGP-4). T.J. Watson Research Center, IBM Corp., Cisco Systems, (1995)
7. Wang, Z., Crowcroft, J.: Quality of service routing for supporting multimedia applications. IEEE Journal on Selected Areas in Communications, vol. 14, no. 7, 9 (1996) 1228-1234

8. Tangmunarunkit, H., Govindan, R., Estrin, D.,Shenker, S.: The impact of routing policy on Internet paths. In Proc. 20th IEEE INFOCOM, 4 (2001)
9. Nichols, K., Carpenter, B.: Definition of Differentiated Services Per Domain Behaviors and Rules for their Specification. RFC 3086, 4 (2001)

Resource Pricing under a Market-Based Reservation Protocol

Jörg H. Lepler[1] and Karsten Neuhoff[2]

[1] Cambridge University, Computer Laboratory, 15 JJ Thomson Avenue
Cambridge CB3 0FD, UK
Joerg.Lepler@cl.cam.ac.uk
[2] Cambridge University, Faculty of Economics, Sidgwick Avenue
Cambridge, CB3 9DD, UK
Karsten.Neuhoff@econ.cam.ac.uk

Abstract. A reservation protocol and pricing model is proposed for the allocation of electronic services, such as computing and communication resources. Whereas centralised allocation mechanisms are frequently analysed and better understood we focus on a decentralised, bilateral market employing a new protocol to enable application agents to request price quotes and to book resource capacity from resource providers. The protocol is implemented in a simulation of competing providers and clients. It is assumed that contracts and prices are confidential, therefore resource providers can only collect information from their own, bilateral trades and from requests for price quotes. They determine the optimal price with a supply function that is extended to continuously update the optimal price quotes with all new information. The simulated equilibrium prices are above competitive levels and deviations from these prices are not profitable. The exercise of market power increases with lower numbers of resource providers and with higher levels of capacity utilisation.

1 Introduction

Individual demand for many applications varies over time and its users benefit from renting variable amounts of a dedicated resource,[1] rather than owning and maintaining machinery sized for peak demand. For example a sports news agency operating live news feeds, tickers and video clips, renting additional computing machinery to cover demand peaks during Olympic games.

Three types of *Resources* can be allocated with the mechanism: First, *Network Bandwidth* connecting focal points of the global infrastructure at a negotiated Quality of Service (QoS) level. Second, *Processor Time* on a server farm and third, *Storage* for long term data deposit. Applications utilising the listed resources include first *Web Services* consisting of http servers, cache proxies

---

[1] The terms *service* and *resource* are both used throughout this document although they can be considered semantically equivalent, as a resource can be viewed as providing a service and vice versa.

B. Stiller et al. (Eds.): QofIS/ICQT 2002, LNCS 2511, pp. 303–314, 2002.

or search engines, second *Media Servers* delivering video/audio streams or relying distributed game content, third *Supercomputing Applications* simulating weather, engineering or scientific problems, and last *Internet Services Providers (ISP)* providing network connectivity.

Market designs can be categorised into centralised auctions and decentralised bilateral trade. Initially B2B sites were expected to move trade towards public, centralised auctions. However, "only few public exchanges are remaining ... and action in B2B turned elsewhere." (Economist 2001). To allow price discrimination and customer specific product offerings companies seem to prefer bilateral trade, potentially facilitated by private exchanges. Therefore we assume that computing resources will be allocated in bilateral negotiations.

To facilitate these negotiations with low transaction costs and short term interactions an automated protocol for resource reservation is developed and implemented in a simulation. The protocol builds on existing support infrastructure and standards under development, including a directory of services, service descriptions and service interactions. The protocol defines the message exchanges between clients wanting to reserve some resource and the providers. The protocol is simulated with several providers offering resources and clients seeking these. Clients make requests to all providers, specifying the time of their usage and desired parameters such as the performance or the Quality of Service level. In order to allow clients to compare offers of different providers, it is not necessary that all providers employ the same negotiation protocol, as long as it also contains the message parts of our protocol. The protocol is simple enough to be an inclusive model for bilateral negotiation. On receiving the request, a provider calculates whether his capacity suffices to satisfy the request and decides on the price to offer the resource.

If the market is competitive, this price falls to variable production costs, when capacity exceeds demand. As variable costs are virtually zero providers would capture little revenue. Market power can allow generators to maintain prices above marginal costs - two mechanisms are usually assumed in economics. First, buyers incur search costs when looking for the best offer and are therefore willing to buy at higher prices rather than to keep on looking for a better price. Second, central market exchanges allow all sellers to simultaneously submit a bid and the auctioneer then clears the market. In Cournot models the bid consists of a quantity the provider offers independent of the price. In a Supply function model the bid consists of a schedule specifying how much capacity the provider is willing to supply at different prices. In both models the auctioneer then determines the price such that demand equals supply. Both Cournot and Supply function models show that prices can be achieved above marginal costs. However, in our model the Internet virtually eliminates search costs and bilateral trade removes the closing time of the auction mechanism. Therefore providers can sell capacity at random times and buyers as well as competitors are never certain whether the provider might not sell additional or cheaper capacity at later stages. Allaz and Vila [1] showed that repeated selling opportunities eliminate market power in a Cournot market and result in a competitive outcome. Does this result transfer to the provision of computer services and will it push prices to short-term marginal costs of zero at times of low demand? Or can we find a pricing strategy that allows providers to maintain prices above marginal costs?

To simulate the answer we extend the supply function model to a bilateral trade setting with continuous trading. Whenever a client asks for capacity, then providers calculate their own supply function and estimate the supply function of their competitors. They imitate the central auctioneer and determine the market-clearing price that would result and offer their capacity at that price to the client. In the absence of any publicly available price information we will refer to this price as estimated market clearing price. Providers collect information on requested bids and the number and prices of successful reservations to update their demand estimation and beliefs about competitors' supply functions.

The model produces an equilibrium if no provider can profitably deviate by undercutting competitors and capturing a bigger market share. In the case of a deviating provider the remaining providers will count fewer reservations, correct their demand estimation downward and bid at lower price. Therefore the deviating provider receives less revenue per reservation, while he obtains more reservations. The net effect has to be simulated by comparing the revenue achieved from selling more slots at lower price with the revenue obtained when no provider deviates. For many parameter choices deviation was initially profitable. Providers can prevent such deviation by changing their updating strategy. Instead of weighting information about demand with the expected error of the observation, providers add additional weight to the observation of successful sales they made. As a result deviation is unprofitable. The analysis suggests that the supply function equilibrium can be transfered to a scenario with bilateral trade and no search costs by using all available information to continuously observe the market.

2   Related Work

Market-based methods have been widely used for control and management of distributed systems as a means of providing a fair allocation of priorities to clients. Work on QoS-network allocations, effects of dynamics of bandwidth allocations and simple bundling of network and server resources can be found in [3, 11]. Danielson [2] used auctions for admission control and pricing for RSVP connections. Internet pricing is similar work [4, 15] providing models for charging network users and enabling differential services based on the willingness to pay as well as passing network congestion feedback on to the user of the network. The research differs from ours, as it does not assume competition between resources, or as in the case of Kelly [7], assumes that the network provider has a monopoly, or as in the case of Kuwabara [12] assume cooperative price adjustments, which would be inhibited by competition authorities. Reiniger [14] describes a market of QoS network bandwidth containing consumers, retailers and wholesalers using cost-benefit (utility) functions to derive pricing on MPEG video streams. How-

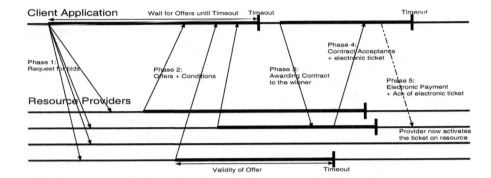

Fig. 1. The Reservation Protocol

ever, their focus is on clients adapting their demand depending on the current load based resource price and therefore congestion is handled gracefully.

Much of research on pricing for competitive markets focused on sales of electronic information products using autonomous agents for negotiation [6, 8]. The market mechanism research applies similar to negotiations and agent bidding strategies, but does not address limited capacities and time dependency.

To compare centralised and decentralised market designs Kirchsteiger, Niederle and Potters [9] studied public versus private exchange markets on the Internet. They conclude that many of the public exchange markets failed and obtained little trading relevance. Traders individually prefer to hide information from their competitors, although they might according to auction theory collectively profit on a public exchange market by maintaining higher prices.

Ogino [13] focused on bilateral interaction and developed a negotiation protocol between clients and network bandwidth providers. Users randomly make requests to providers. Providers have a fixed price schedule for each service level and only decide whether to provide bandwidth or whether to wait for more lucrative requests. Ogino shows that if clients are price sensitive, then providers that are given a lower price schedule increase their market share and profits. In contrast we allow providers to charge prices responding to market conditions - providers continuously assess demand and competitors' behaviour to determine the optimal price for their bids. The equilibrium prices stay above competitive levels.

3   Allocation Protocol and Language

Figure 1 shows the message sequence between one client and a few resource providers leading to a resource reservation. This protocol is based on a protocol described by Ogino [13], which achieves competition among providers and leads to balanced loads. It is extended to address economic considerations and to enable differentiation between clients. The following messages are exchanged:

First, the client application sends a *Request* message to the list of possible resource providers including the attributes of the resource usage (type of service, QoS-level, etc.). Second, the resource provider replies with an *Offer* message considering the current market situation as well as previous negotiation results. Third, the client chooses the best offer among the bids that satisfy his required specifications and notifies the successful provider with the *Award* message. On receipt of the award message from the client, the provider issues and sends an electronic *Ticket* message to the client. Finally, the client sends the electronic *Payment* message to the resource provider, acknowledging the receipt of the electronic ticket. The resource provider then activates the issued ticket and confirms the status of the reservation entry in his booking table.

The aim of this protocol is to provide a realistic framework for bilateral negotiations, taking into account issues such as accountability and resilience against fraud. The choice of short timeouts allows resource providers to protect themselves from malicious clients who attempt to block a provider's schedule with fake requests. Bilateral negotiation protocols have a practical advantage over centralised designs: A centralised market mechanism like an auction would require the market participants to agree on essential service parameters, where bilateral negotiation can be individualised. Further, as pointed out in the related work (section 2), economic comparison of public versus private exchanges [9] indicates the preference of sellers for private exchanges.

4   Economic Framework

Investment in computer resources depends on the expected market price for such resources. In a long-term equilibrium the average price equals total cost of providing the service. Higher prices result in capacity expansion or entry of additional providers to offer the service whilst lower prices would result in reduced investment. In the short term, prices can differ from total cost, as telecommunication bandwidth providers discovered in 2002. We simulate prices that would evolve in short-term markets. Collusion could enforce higher prices by punishing deviations in subsequent periods, but is illegal and not part of the model.

In a perfectly competitive world, which is characterised by a large number of resource providers, and supply exceeding demand, providers offer at marginal cost, which for electronic services is close to $\emptyset$. If one of hundreds of resource providers would withhold capacity it would have no effect on the market price and therefore not increase profits on his remaining capacity. If demand exceeds total capacity, the price will rise until they once again match each other. The resulting allocation is efficient because consumers that are willing to pay the most obtain the resources and resources are only unused if no further demand exists even at price zero.

However, if the number of resource providers is small, then they can exercise market power and profitably maintain prices above competitive levels by restricting output below competitive levels. Typically, Cournot-Nash models are used to model market power. They define an equilibrium by a set of output choices of all generators such that, given other providers' output, each generator chooses the profit maximising output. In the continuous bilateral negotiations model the assumption that providers take the output of other providers as fixed is no longer appropriate. Therefore, we model the strategy space of providers based on a supply function model introduced by Klemperer and Mayer [10] and subsequently applied to the electricity market by Green and Newbery [5]. Supply function equilibria are based on a setting of producers bidding a supply function into a wholesale market, which specifies how much output the producer is willing to provide at any given price. The auctioneer aggregates all supply functions and matches demand with supply, thereby determining the market clearing price as well as the output of all providers. Klemperer showed that market equilibria exist if providers specify their supply function to maximise profits whilst taking other providers' supply functions as fixed. These traditional supply function equilibria require that providers can commit to the supply functions they submitted. Decentralised reservation mechanisms do not have such a commitment device. Providers continuously receive requests for slots and can deviate from a suggested price at any time. Instead of the commitment achieved through a centralised auction place in a traditional supply function equilibrium, we introduce continuous updating about market demand. If a provider reduces price below the price suggested by the supply function then other providers receive fewer contracts. When updating their beliefs the other providers assume lower overall demand and as a reaction reduce their price. Total price level falls and profits of all providers are reduced.

We define a supply function equilibria in the bilateral trade setting by two criteria. First, given competitors' supply functions and assuming that competitors bid according to their supply function, the provider's supply function is profit maximising. Second, given the supply functions and the mechanisms providers use to update their beliefs about demand, it is most profitable for a provider to bid according to her supply function. Therefore, we have to analyse whether it suffices to update beliefs about total market demand to prevent deviations. The simulation shows, that for small numbers of providers and higher demand faithful updating suffices to make deviations unprofitable. For other parameter choices, providers have to be suspicious and put excessive weight on the number of reservations they count, to ensure deviations from the supply function equilibrium are unprofitable.

Clients Clients request computational resources with probability R. In equilibrium, price does not change over the bidding period, therefore the time when clients ask for quotes for resources is independent of the strategies of resource providers and without loss of generality can be assumed to be uniformly distributed on an interval $[0, T]$. The reservation price of any client requesting computational resources is uniformly distributed between 0 and P. Therefore demand expected at the beginning of the bidding period at price p is $N \cdot R \cdot (1 - \frac{p}{P})$ with $p \in [0, P]$.

Strategy of Resource Providers All M symmetric resource providers independently develop a strategy $q(p)$ which determines the capacity q they provide at a given market price p. Each resource provider subsequently calculates the estimated market clearing price, balancing supply $M\,q(p)$ and demand based on his estimation of demand parameter r.

A resource provider chooses his strategy to maximise expected profits. Profits equal the product of price and capacity sold. Capacity sold equals total demand at price p minus the capacity provided by the $M-1$ other resource providers at that price. All costs are assumed fixed in the short term irrespective of utilisation level, and do therefore not appear in the profit function:

$$\max_{q(p)} E_r\left[\left(N\,r\,(1-\frac{p}{P})-(M-1)\,q(p)\right)p\right]. \tag{1}$$

When determining their supply function resource providers assume demand function and supply function $q(p)$ of other resource providers are given. Differentiating equation (1) with respect to p and using the market clearing condition $M\,q(p)=N\,r\,(1-\frac{p}{P})$ with symmetry among providers gives: $\left(-N\frac{r}{P}-(M-1)\frac{d\,q(p)}{d\,p}\right)p+q(p)=0$. The general solution for $q(p)$ to satisfy this differential equation is:

$$q(p)=A\,p^{\frac{1}{M-1}}-N\,\frac{r}{P\,(M-2)}\,p. \tag{2}$$

We follow Klemperer to determine the supply function, such that the supply schedule $q(p)$ is between the competitive schedule as a lower bound and the Cournot oligopoly schedule as an upper bound. Providers can choose any schedule within this range. Subsequent deviations will not be profitable. We assume that providors choose the most profitable schedule, the schedule that crosses the Cournot schedule for individual providers output at the capacity K of each provider.

With the calculated A, the supply function (2) is complete and market participants can solve for the equilibrium price $p \in [0; P]$, based on their estimation of r: $M\,A\,p^{\frac{1}{M-1}}-\frac{2\,N\,r}{P\,(M-2)}\,p-N\,r=0$. We did not obtain an analytic solution. However existence and uniqueness is guaranteed, because the left hand side is negative for $p=0$, strictly monotonically increasing for $p>0$ and is positive for p towards P.

Updating Providers use three types of information to update beliefs about r at any time $t \in [0,T]$ in the bidding period. Two types are identical for all providers, namely the ex ante probability R and the number of requests for quotes sent out by clients. Each provider furthermore privately counts how many of his offers have been accepted by time t. All three observations are weighted with the inverse of their variance to form the updated r. To simplify

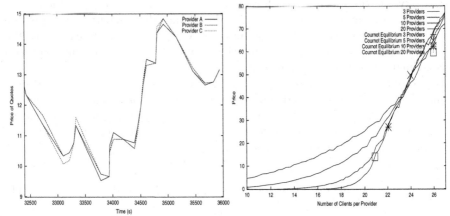

Fig. 2. Price quotes of three providers and 15 clients per provider over one example period.

Fig. 3. Scaling Market Size: Equilibrium price as a function of number of clients per provider for 3,5,10 and 20 providers.

the calculation the calculation of variances is based on R not on r. To deter deviation, additional weight can be put on the number of successfully booked resources. This is implemented by scaling down the variance of resources booked. The description of the detailed updating process is available from the authors.

5 Analysis

We simulate the pricing function using the request protocol and evaluate the pricing function under deviation from a competing provider.

5.1 Competition with Symmetric Pricing Functions

M servers offer the resource and have ten identical slots to satisfy client requests. N clients have a random valuation between $\emptyset$ and 300 for each of these slots. Users ask for price quotes at randomly chosen points during the reservation period. Scaling to any other period with probability $R = 0.5$.

Figure 2 shows the prices quoted by three providers A, B and C over one independent example period. Clients make requests at random times, prompting the provider to reply immediately with a quote based on the conditions at the current time. Providers continuously update their assumptions about the realised demand r (as described in Section 4), therefore quoted prices fluctuate between high and low areas within a bidding period as can be seen in figure 2. The clients collect all replies and select the cheapest offer.

Figure 3 displays the average price quoted by providers for different numbers of competing providers dependent on demand, as given by the number of clients

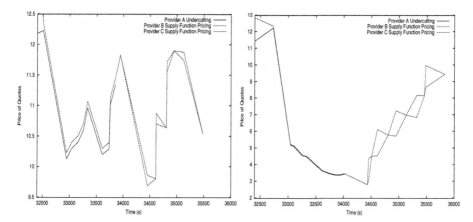

Fig. 4. Price quotes with one seller undercutting (solid) the competitors (dashed).

Fig. 5. Prices with undercutter (solid) and increased weight on accepted bookings.

per provider. The market clearing price under perfect competition would be $\emptyset$ for less than 20 clients per server. The fewer servers compete, the more market power they exercise, raising the price. If the number of clients per server is above 25, then the Cournot output choice, which gives the upper limit of exercise of market power, is above the available capacity. Providers offer all capacity and return to the competitive price.

5.2 Competition with Undercutting

We test the robustness of this bidding strategies against deviation of a resource provider. The bidding strategy is not robust if deviation is profitable. The deviating resource providor calculates which price other providors will quote by keeping the counter for succesfull bids at $\emptyset$ ($\#Bid_{acpt} := \emptyset$) . Then he slightly reduces the price to undercut all offers.

Is this undercutting strategy profitable? Figure 4 shows the same bidding period as in Figure 2, with resource seller A applying the undercutting strategy. A is selling all its resource capacity before any other provider makes the first sale. The prices drop with the undercutting because all sellers assume that sales are evenly distributed among them. If a server is not selling his predicted share of the demand, this server drops the price. This is the threat to the undercutting strategy: If the undercutting seller is making less profit than not by undercutting, then the supply function is an equilibrium strategy.[2]

---

[2] There seems to be a second threat to the stability of the bidding strategies. If a server undercuts at the last offer at the end of the bidding period, then he will not feel the negative impact of lower prices on subsequent bids. However, as clients bid at random times servers never know how many more requests they will receive. Future work will focus on evaluating such deviation strategies.

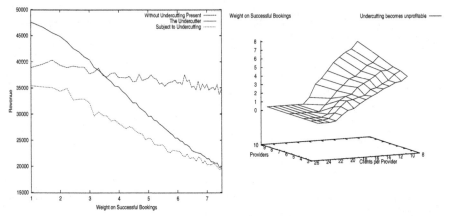

Fig. 6. Profitability of Undercutting (solid) as a function of weight on successful bookings.

Fig. 7. Weights required on successful bookings to prevent undercutting as a function of number of providers and clients.

In the example price only fell slowly and undercutting was therefore profitable. To prevent deviations resource providers put additional weight on the observation of accepted bids. Figure 5 shows the price development for the same hour as in Figure 4, this time with A undercutting and all servers applying the weight $w_{v3} = 5$. Now, the prices drop very quickly to low levels. The undercutting seller is making on average about 20% less profit than if not deviating.[3]

The disadvantage of increasing the weight w is that it reduces the profit for all sellers even if there is no undercutting seller present. For very high values of $w > 7$ this loss due to the increased weight amounts to about 10%. However, it suffices for providers to decrease profits of the undercutter, such that they are below the profits he would make when following the supply function strategy. The example in Figure 6 shows undercutting becomes unprofitable for any $w > 3.3$. The lost profits due to the increased weight amounts in this case to about $3-5\%$. One would expect, that an unbiased random effect should not have a net impact on the average price. However, as only the lowest bid is accepted, the selection bias reduces the average price. Furthermore, expected revenue is concave in the bid price, therefore the revenue at average price is higher than the average of revenues at different prices.

Figure 7 maps the weight w at the breaking point when undercutting becomes unprofitable as a function of number of clients and resource providers. The number of clients per server is an indicator of total demand. The weight w needs to be high in cases of low demand. Conversely deviating is unprofitable

---

[3] We chose a simple undercutting strategy and can therefore not guarantee that a complex strategy of undercutting might be more profitable and requiring a higher factor of sensitivity w to prevent deviations.

with high demand because lowered prices reduce revenue more than is gained by a small increase in capacity utilisation.

Figure 7 illustrates a further phenomenon: In the case of low demand a scenario with more servers requires greater weight w than fewer servers. On the other hand in the case of higher demand - just short of exceeding resources - fewer servers require higher steering weight w than more servers. This is difference is due to two factors:

First, in the case of higher demand, the prices with fewer servers are higher due to the resource providers shortening the supply, as described in section 5.1. Higher prices and a shortened supply imply that undercutting becomes more profitable. Thus, the other resource sellers need to react stronger to prevent undercutting. Secondly, in the case of low demand, the limits of preventing undercutting from being successful are reached. The more resource sellers are participating in the market, the longer it takes for the participants to detect the undercutting because they are not getting their share of the market.

6   Conclusions

We presented a protocol and bidding strategy for reservation of computational resources. The protocol facilitates a market to the benefit of resource providers and clients, it allows for a decentralised market based on bilateral negotiation and can be implemented to cover security issues.

We demonstrated that the presented bidding strategy enables resource providers to exercise market power and to keep prices above perfect competitive level. Particularly, we showed how development of prices through updating during bidding periods suffices to prevent simple deviation strategies. We do not require collusion using punishment strategies through the prices charged for other periods of resource utilisation to maintain prices above competitive levels.

Further work will focus on complex deviation strategies, applying reinforcement learning to search for undercutting strategies. So far we have not addressed the effects of demand that is flexible over time. We intend to include elastic client behaviour in our future work.

Acknowledgements We are grateful to David Newbery, Martin Richards, Tanga McDaniel and several anonymous referees for helpful comments, and the Gottlieb Daimler Foundation and Economics and Social Research Council (award R42200034307) for financial support.

References

[1] Blaise Allaz and Jean-Luc Vila. Cournot Competition, Forward Markets and Efficiency. In *Journal of Economic Theory*, pages 1–16, issue 59, 1993.

[2] Ketil Danielson and Judith Molka-Danielsen. Admissions and Preemption in Auction-based Reservation Networks. In *ISQE Workshop, MIT Boston*, 1999.

[3] D. F. Ferguson, C. Nikolau, and Y. Yemini. An Economy for Flow Control in Computer Systems. In *Proceedings of the INFOCOM*, 1990.

[4] P. Fishburn and A. Odlyzko. Dynamic Behavior of Differential Pricing and QoS Options for the Internet. In *ACM Conf. on Info. and Comp. Economies*, 1998.

[5] Richard J. Green. and David M. Newbery. Competition in the British Electricity Spot Market. In *The Journal of Political Economy*, pages 100 (5), 929–953, 1992.

[6] R. H. Guttman, A. G. Moukas, and P. Maes. Agent-mediated Electronic Commerce: A Survey. Technical report, 1998.

[7] Frank P. Kelly. On tariffs, policing and admission control for multiservice networks. In *Oper.Res.Lett, vol. 15, no. 1*, 1994.

[8] J. Kephart and A. Greenwald. Shopbot economics. In *5th Europ. Conf. on Symbolic and Quantitative Approaches to Reasoning with Uncertainty, UCL, London*, 1999.

[9] Georg Kirchsteiger, Muriel Niederle, and Jan Potters. Public versus Private Exchanges. In *Tilburg, CentER Discussion paper*, 2001.

[10] Paul D. Klemperer and Margaret A. Meyer. Supply Function Equilibria in Oligopoliy under Uncertainty. In *Econometrica*, pages 57 (6), 1243–1277, 1989.

[11] S. Lalis, C. Nikolaou, D. Papadakis, and M. Marazakis. Market-driven Service Allocation in a QoS-capable Environment. In *ACM Int. Conf. on Info. and Computation Economies (ICE-98)*, 1998.

[12] Y. Nishibe, K. Kuwabara, T. Suda, and T. Ishida. Distributed channel allocation in ATM networks. In *Proceedings of GLOBECOM '93*, 1993.

[13] Nagao Ogino. Connection Establishment Protocol Based on Mutual Selection by Users and Network Providers. In *ACM Int. Conf. on Information and Computation Economies (ICE-98)*, 1998.

[14] D. Reininger, D. Raychaudhuri, and M. Ott. Market Based Bandwidth Allocation Policies for QoS Control in Broadband Networks. In *ACM Int. Conf. on Info. and Computation Economies*, 1998.

[15] J. Sairamesh, D. F. Ferguson, and Y. Yemini. An Approach to Pricing, Optimal Allocation and Quality of Service Provisioning In High-Speed Packet Networks. In *14th Joint Conf. of the IEEE Computer and Communication Soc. (Vol. 3)*, 1995.

The Economic Impact of Network Pricing Intervals*

Errin W. Fulp[1] and Douglas S. Reeves[2]

[1] Department of Computer Science, Wake Forest University,
P.O. Box 7311, Winston-Salem N.C. 27109-7311 USA
fulp@wfu.edu
[2] Department of Computer Science *and* Department of Electrical and Computer
Engineering, N.C. State University, Box 7534, Raleigh N.C. 27695 USA
reeves@eos.ncsu.edu

Abstract. Interval pricing can provide an effective means of congestion control as well as revenue generation. Using this method, prices are
fixed over intervals of time, providing adaptability and predictability. An
important issue is the interval duration associated with price updates.
While previous research has discussed the effect of interval lengths on
congestion control, this paper investigates the economic impact of price
interval duration. Smaller intervals yield higher profits since prices are
more responsive to changing demands. However, experimental results indicate only a modest profit gain (no more than 5%) is achieved when
smaller intervals are used as opposed to larger intervals (for example
100 times longer). Given users preferences toward fewer price changes,
smaller price intervals may hold few economic benefits.

1 Introduction

A growing number of applications (e.g., multimedia-oriented) require certain
network resources for their proper operation. Network resources such as bandwidth of each physical link, buffer space and processing time at each node, are
finite and must be allocated in a fair and cost-effective manner. It has been
demonstrated that pricing is an effective method for achieving fair allocations
as well as revenue generation [1, 2, 3, 4, 5, 6, 7]. Furthermore, pricing resources
prevents users from over allocating resources or requesting service classes more
stringent than necessary.

Most pricing-based methods rely on the basic microeconomic principle of
supply and demand to determine the appropriate resource price. For example,
price increases can be used to lower demand as it approaches supply (capacity).
Conversely, decreasing prices can be used to encourage usage. The objective is to

---

* This work was supported by AFOSR contract F49620-99-1-0264 and DARPA contract F30602-99-1-0540. The views and conclusions contained herein are those of the
authors and should not be interpreted as necessarily representing the official policies or endorsements, either expressed or implied, of the AFOSR, DARPA, or the
U.S. Government.

B. Stiller et al. (Eds.): QofIS/ICQT 2002, LNCS 2511, pp. 315–324, 2002.

price resources so demand equals supply, at this point allocations are provably fair and profit is maximized [8]. Given a microeconomic-model of user demands, the appropriate price and supply can be determined. However, given the dynamic nature of computer networks (changing user demands), determining when prices should change (time-scale) remains an important issue.

Users prefer a price that is predictable, while service providers prefer a price that can change based on current congestion [9, 10, 7]. For example, spot market prices are updated over short periods of time to reflect congestion [11]. While this method does provide fair allocations under dynamic conditions, users can not accurately predict the cost of their sessions due to possible price fluctuations. In general, most users dislike the uncertainty and perceived complexity associated with a dynamic or spot price, even though it may result in lower costs [10]. In contrast, fixed prices provide predictable costs; however, there is no incentive for the user to curtail consumption during peak (congested) periods. Service providers typically prefer a price structure that reflects actual usage; thus, providing more control. As a compromise, prices can be based on slowly varying parameters such as Time of Day (ToD) statistics. As noted in [12, 13, 5, 9], the aggregate demand for bandwidth changes considerably during certain periods of the day. This is depicted in figure 1, which shows the aggregate traffic arriving and departing Wake Forest University campus during a week. Peak demand is evident during business hours, while network usage decreases at night. Given a ToD based pricing structure, an important question is determining the appropriate interval length. The effect of ToD duration (pricing interval) on congestion control has been discussed (smaller intervals provide better congestion control [5, 7]); however, the specific impact on profit has not.

This paper investigates the effect of pricing intervals on profit. For example, given a service provider and a set of customers, this paper seeks to determine the economic advantages of different sized pricing intervals. Having many smaller pricing intervals can result in higher profits. However, experiments indicate the profit increase is less than 5% on average as opposed to implementing longer intervals. As a result, the economic benefit of having smaller duration intervals is limited considering users prefer flat and less complex pricing structures.

The remainder of this paper is structured as follows. Section 2 describes the network pricing model used, consisting of individual users and a service provider. Optimal strategies for bandwidth provisioning and allocation (pricing) are then presented in section 3. In section 4, the economic impact on different sized pricing intervals is discussed and investigated experimentally. Finally, section 5 provides a summary of network pricing intervals and discusses some areas of future research.

2   Network Pricing Model

The pricing model consists of a network service provider and a set of users (customers). Users require bandwidth for their network applications. A user may start a session at any time, request different levels of QoS, and have varying

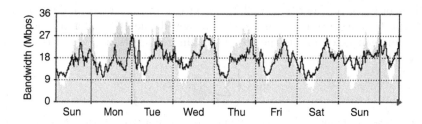

Fig. 1. Aggregate bandwidth usage for Wake Forest University. The dark line represents the incoming traffic, while the shaded area is traffic departing campus.

session lengths. Furthermore, users desire immediate network access (minimal reservation delay). In contrast, service providers own large amounts of bandwidth (or rights to bandwidth) and create connections across their network [14]. A connection would serve multiple users, and persists over a long period of time[1]. These large aggregate connections provide scalability; however, single-user connections (e.g., using the ReSource reserVation Protocol, RSVP [16]) are also possible. Once a connection is established (provisioned), portions of the connection are sold (allocated) to individual users at a price.

The price of bandwidth (charged to users) will be usage-based, where the user cost depends on the current price and the amount consumed. As previously mentioned, an important issue is the time-scale associated with the price. For example, prices could remain fixed for long periods of time or continually change based on current congestion levels. As a compromise, this model will use interval pricing. A day will be divided into T equal length periods of time, where $t = 1, ..., T$. To provide predictability, these prices (next day) are known a priori by the users (prior pricing [7]) via a price-schedule $\{p_t\}$, where p_t is the price of bandwidth during the t ToD period. Given the price schedule, the interval durations, and the required bandwidth, the user can predict the cost of a session. The bandwidth of a connection is sold on a first come first serve basis; no reservations are allowed. If the amount is not available at the beginning of the session, the user is blocked. However, users who can not afford p_t are **not** considered blocked.

The service provider is responsible for establishing connections and allocating portions of the connection to individual users. Within this model, acquiring resources for a connection is provisioning, while selling portions of the connection will be referred to as allocation. Primary goals of the service provider will be profit maximization and minimizing the blocking probability.

3   Optimal Resource Provisioning and Allocation

A service provider must manage several connections simultaneously, where each connection could represent a unique pair of ingress and egress routers, a QoS

---

[1] Analogous to an Internet DiffServ connection across a network domain [15].

class, or both. As described in [8], the service provider is interested in maximizing the profit of each connection, which is achieved when the difference between the revenue generated minus the cost is maximized. This is given in the following formula,

$$\max \left\{ \sum_{t=1}^{T} [r(x_t) - c(s)] \right\} \tag{1}$$

The revenue generated by the connection during ToD period t is $r(x_t)$ and is based on x_t which is the aggregate user demand. Note the profit maximization is over T consecutive ToD periods (which could represent a DiffServ SLA term). Viewing this as an optimization problem, the first order conditions are

$$\sum_{t=1}^{T} \frac{\partial r(x_t)}{\partial x_t} = T \cdot \frac{\partial c(s)}{\partial s} \tag{2}$$

Note the connection supply, s, is constant for each ToD period. The left-hand side of equation 2 is also referred to as the marginal revenue, which is the additional revenue obtained if the service provider is able to sell one more unit of bandwidth. The right-hand side of equation 2 is referred to as the marginal cost, which is the additional cost incurred. A solution to the optimization problem exists, if the cost and revenue functions are continuous and convex. Therefore, to determine the appropriate provisioning amounts and prices, these functions must identified.

The Cobb-Douglas demand function will be used to model aggregate user demand (multiple users using the same connection). The Cobb-Douglas demand function is commonly used in economics because it is continuous, convex, and has a constant elasticity. A constant elasticity assumes users respond to proportional instead of absolute changes in price, which is more realistic. Therefore, this demand function is popular for empirical work. For example, the Cobb-Douglas demand function has been used for describing Internet demand in the INDEX Project [17]. The Cobb-Douglas function has the following form,

$$x_t(p_t) = \beta_t \cdot p_t^{-\alpha_t} \tag{3}$$

where p_t is the bandwidth price during ToD t and the approximate aggregate wealth of users is denoted by β_t. The elasticity (own-price) during ToD t is α_t, which represents the percent change in demand in response to a percent change in the price. Due to the dynamic nature users, demand is expected to change over time. These changes may reflect variable application demands, ToD trends, pricing, or the introduction of new technology. For this reason, demand prediction and estimation must be employed [4, 12], where the demand curve parameters (α_t and β_t) are estimated using previous ToD measurements. During a ToD the price is set and the demands are recorded (demand during a ToD will vary based on the number of users and their applications). This results in a set of demands for each ToD price. For this paper, the highest demand observed for a ToD price was used for parameter estimation, which resulted in a conservative demand curve. If the estimation is correct, demand should equal supply, which results in a zero blocking probability.

Given the estimated aggregate demand function, the revenue earned is the price multiplied by the demand,

$$p_t \cdot x_t(p_t) = \left(\frac{\beta_t}{x_t(p_t)}\right)^{\frac{1}{\alpha_t}} \cdot x_t(p_t) = \beta_t^{\frac{1}{\alpha_t}} \cdot [x_t(p_t)]^{1-\frac{1}{\alpha_t}} \tag{4}$$

Taking the derivative of equation 4 with respect to demand yields the marginal revenue for ToD period t. Similarly, taking the derivative of the cost function yields the marginal cost. Substituting these values into equation 2 results in a system of equations that can be solved for s which is the appropriate amount to provision. Since the marginal equations (revenue and possibly costs) are nonlinear, a direct solution can not be found [8]. For this reason, gradient methods (e.g. Newton) can be used to determine the optimal provisioning amounts [18]. Due to the time typically associated with negotiating (or establishing) a connection (SLA) [8], calculations can be performed off-line; therefore, convergence time is not critical.

Although the technique presented in this section does determine the optimal prices and connection supply, it does not (like many others) address the appropriate price-interval length. Since price-interval length can impact profit, its economic effect must be considered when managing network resources.

4   Pricing Intervals

As described in the introduction and depicted in figure 2, the duration of a price interval can range from extremely large to very small. Smaller intervals provide greater congestion control since prices can quickly adjust based on user demand [10, 7]. This flexibility also increases profits, since prices can be set to encourage usage. Consider the allocated bandwidth for a single price interval given in figure 2(a). Near the end of this graph, the profit decreases due to a reduction in demand. Using shorter intervals, prices are lowered during this time as seen in figure 2(c). This price reduction encourages more usage and increases profits. Even using seven intervals during a day, seen in figure 2(b), results in higher profits than a single price. Of course, once users have purchased their desired maximum bandwidth, further price reductions will not increase demand. Ideally, a service provider wants to update the price whenever the aggregate user demand changes. Changing the price at a rate faster than the change in demand would not increase profits. Although smaller price intervals are advantageous to the service provider, users prefer a simpler price structure. Generally, users are adverse to the possibility of multiple price changes during a session, even if it could result in a lower cost [7]. As a result, a service provider could lose customers, which would result in lower profits, if rival service providers offer a simpler pricing structure (fewer intervals).

Determining the appropriate number of intervals in a day can be viewed as an optimization problem. Let $g(T)$ represent the profit (revenue minus the

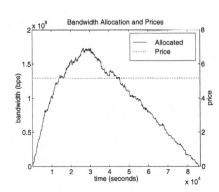

(a) One interval.

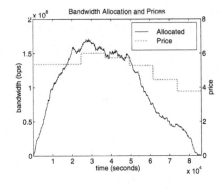

(b) Seven intervals.

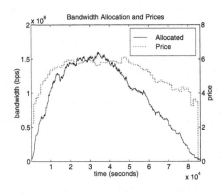

(c) Seventy one intervals.

Fig. 2. Effect of interval duration on aggregate user demand.

actual connection cost) obtained using T intervals, which can be determined using equation 1. In addition, let $l(T)$ represent the loss in revenue due to the number of intervals in a day[2]. We will assume $l(T)$ increases as the number of intervals increases, which represents a higher inconvenience or complexity value users associate with more pricing intervals. Furthermore, let $l(1) = 0$ indicating users prefer a flat price. Under these circumstances the service provider seeks to determine T such that,

$$\max_{T \geq 1} \{g(T) - l(T)\} \qquad (5)$$

A solution for T can be found if these functions are continuous and convex. The value of $g(T)$ can be determined using equation 1 and demand estimation techniques. Determining $l(T)$ is difficult, since it relies on user feedback on pric-

---

[2] Again, equation 1 only considers the cost of the connection in terms of resources required to support it.

ing intervals preferences. Although we can assume the function increases as T increases (users prefer flat prices), realistic data is required to determine the actual properties of this function. We can gain insight on the economic effect using simulation by measuring the profit associated with different price-interval lengths. If there is relatively little monetary gain using smaller intervals, the service provider should implement a simpler user-friendly price structure.

4.1   Experimental Results

In this section, the effect of price interval duration on service provider profit is investigated using simulation. For each simulation, a random number of users, uniformly distributed between 200 and 500, interacted with a single service provider during a day. Users started their sessions at random times using a Poisson distribution with mean equal to 9:00am. The duration of a session was then uniformly distributed between 0 and 12 hours. This resulted in a high aggregate demand during the midday, simulating peak hours [13]. Users had an elasticity α uniformly distributed between 1.75 and 2.5 (consistent with the IN-DEX project [17]), and a wealth β uniformly distributed between 1×10^8 and 2×10^8. The minimum demand of each user was uniformly distributed between 0.5 Mbps and 2 Mbps to represent a variety of traffic. Furthermore, users were considered independent and did not share connections. The service provider used the optimization techniques described in the previous section (including demand estimation) to provision and price bandwidth.

For each experiment, we are interested in measuring the percent change in profit as the number of intervals increases during a day[3]. A single experiment consisted of six simulations, each simulating a different number of pricing intervals. User arrivals and demands were randomly generated at the beginning of each experiment and were used for the six simulations. For each experiment, the first simulation was performed to estimate demand curve parameters and to determine the smallest aggregate demand interval. An aggregate demand interval is the amount of time between successive demand changes. Changing prices more frequently than the shortest aggregate demand interval should not increase profits, as described in the beginning of this section. Let T_* represent the number of pricing intervals based on the shortest aggregate demand interval. Five additional simulations were then performed, where the number of intervals were equal to 1, $T_*/100$, $T_*/10$, T_*, and $2 \cdot T_*$. Note T_* was greater than 100 for every experiment. Finally, 5000 independent experiments were conducted (30000 simulations total) to provide averages and 95% confidence intervals.

The experimental results are shown in figure 3, which depicts the percent change in profit as the number of intervals increases. Note, the percent change in profit is compared to the profit obtained from a single interval. The parameter γ is the normalized number of intervals used for a simulation (normalized by the maximum number of intervals required, T_*, for the experiment); therefore, γ is calculated using the formula

---

[3] For a given day, we assume all intervals are equal in duration.

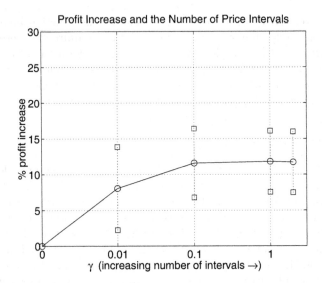

Fig. 3. Simulation results showing the profit change as the number of pricing intervals in a day increases. Percent change in profit is based on the profit of a single interval ($\gamma = 0$). For each data point, the 95% confidence intervals are shown.

$$\gamma = \frac{T}{T_*}$$

One interval was represented by $\gamma = 0$. As expected, profit increased as the number of intervals increased. Initially, the profit increased rapidly. At $\gamma = 0.01$ the percent increase was 8.1, while the percent profit increase was 11.6 at $\gamma = 0.1$ (a difference of 3.5) . However, at $\gamma = 1$ the percent profit change was 11.8, which is negligible as compared to the profit obtained when $\gamma = 0.1$. Using more than T_* intervals ($\gamma > 1$) did not increase profits as compared to T_* intervals ($\gamma = 1$), which was expected. Furthermore, the standard deviation of the percent-average-profit-change reduced as the number of intervals increased, from 5.8 at $\gamma = 0.01$ to 4.3 at $\gamma = 2$. No users were blocked in any of the simulations performed. In summary, a service provider can increase profits by using more intervals; however, there is limited monetary benefit associated with using intervals lengths that approach the smallest change in aggregate user demand.

5   Conclusions

Pricing has been demonstrated as an effective method for allocating network resources as well as revenue generation. Pricing resources also prevents users from over allocating resources or requesting service classes more stringent than necessary. The objective is to price resources so demand equals supply, at this point allocations are provably fair and profit is maximized. However, given the dynamic nature of computer networks (changing user demands), determining when prices should change (time-scale) is problematic. Prices could remain fixed over long periods of time providing predictable network usage costs, but limited congestion control. In contrast, prices could continually change based on demand yielding effective congestion control, but no cost predictability. Interval pricing is a compromise between these two extremes, where prices remain fixed over a certain time span. This method can provide predictability as well as congestion control; however, determining the appropriate interval length is difficult. The impact of interval length on congestion control has been addressed, indicating shorter intervals provide better control. This paper investigated the economic impact of price interval duration. Shorter intervals can provide higher profits; however, experimental results indicated these gains are modest. For example, experiments conducted indicate a profit increase no larger than 5% was achieved when smaller intervals were used as opposed to larger intervals (for example 100 times longer). Therefore, given users preferences toward fewer price changes, smaller price intervals hold few economic benefits.

Future areas of research include determining user price-interval preferences and investigating variable length price-intervals. Users are adverse to multiple price changes during a session; however their preferences must be quantified to solve the optimization problem introduced in this paper. Furthermore, most interval pricing techniques require all intervals to be equal length. Using different sized intervals during the day may increase profits; yet, still provide a user friendly price structure. Both items are integral in the development of interval pricing.

References

[1] Anerousis, N., Lazar, A.A.: A framework for pricing virtual circuit and virtual path services in atm networks. ITC-15 (1997) 791 – 802
[2] Courcoubetis, C., Siris, V.A., Stamoulis, G.D.: Integration of pricing and flow control for available bit rate services in atm networks. In: Proceedings of the IEEE GLOBECOM. (1996) 644 – 648
[3] Ferguson, D.F., Nikolaou, C., Sairamesh, J., Yemini, Y.: Economic models for allocating resources in computer systems. In Clearwater, S., ed.: Market Based Control of Distributed Systems. World Scientific Press (1996)
[4] Kelly, F., Maulloo, A.K., Tan, D.K.H.: Rate control for communication networks: Shadow prices, proportional fairness and stability. Journal of the Operational Research Society 49 (1998) 237 – 252
[5] Paschalidis, I.C., Tsitsiklis, J.N.: Congestion-Dependent pricing of network services. IEEE/ACM Transactions on Networking 8 (2000) 171–184
[6] Wang, X., Schulzrinne, H.: Pricing network resources for adaptive applications in a differentiated services network. In: Proceedings of the IEEE INFOCOM. (2001)
[7] Yuksel, M., Kalyanaraman, S.: Effect of pricing intervals on the congestion-sensitivity of network service prices. In: Proceedings of the IEEE INFOCOM. (2001)
[8] Fulp, E.W., Reeves, D.S.: Optimal provisioning and pricing of internet differentiated services in hierarchical markets. In: Proceedings of the IEEE International Conference on Networking. (2001)

[9] Reichl, P., Flury, P., Gerke, J., Stiller, B.: How to overcome the feasibility problem for tariffing internet services: The cumulus pricing scheme. In: Proceedings of the ICC. (2001)

[10] Shih, J.S., Katz, R.H., Joseph, A.D.: Pricing experiments for a computer-telephony-service usage allocation. In: Proceedings of the IEEE Globecom. (2001)

[11] Fulp, E.W., Ott, M., Reininger, D., Reeves, D.S.: Paying for qos: An optimal distributed algorithm for pricing network resources. In: Proceedings of the IEEE Sixth International Workshop on Quality of Service. (1998) 75 – 84

[12] Morris, R., Lin, D.: Variance of aggregated web traffic. In: Proceedings of the IEEE INFOCOM. (2000)

[13] Odlyzko, A.: The economics of the internet: Utility, utilization, pricing, and quality of service. Technical Report 99-08, DIMACS (1999)

[14] Nichols, K., Jacobson, V., Zhang, L.: A two-bit differentiated services architecture for the internet. Internet Draft http://ds.internic.net/internet-drafts/draft-nichols-diff-svc-arch-00.txt (1997)

[15] Blake, S., Black, D., Carlson, M., Davies, E., Wang, Z., Weiss, W.: An architecture for differentiated services. IETF RFC 2475 (1998)

[16] Braden, R., Zhang, L., Berson, S., Herzog, S., Jamin, S.: Resource reservation protocol (rsvp) - version 1 functional specifications. IETF RFC 2205 (1997)

[17] Varian, H.R.: Estimating the demand for bandwidth. Available through http://www.INDEX.Berkeley.EDU/public/index.phtml (1999)

[18] Yakowitz, S., Szidarovszky, F.: An Introduction to Numerical Computations. second edn. Macmillan (1989)

Pricing and Resource Provisioning for Delivering E-content On-Demand with Multiple Levels-of-Service

Srinivasan Jagannathan and Kevin C. Almeroth

Department of Computer Science, University of California, Santa Barbara,
CA 93106-5110
{jsrini,almeroth}@cs.ucsb.edu

Abstract. Businesses selling multimedia rich software or *e-content* are growing in the Internet. The e-content can be downloaded or streamed immediately after an on-line transaction. Since Internet connection speeds are variable, ranging from dial-up access speeds to broadband speeds, a content provider may provide different *levels-of-service (LoS)* for the same content. If a provider offers service at different LoS, for example at 56 kbps and 128 kbps, how should the price of the service be set such that the provider makes the most money? In addition, how should the server resources be provisioned among the different service levels? In this paper, we address such pricing and resource provisioning issues for selling e-content at multiple service levels.

1   Introduction

Available bandwidth and usage have increased in the Internet. Use of the Internet to purchase goods and services is also increasing. At the same time, the multimedia capabilities of computers are improving while remaining affordable. Together, these trends have spawned services offering video-on-demand, downloadable CDs, etc. In these services, customers should have adequate bandwidth to receive content and multimedia-capable computers to view it. While such services are growing, all users are not connected to the Internet at the same speed. Some users connect at dial-up speeds and others at broadband speeds. To accommodate this heterogeneity, a content provider may serve the same content at different quality levels. For instance, many web sites offer the same streaming content at two quality levels: 56 kbps and 128kbps. Since requests for different levels-of-service (LoS) consume different amount of server resources and are presumably quoted a different price, one must examine how to provision resources for each LoS. In this paper, we seek to answer the following questions: 1) how should a content-provider price content at each LoS, and 2) how should a content provider allocate resources for each LoS.

Our work builds on our earlier research for pricing on-demand delivery of e-content when there is single LoS [1, 2, 3, 4]. In our earlier work [5], we compared a number of simple pricing schemes using simulations. These pricing schemes

B. Stiller et al. (Eds.): QofIS/ICQT 2002, LNCS 2511, pp. 325–336, 2002.

could be classified as being *static* or *dynamic*. In a static pricing scheme, the price of the content does not change frequently. In a dynamic pricing scheme, the price may vary on much smaller time scales based on factors like current server load, request arrival rate etc. Based on our simulations, we believe that there exist fixed prices which generate very high revenues but, finding these prices is non-trivial. We formulated a dynamic pricing scheme called HYBRID which not only generated consistently high revenues across a range of simulation scenarios and customer populations, but also reduced the number of requests rejected due to lack of server resources. In this paper, we primarily focus on extending the HYBRID pricing scheme to systems with multiple levels-of-service (multi-LoS systems). We validate our work through simulations.

There are two challenges in designing pricing schemes for multi-LoS systems. First, it is difficult to quantify the "capacity" of the system. For instance, consider a system where resources are quantified in terms of channels. Consider a system with 100 channels and two LoS. Suppose that for a lower LoS one channel is allocated, and for a higher LoS two channels are allocated. Then the system can accomodate 100 low LoS requests or 50 high LoS requests. The actual number of requests that the system serves will vary with the relative fraction of high vs low LoS requests. Moreover, when there are more requests than the system can serve, it is difficult to decide which requests to satisfy. For instance, if it is known that customers are willing to pay at least $5 for the lower LoS and $7 for the higher LoS, accepting low LoS requests will increase the revenue when resources are constrained. However, since how much customers are willing to pay is not known, deciding which requests to serve is difficult.

The other challenge in pricing multi-LoS systems is in understanding customer behavior. For customers with high bandwidth connections, the choice of LoS depends not only on the LoS actually desired but also on how other LoSs for that content are priced. For instance, suppose that a customer with a high bandwidth connection is willing to pay $9 for a low LoS. If the desired LoS is priced at $6 and the higher LoS at $8, then the customer may choose the higher LoS. Though this increases the revenue by $2, it may prevent another low LoS request from being satisfied. The system loses $4 in this case. In this paper, we make a simplifying assumption that a customer's choice of LoS is independent of the price for other LoSs. This is a reasonable assumption because in the Internet today, content is typically served at a LoS where there is a perceptible difference in quality between the LoS. Customers with high bandwidth connections may typically not purchase content at low LoS. We shall address the general problem where choice of LoS is correlated with price in future work.

We now briefly survey related work. Basu and Little[6], have formulated models for VoD and pricing issues related to them. Mackie-Mason et al.[7] investigate adaptation to changes in consumer variables for an information goods market. Sairamesh and Kephart [8] discuss competition and price wars in information goods markets. Their analysis assumes that each competitor sells at a different LoS. All the above do not consider distribution constraints of the content provider. Wolf et al. [9] study how to maximize profits when broadcasting digital goods. When resources are constrained, they schedule the delivery at a later time, and pay a penalty for late delivery by charging a lower price. Chan and Tobagi [10] design scheduling schemes for batched delivery of video-on-demand, when the fixed price for the content is known. Their work does not consider multiple levels-of-service. To the best of our knowledge, though there has been considerable work on connectivity pricing, there has been very little work on content pricing.

The rest of the paper is organized as follows. Section 2 describes a formulation for revenue earned in multi-LoS systems. Section 3 describes our HYBRID pricing scheme and two other dynamic pricing schemes adapted from the work by Sairamesh and Kephart [8]. Section 4 discusses the simulation framework and the experiments we perform. Results are presented in Section 5. We conclude the paper in Section 6.

2   Revenue Model and Resource Provisioning

We consider a system where requests are satisfied if resources are available and the customer agrees to pay the quoted price. We assume that all server resources can be quantified and mapped to a real number. One approach to doing this is to consider the bottleneck resource at the server as the indicator of system resources. For example, if bandwidth is the bottleneck, then the total available bandwidth is modelled as the system capacity. For the purposes of this paper, we shall assume that available connection bandwidth of the content provider is the measure of system resources. When a request is served, some of the connection bandwidth is allocated to that request[1]. Requests are processed on a First-Come-First-Served basis. If there is insufficient bandwidth available when a request arrives, then the request is rejected. In our model, we assume that once the content provider makes the initial infrastructural investment, there are either negligible or fixed costs in maintaining the resources (caches, servers, bandwidth etc.), i.e., there are no additional costs based on number of requests served. This is a reasonable assumption because servers incur fixed costs and bandwidth can be bought at a flat monthly rate. If maintenance costs are negligible or fixed, profit maximization is equivalent to revenue maximization. We also assume that the market is monopolistic, i.e., there is no other entity selling the same content. This is a realistic assumption in many scenarios where the content owner personally sells the content or has licensed it to a single distributor.

Table 1 presents the symbols we have used in our analysis. Since we assume that a customer's choice of LoS is independent of price, we can treat the same content at different LoS as different products. Consider an arbitrary customer who wants to purchase content $p_{i,j}$. We denote his/her decision to purchase the service by the random variable $\Upsilon_{i,j}$ which can take two values, 1 for accept and 0 for reject. Let $E[\Upsilon_{i,j} \mid \psi_{i,j}]$ denote the expectation of the decision to purchase content $p_{i,j}$ when the price is $\psi_{i,j}$. The expectation of revenue per unit time is given by: $\mathcal{R} = \sum_{i=1}^{m} \sum_{j=1}^{L} \lambda_{i,j} \psi_{i,j} E[\Upsilon_{i,j} \mid \psi_{i,j}]$

---

[1] This does not imply that network resources are reserved.

Notation	Description
m	Number of products
L	Number of levels of service
$\mathcal{B}$	Total system resources
b_j	Resources provisioned for j^{th} LoS
l_j	Resources for serving a request at j^{th} LoS
$p_{i,j}$	i^{th} product at j^{th} LoS
$\Upsilon_{i,j}$	Decision to purchase $p_{i,j}$ (0 or 1)
$\psi_{i,j}$	Price of $p_{i,j}$
$\lambda_{i,j}$	Request arrival rate for $p_{i,j}$
$\mathcal{R}$	Total revenue per unit time
d	Mean service time
ρ	System Utilization

Table 1. Symbols used.

To model resource constraints, we use the notion of system utilization. System utilization, ρ, is defined as the ratio of the number of requests entering the system per unit time to the number of serviced requests exiting the system per unit time. In a stable system, this ratio must be less than or equal to 1. Therefore, we impose an additional constraint that the predicted system utilization should be less than or equal to 1. System utilization is easily defined when there is a single LoS. If l is the resources consumed by a request at this LoS, the system utilization can be computed as $\frac{dl}{\mathcal{B}} \sum_{i=1}^{m} \lambda_i E[\Upsilon_i \mid \psi_i]$. However, with multiple LoS, and requests at each LoS consuming different amount of resources, it is not possible to quantify the number of serviced requests exiting the system. We therefore take a different approach. Suppose that the system resources are partitioned into $\langle b_1, b_2, ..., b_L \rangle$, where b_j is the resource provisioned for level j. Then, we can impose the system utilization constraint independently for each LoS. We solve an independent constrained maximization problem for each LoS. The total revenue earned critically depends on how the resources are partitioned for each level. Notice that resources consumed by requests for level j will be less than or equal to $\sum_{i=1}^{m} l_j \lambda_{i,j}$. Based on this, we provision resources as follows:

$b_j = \frac{\sum_{i=1}^{m} l_j \lambda_{i,j}}{\sum_{j=1}^{L} \sum_{i=1}^{m} l_j \lambda_{i,j}}$ The revenue maximization problem is then given by:

- Maximize: $\sum_{j=1}^{L} \mathcal{R}_j$ where $\mathcal{R}_j = \sum_{i=1}^{m} \lambda_{i,j} \psi_{i,j} E[\Upsilon_{i,j} \mid \psi_{i,j}]$
- Subject to:
  - $\psi_{i,j} \geq 0, 1 \leq i \leq m, 1 \leq j \leq L$
  - $\rho_j \leq 1, 1 \leq j \leq L$, where $\rho_j = \frac{dl_j}{b_j} \sum_{i=1}^{m} \lambda_{i,j} E[\Upsilon_{i,j} \mid \psi_{i,j}]$

As can be observed, the revenue model relies on knowledge of the request arrival rate and the expectation of the decision to purchase, given the price. The request arrival rate can be monitored. However the expectation of the decision to purchase is not known. In the next section we outline the HYBRID scheme which estimates the expectation of the decision to purchase.

3   Dynamic Pricing Algorithms

In this section, we briefly describe the HYBRID pricing scheme. The HYBRID algorithm is based on the premise that customers are rational human beings. We are interested in the fraction of requests that will result in successful transactions. For a rational customer population, it can be argued that this fraction is a non-increasing function of the quoted price. For a price x, let $f(x)$ denote the fraction of customers who will accept the price. Let x_{low} be a price below which $f(x)$ is exceptionally high, say more than t_h and let x_{high} be a price above which $f(x)$ is exceptionally low, say below t_l. Then $f(x)$ can be approximated in the domain $[x_{low}, x_{high}]$ using some non-increasing function. We propose a family of decreasing functions which depend on a parameter δ described as follows.

$$f(x) = \begin{cases} t_h & , & 0 \le x < x_{low} \\ (t_h - t_l)\left[1 - \left(\frac{x - x_{low}}{x_{high} - x_{low}}\right)^{\delta}\right] + t_l & , & x_{low} \le x \le x_{high} \\ t_l & , & x > x_{high} \end{cases} \qquad (1)$$

By experimenting with different prices to observe the fraction of customers who accept the price, and using statistical methods like least squared errors, one can estimate the parameter δ, and the threshold prices x_{low} and x_{high}. Notice that $f(x)$ is also the expectation of the decision to purchase, given price x. Once all the parameters are known, the content provider can predict the customer behavior and thereby choose a price[2] using the optimization problem described in the previous section. In HYBRID, the customer reaction is continuously monitored, and the price is varied at regular intervals[3]. The details of this algorithm are presented in our earlier work [5].

After performing simulations with the scheme described above, we observed that while the revenue earned was high, the number of requests rejected due to lack of resources was also high. This was mainly because when the algorithm experimented with low prices, more customers accepted the service than could be accommodated by the server. We therefore modified the algorithm as follows. Whenever the server load increased beyond a certain threshold, an exponentially increasing price was quoted. Suppose that x is the fraction of available resources that have been allocated to satisfy requests. Let L and H be the lowest price and highest price that the content provider decides to quote to customers. Then, if x is greater than a threshold, the price quoted to a customer, irrespective of the content requested, is given by: $(L-1) + (H - L + 1)^x$. This modified algorithm, generated consistently high revenues while at the same time minimizing the number of requests rejected due to lack of resources.

Other Dynamic Pricing Algorithms: We present two dynamic pricing algorithms adapted from the work of Sairamesh and Kephart [8]. These algorithms were observed to converge to the game-theoretic optimal price in a competitive

---

[2] The price so obtained may not be the global optimum.
[3] Temporal price variations are an inherent feature in many commodity markets.

market in the simulations performed by Sairamesh and Kephart. We chose these algorithms for the purposes of evaluation and comparison with our algorithm.

Let L and H be the lowest and highest prices respectively that the content provider decides to quote. In the Trial-and-Error-Pricing (TEP) algorithm, an initial price is chosen at random in the range $[L, H]$. At regular intervals, with a small probability (called small jump probability) a random price increment is chosen from a standard normal distribution having very small σ. After the price change, revenue earned in the next interval is monitored. If revenue earned per request is lower than before, the old price is restored. In addition, with a very small probability (called big jump probability), a new price is chosen at random. The big jump probability is much smaller than the small jump probability.

The Derivative-Following-Pricing (DFP) algorithm is similar to the TEP algorithm. An initial price in the range $[L, H]$ is chosen at random. At regular intervals, the price is varied by a random step size. If in the next interval, revenue per customer increases, then the next increment is chosen in the same direction, i.e., price is increased. If however, the revenue decreases, then the direction of increment is reversed, i.e., the price is decreased. At all times, the price is kept in the range $[L, H]$.

4   Simulations

We performed simulations to evaluate our pricing algorithm. Our implementation was designed to model a content delivery system. All our simulations are averaged over five runs with different seed values for the random number generator. We describe the components of our simulation below.

System Description: We performed simulations with two different systems, one with a T3 (45 Mbps) outgoing link and the other with OC-3 (155 Mbps) outgoing link. We assumed that there are enough servers to accommodate all the incoming requests. Therefore, outgoing bandwidth is the bottleneck resource. In our system, customers could choose from one of two LoSs: 56kbps or 128kbps. We chose request service times from a uniform distribution between 90 and 110 minutes.

Customer Choice of Products: In all our simulations we assume that there are 100 products for the customer to choose from. Customer choice of the products was assumed to follow a Zipf-like distribution with zipf-exponent. In a Zipf-like distribution, the ith popular product in a group of m products is requested with probability $\frac{\frac{1}{i^\theta}}{\sum_{j=1}^{m} \frac{1}{j^\theta}}$.

Customer Valuation Model: We assume that the products are partitioned into classes. A customer's valuation for a product is drawn from a probability distribution which is common for all products in a class. If the quoted price is less than this valuation, the customer accepts the price. We assume that the content provider knows how the products have been partitioned, but has no knowledge about the probability distribution. This is a reasonable assumption because, in real-life, the content provider can partition products into "New", and "Old" classes.

Since humans typically think in terms of discrete values, we chose three possible discrete probability distributions for modelling customer valuations: Uniform, Bipolar, and Zipf. We also chose one continuous distribution–Normal. We ignored negative values drawn from this distribution. In real life, customer valuations may not conform to any of these distributions. But in the absence of real life data, our objective was to test the robustness of the pricing algorithms over a range of "feasible" customer behavior patterns.

In all our simulations, our unit of currency is dimes (10 dimes = \$1). We performed simulations with numerous customer valuations. We present results for valuations corresponding to prices charged in movie theaters[4]. We chose two classes of products. For simplicity of labelling, we shall refer to these classes as "Old", and "New". For Uniform distribution, and 56kbps LoS, valuations were in the range $[15, 25]$ and $[25, 45]$ dimes for Old and New movies respectively. For 128 kbps LoS, valuations were in the range $[50, 70]$ and $[60, 90]$ dimes respectively. For Zipf distribution, and 56 kbps LoS, valuations were in the range $[20, 35]$ and $[30, 45]$ dimes for Old and New movies respectively. For 128 kbps LoS, valuations were in the range $[45, 60]$ and $[70, 99]$ dimes respectively. For Bipolar distribution, and 56 kbps LoS, valuations drawn from $\{15, 32\}$, $\{30, 38\}$ dimes respectively. For the 128 kbps LoS, valuations were drawn from $\{50, 67\}$ and $\{70, 87\}$ dimes respectively. For Normal distribution, the $\langle \mu, \sigma \rangle$ of the distributions were: $\langle 20, 5 \rangle$ for (Old, 56 kbps LoS), $\langle 35, 5 \rangle$ for (New, 56 kbps LoS), $\langle 50, 10 \rangle$ for (Old, 128 kbps LoS), and $\langle 80, 10 \rangle$ for (New, 128 kbps LoS).

Pricing Policy: We assume that the content-provider will charge at least \$1 and not more than \$10 for serving the content. We simulated all three pricing algorithms described in the previous section. For the Trial-and-Error-Pricing algorithm, we set small jump probability to be 0.05 and big jump probability to be 0.001 as mentioned by Sairamesh and Kephart [8]. For the Derivative-Following-Pricing algorithm, price increments were chosen from a uniform distribution in the range $[0, 10]$. In case of the HYBRID algorithm, we chose a server load threshold of 0.75. When current server load exceeded this threshold, exponentially increasing prices were charged.

Request Arrival Process: We obtained hourly logs from a Content Delivery Network[5] for all content requests between the dates December 28, 2000 and September 8, 2001. The data we obtained consisted of the number of connections and the number of bits transmitted per second during each hour of the observation period. The data was collected over streaming servers across the United States. We obtained data for Real, Windows Media and Quicktime connections. Of these, we could only use data from Quicktime connections because of differences in the way Real and Windows Media count the number of connections. Assuming that requests could only be of two types: 56 kbps or 128 kbps, we estimated the relative fraction of requests for each level of service. We then scaled the data we obtained so that it matched the total bits transmitted by

---

[4] We have observed theaters charging anywhere in the range of \$2.50 to \$8.50 for movies.

[5] We have withheld the name of the CDN upon request.

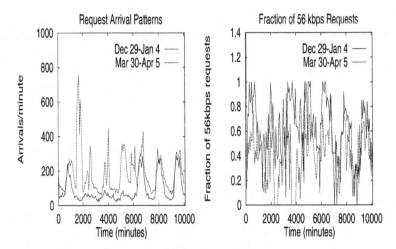

Fig. 1. Request arrival process.

Real, Windows Media and Quicktime put together. We chose two representative weeks for running our simulations: December 29 to January 4, and March 30 to April 5. The first week is during peak holiday season. The second week is during a normal week during the year. The request arrival rates and the fraction of 56 kbps requests for both the weeks is shown in Figure 1. Notice that the request arrivals are lower for the peak holiday season. This could possibly be because the CDN usage could have grown over the five months in between. Also, notice the the fraction of requests for 56 kbps service varies a lot in both the weeks.

Metrics: We use two metrics in our simulations: (1) revenue earned, and (2) percentage of requests denied service because they could not be scheduled due to lack of resources. The higher the revenue earned by a pricing algorithm, the better the performance. Ideally, we would like to compare the revenues earned by each algorithm with the predicted maximum expectation of revenue, computed using complete knowledge of system and customer parameters. However, as we shown in our earlier work [5], the revenue maximization problem is intractable when customer valuations conform to discrete probability distributions. Even for the normal distribution, in a system with 100 products and a variety of request arrival rates, it is difficult to compute the globally optimal revenue. We therefore only compare the revenues earned by the different algorithms in different scenarios. Thus, it is quite likely that even though one algorithm performs very well, the revenue earned using that algorithm may be very far from the global optimum. Our focus therefore has been to ascertain if one of the algorithms performs *consistently* well in comparison to the other algorithms across the different customer valuation and system load profiles. Our other metric, the fraction of denied requests, in very important for a commercial system for two reasons. First, a high percentage of denied requests indicates that the content-provider is not living up to service guarantees. Second, it indicates that the content provider is unable to manage available resources efficiently.

5   Results

We now present our simulation results. We simulated two situations: (1) there is no change in the customer behavior. However, the request arrivals are dynamic, and (2) there is a change in the customer behavior during peak hours.

To better illustrate the performance of each pricing algorithms, across different system and customer profiles, we present the results in tabular form. Each entry in the table is an ordered pair $\langle \mathcal{R}, r \rangle$. $\mathcal{R}$ is the mean revenue earned by that pricing algorithm over a number of simulations and r is the mean request denial rate. The mean revenue is rounded to the nearest 1000. By denial rate we mean the fraction of requests that could not be served (due to lack of resources) even though the customer agreed to the price. Note that it will not be appropriate to compare revenues within a column because, the customer valuations are different for different distributions.

December 29 2000

	T3 link			OC-3 link		
	HYBRID	TEP	DFP	HYBRID	TEP	DFP
Uniform	⟨ 469, 0.03 ⟩	⟨ 268, 0.49 ⟩	⟨ 356, 0.69 ⟩	⟨ 1273, 0.00 ⟩	⟨ 846, 0.35 ⟩	⟨ 1151, 0.53 ⟩
Normal	⟨ 472, 0.02 ⟩	⟨ 229, 0.39 ⟩	⟨ 300, 0.62 ⟩	⟨ 1207, 0.00 ⟩	⟨ 713, 0.26 ⟩	⟨ 967, 0.47 ⟩
Bipolar	⟨ 458, 0.05 ⟩	⟨ 250, 0.49 ⟩	⟨ 388, 0.39 ⟩	⟨ 1399, 0.01 ⟩	⟨ 790, 0.35 ⟩	⟨ 1220, 0.25 ⟩
Zipf	⟨ 310, 0.01 ⟩	⟨ 130, 0.33 ⟩	⟨ 268, 0.53 ⟩	⟨ 947, 0.00 ⟩	⟨ 407, 0.21 ⟩	⟨ 866, 0.40 ⟩

March 31 2001

	T3 link			OC-3 link		
	HYBRID	TEP	DFP	HYBRID	TEP	DFP
Uniform	⟨ 454, 0.03 ⟩	⟨ 219, 0.55 ⟩	⟨ 310, 0.70 ⟩	⟨ 1406, 0.01 ⟩	⟨ 652, 0.48 ⟩	⟨ 966, 0.62 ⟩
Normal	⟨ 485, 0.04 ⟩	⟨ 183, 0.50 ⟩	⟨ 260, 0.71 ⟩	⟨ 1461, 0.02 ⟩	⟨ 559, 0.43 ⟩	⟨ 850, 0.63 ⟩
Bipolar	⟨ 440, 0.04 ⟩	⟨ 208, 0.54 ⟩	⟨ 306, 0.60 ⟩	⟨ 1372, 0.02 ⟩	⟨ 627, 0.46 ⟩	⟨ 979, 0.52 ⟩
Zipf	⟨ 319, 0.00 ⟩	⟨ 125, 0.46 ⟩	⟨ 231, 0.55 ⟩	⟨ 1025, 0.00 ⟩	⟨ 416, 0.39 ⟩	⟨ 781, 0.48 ⟩

April 3 2001

	T3 link			OC-3 link		
	HYBRID	TEP	DFP	HYBRID	TEP	DFP
Uniform	⟨ 428, 0.02 ⟩	⟨ 179, 0.54 ⟩	⟨ 265, 0.65 ⟩	⟨ 1264, 0.01 ⟩	⟨ 550, 0.40 ⟩	⟨ 852, 0.51 ⟩
Normal	⟨ 459, 0.04 ⟩	⟨ 163, 0.50 ⟩	⟨ 257, 0.60 ⟩	⟨ 1331, 0.01 ⟩	⟨ 513, 0.37 ⟩	⟨ 862, 0.47 ⟩
Bipolar	⟨ 435, 0.04 ⟩	⟨ 171, 0.52 ⟩	⟨ 278, 0.59 ⟩	⟨ 1321, 0.01 ⟩	⟨ 533, 0.39 ⟩	⟨ 895, 0.45 ⟩
Zipf	⟨ 325, 0.03 ⟩	⟨ 119, 0.46 ⟩	⟨ 229, 0.57 ⟩	⟨ 1033, 0.00 ⟩	⟨ 400, 0.34 ⟩	⟨ 758, 0.44 ⟩

Fig. 2. ⟨Revenue, denial-rate⟩ of algorithms with no changes in customer behavior.

Simulation 1: Since the performance was consistent through all the days of the week, we only present results for single days during the chosen weeks. We presents for Dec 29 2000 (Friday, holiday season), March 31 2001 (weekend) and

March 2001

	T3 link			OC-3 link		
	HYBRID	TEP	DFP	HYBRID	TEP	DFP
Uniform	⟨ 473, 0.04 ⟩	⟨ 240, 0.60 ⟩	⟨ 318, 0.73 ⟩	⟨ 1469, 0.02 ⟩	⟨ 711, 0.52 ⟩	⟨ 995, 0.65 ⟩
Normal	⟨ 497, 0.06 ⟩	⟨ 201, 0.55 ⟩	⟨ 270, 0.74 ⟩	⟨ 1533, 0.03 ⟩	⟨ 617, 0.48 ⟩	⟨ 873, 0.66 ⟩
Bipolar	⟨ 458, 0.05 ⟩	⟨ 242, 0.60 ⟩	⟨ 314, 0.65 ⟩	⟨ 1424, 0.03 ⟩	⟨ 715, 0.52 ⟩	⟨ 1006, 0.57 ⟩
Zipf	⟨ 340, 0.02 ⟩	⟨ 158, 0.53 ⟩	⟨ 253, 0.63 ⟩	⟨ 1097, 0.01 ⟩	⟨ 509, 0.45 ⟩	⟨ 839, 0.55 ⟩

Table 2. ⟨Revenue, denial-rate⟩ of algorithms with changes in customer behavior.

April 3 2001 (weekday). Table 2 presents the results for the first set of simulations. In Table 2 we observe that on average, the HYBRID algorithm earns 75% to 150% more than TEP and 10% to 87% more than DFP across all the workloads and for both the system capacities. However, there were some simulations in which the TEP algorithm earned comparable revenues. This happened when the initial price chosen was close to the ideal fixed price. Even in those simulations, the HYBRID algorithm earned as much or slightly more revenue. The DFP algorithm outperforms the TEP algorithm in all the cases shown here, mainly because it learns the customer behavior better than TEP. We also observe that the service denial rate is very high for both TEP (0.21 to 0.55) and DFP (0.25 to 0.70) algorithms. This is because, they charge a low price and cannot accommodate all the requests. Such high service denial rates would be unacceptable in a commercial content delivery system. We also note that the revenues with OC-3 link are higher than with the T3 link. This is clearly because the system can accommodate more requests. Note that the revenues during the weekend (March 31) are in general more than the revenues during the weekday (April 3) for all the algorithms. This is because of the differences in the request arrival pattern.

Simulation 2: In the second set of simulations, we varied the customer valuation during peak hours. For the results presented in this paper, all the customer valuations were increased by 10-20 dimes during peak hours. Table 2 presents the results. Since the results are similar to the first set, we present results only for March 31 due to space constraints. The HYBRID algorithm consistently generates high revenues in comparison to the other algorithms across all customer distributions and resource constraints, mainly because it learns the customer behavior by experimenting with different prices. All the results appear consistently similar to the results in the first set of simulations. All revenues are marginally higher than in the first set because the customer valuations are higher during the peak hours. The revenues are however not significantly higher because the customer valuations did not increase significantly and moreover both the systems did not have enough capacity to satisfy all the requests. We also observe that the service denial rate is higher for all the algorithms. This is because, customers have more money to spend during peak hours, and therefore accept the quoted price more often. Note that the increase in service denial rate is higher in case of DFP (around 0.03 to 0.08) and TEP (0.04 to 0.07) than in case of HYBRID (around 0.01 to 0.02).

The reason why DFP and TEP do not perform well is that the algorithms do no consider resource constraints. They were primarily designed for a scenario where e-content can be delivered at leisure. We also ran other simulations to evaluate our choice of parameters for the algorithms. We only present a summary of our findings due to reasons of space. We observed that performance of the TEP and DFP did not vary when we changed the interval after which prices are reassessed. This was because the jumping probabilities for TEP are very small, and in case of DFP, the algorithm itself is independent of the interval. In case of the HYBRID algorithm however, we observed that a small interval generates higher revenue but increases service denial rate. We found that an interval of 45 minutes was ideal in terms of revenue as well as service denial rate. The results presented in this paper used a 45 minute interval for all the algorithms. We also observed that by increasing the big jump and small jump probability, the performace of the TEP was more erratic. The revenue did not increase or decrease in a consistent way with increasing jumping probability.

6   Conclusions

In this paper we developed an approach for pricing delivery of e-content in a system with multiple LoS. The pricing scheme, called HYBRID, was dynamic because the price varied with time. The pricing scheme was based on oberving customer reactions to price and provisioning of resources among the different levels of service. Resources were dynamically provisioned based on the amount of resources that requests for each LoS could consume. We compared the performance of this scheme with two other simplistic dynamic pricing schemes adapted from work by other researchers. We performed simulations using semi-realistic data to evaluate the performance of the algorithms. We observed that the HYBIRD pricing scheme consistently generates high revenues across a range of customer and system profiles. We also observed that the HYBRID pricing scheme reduced the number of customers rejected service due to resource constraints mainly by charging high prices at times of peak load. We observed that the two other dynamic pricing schemes failed to generate higher revenues mainly because they do not consider resource restrictions for content delivery.

In this work, we assumed that the customer's choice of LoS is independent of the price of the content. Such an assumption is valid in today's Internet where there is a big difference in quality between content avaiable at dial-up speeds and content available at broadband speeds. Such an assumption may not be valid in the future where more customers will have broadband connectivity, and content providers will possibly provide content at a range of quality levels, each marginally different from the other. Pricing mechanisms for such markets is an avenue for future research.

References

[1] S. Jagannathan and K. C. Almeroth, "Price issues in delivering e-content on-demand," *ACM Sigecom Exchanges*, vol. 3, May 2002.

[2] S. Jagannathan, J. Nayak, K. Almeroth, and M. Hofmann, "E-content pricing: Analysis and simulation," tech. rep., University of California Santa Barbara, November 2001. available at
http://www.nmsl.cs.ucsb.edu/papers/ECONTENTPRC.ps.gz.

[3] S. Jagannathan and K. C. Almeroth, "An adaptive pricing scheme for content delivery systems," in *Global Internet Symposium*, (San Antonio, Texas, USA), November 2001.

[4] S. Jagannathan and K. C. Almeroth, "The dynamics of price, revenue and system utilization," in *Management of Multimedia Networks and Services*, (Chicago, Illinois, USA), October 2001.

[5] S. Jagannathan, J. Nayak, K. Almeroth, and M. Hofmann, "On pricing algorithms for batched content delivery systems," tech. rep., University of California Santa Barbara, January 2002. available at
http://www.nmsl.cs.ucsb.edu/papers/BatchingPrc.ps.gz.

[6] P. Basu and T. Little, "Pricing considerations in video-on-demand systems," in *ACM Multimedia Conference*, November 2000.

[7] J. Mackie-Mason, C. H. Brooks, R. Das, J. O. Kephart, R. S. Gazzale, and E. Durfee, "Information bundling in a dynamic environment," in *Proceedings of the IJCAI-01 Workshop on Economic Agents, Models, and Mechanisms*, August 2001.

[8] J. Sairamesh and J. Kephart, "Price dynamics of vertically differentiated information markets," in *International Conference on Information and Computation Economies*, 1998.

[9] J. Wolf, M. Squillante, J. Turek, and P. Yu, "Scheduling algorithms for broadcast delivery of digital products," *IEEE Transactions on Knowledge and Data Engineering*, 2000.

[10] S. Chan and F. Tobagi, "On achieving profit in providing near video-on-demand services," in *Proceedings of the 1999 IEEE International Conference on Communications (ICC'99)*, June 1999.

Providing Authentication & Authorization Mechanisms for Active Service Charging[1]

Marcelo Bagnulo[1], Bernardo Alarcos[2], María Calderón[1], Marifeli Sedano[2]

[1] Departamento de Ingeniería Telemática. Universidad Carlos III de Madrid.
Av. Universidad 30 - 28911 LEGANES (MADRID)
{marcelo,maria}@it.uc3m.es
[2] Área de Ingeniería Telemática, Universidad de Alcalá de Henares
28871 Alcalá de Henares (MADRID)
{bernardo,marifeli}@aut.alcala.es

Abstract. Active network technology enables fast deployment of new network services tailored to the specific needs of end users, among others features. Nevertheless proper charging for these new added value services require suitable authentication and authorization mechanisms. In this article we describe a security architecture for SARA (Simple Active Router-Assistant) architecture, an active network platform deployed in the context of the IST-GCAP project. The proposed solution provides all the required security features, and it also grants proper scalability of the overall system, by using a distributed key-generation algorithm.

1. Introduction

Network services provision versatility has been dramatically improved by the introduction of active networking [1]. This technology provides network nodes with dynamic programmability capabilities, enabling the provision of customized services in a per customer basis, thus allowing clients to select specific services to be used when coursing its traffic. Since these services provide an added value to the users and its provision can imply additional costs to the operator, new charging mechanisms compatibles with this new service dynamic must be deployed as well. This charging mechanisms need that proper authentication and authorization guarantees be provided by the active network security architecture. However, heavy security can preclude deployment in real scenarios because of the imposed overhead in terms of processing, bandwidth and/or latency. So, in order to achieve a deployable active network architecture, the security solution must not only provide the guarantees needed for the charging system but it must also grant the scalability of the system. In this article, we will present the authentication and authorization mechanisms needed to provide an suitable charging system for active networks based on the SARA platform which can fulfill both requirements thanks to a distributed key-generation algorithm and to architectural features of the SARA platform.

---

[1] This work has been funded by CICYT under project AURAS.

B. Stiller et al. (Eds.): QofIS/ICQT 2002, LNCS 2511, pp. 337-346, 2002.
© Springer-Verlag Berlin Heidelberg 2002

The remainder of this article is structured as follows: in section 2 an introduction to SARA is presented, along with a description of its active packet exchanges. In section 3, the security solution requirements are detailed, including threats assessment and scalability requirements. Next, in section 4, the security architecture is described. In section 5, related work is presented and finally, section 6 is devoted to conclusions.

2. About Active Networks

There is clear trend towards extending the set of functions that network routers support beyond the traditional forwarding service. Active network technology aims to allow intermediate routers to perform computations up to the application layer and therefore making network more intelligent. Besides, this technology supports the deployment and execution on-the-fly of new active services, without interrupting the network operation. In this way, an active network is in the position to offer dynamically customized network services to customers/users.

This dynamic network programmability can be conceived by two different approaches. Some active networks platforms follow a discrete approach. This mean, packets don't include the code to be executed in the active routers, but exist a separate mechanism for injecting programs intro an active router. Usually this download is done from a server code or other system with the responsibility of storing the code. Others follow a integrated approach, and packets (called capsules) include not only user data but the code for the forwarding of the own packet as well. This code is then executed at the routers, or switches, as the packet propagates through the network.

Potential advantages of active networking include the opening up of the network to third parties, the easy introduction of sophisticated and unanticipated network services, and significant speedups in the deployment of such services.

2.1. About SARA

SARA (Simple Active Router Assistant) [2] is an active router prototype developed in the context of the IST project GCAP [3]. It is based on the router-assistant paradigm, meaning that active code does not run directly on the router processor but on a different device, called assistant, which is directly attached to the router through a high-speed LAN. Hence, the router only has to identify and divert active packets to its assistant. Active packets are identified by the router alert option, enabling active router location transparency, since active packets need not to be addressed to the active router in order to be processed by it. After requested processing is performed by the assistant, the packets are returned to the router in order to be forwarded. The active code needed to process active packets is dynamically downloaded from Code Servers when it is not locally available in the assistant. In this way safety is checked in advance, since only registered harmless-proofed code is allowed to run on the network. Thus the presumed target scenario is one in which a central administration provides active services loaded on the fly from a choice of known applications that have been provided by the customer or network manager.

SARA is available in two platforms: One fully based on linux [2] (playing both roles: router and assistant as a development scenario) and a hybrid platform where the router used is an Ericsson-Telebit AXI462 running a kernel adapted to work with an active assistant.

2.1.1. SARA Packet Exchanges

We will next introduce the packet exchanges performed, so we can detect the authentication and authorization requirements.

2.1.1.1. Elements involved in the packet exchange
Source: User terminal that generates traffic and uses the active features of the network.
Destination: It is the terminal that *Source* addresses its traffic to.
Active network operator: provides network services and additional active services. There are two main elements in the active network:
  Active Router: It is an active router (router plus assistant) capable of processing active packets. It is also able of obtaining the active code needed.
  Code Server: It is the active code repository that serves the *Active routers*.

2.1.1.2. Packet exchange description

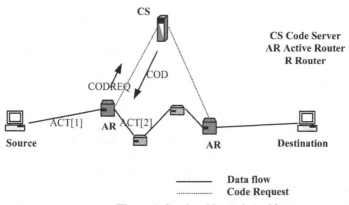

Figure 1. Services Network Architecture.

When *Source* needs special active processing for a flow of packets between itself and *Destination*, it must send packets (ACT[1] in the Figure 1), addressed to *Destination*, containing the Router Alert option and the identification of the active code that it desires to be executed. When this packet reaches the first *Active Router,* it is inspected and the identification of the active code is extracted. If the Active Code is locally available at the *Active Router*, it performs the requested process and then forwards the packet (ACT[2] in the Figure 1). If the needed active code is not locally available the *Active Router* requests it to the *Code Server* (CORREQ in the Figure 1). The *Code*

Server then sends the requested code to the *Active Router* ([COD] in the Figure 1), which now processes the packet and forwards it to the next hop. The same procedure is executed by all the *Active Routers* along the path, until the packet reaches *Destination*, where the packet is received. Next active packets of this flow will presumably follow the same path, so the *Active Routers* will be capable of processing them without needing to request the code from *Code Servers* again.

3. Security Architecture Requirements

In this section we will present the different requirements imposed to the security architecture. Charging system imposes the need for authentication and authorizations of , i.e. it must be possible to verify that the user that is requesting the code (*Source*) is authorized to executing it at this moment. In addition, the service request must be provided in a non repudiable fashion, since it is considered as an asset when charging is involved. However, it must be noted that non repudiation features are usually expensive, because they require the usage of public key cryptography. Besides, other security issues impose additional requirements that will be presented next. Finally we will describe other general requirements, specially emphasizing in scalability aspects that also have great impact in the final solution.

3.1. Additional Security Requirements

In order to perform an exhaustive analysis, we will retrieve the security requirements from each elements' perspective.

From the *Active Router's* perspective, it is relevant that the active code loaded into the routers is provided by an authorized *Code Server* and not from an unauthorized one. Besides, the code integrity must be preserved while it is transmitted from the *Code Server* to the *Active Router*.

From the *Code Server's* perspective, it must be able to authenticate *Active Routers* that are requesting active code, since not all the code will be available to all routers. Furthermore, the security solution must provide confidential code transfer, in order to prevent unauthorized parts to inspect the delivered code.

From the *Source's* perspective, it must be able to be certain no other user is requesting active services on its behalf, so that it is only charged for the services that it has requested. It must also be the only one capable to control its active services, meaning that no other user is capable of introduce new active packets or modify active packets sent by *Source*, interfering with the requested active services.

From the *Destination's* perspective, there are no requirements since it does not demand active services from the network, it just receives packets sent using them. In case that *Destination* would be interest in answering this packets using also active services, it would become *Source* and *Source's* requirements would apply. It should be noted that end-to-end security is out of the scope of this security solution.

3.2. Other General Requirements

Zero user knowledge at the *Active Routers:* In order to build a manageable solution, user management must not be performed on each and every *Active Router.* A central database containing all the users information, including access rights would be the preferred solution.

Active Router transparency: It must not be required that *Source* be aware of which *Active Routers* are in the path used by the network to transport packet towards destination. This means that active packets sent by *Source* must not be dependent of which *Active Routers* are addressed to.

4. Security Architecture

4.1. *Source* Authorization

4.1.1. Authorization Paradigm
The key feature that must be provided by the security architecture is authorization i.e. *Sources* must be authorized to execute the solicited code on *Active Routers*. There are two authorization paradigms that can be used: authorization based in access control lists or authorization based in credentials. The first paradigm is based on the existence of an access control list (locally available or in a remote location) that must be queried every time an *Active Router* receives an active packet sent by *Source*, in order to validate the *Source's* permissions. In this case the identity of requesting part must be authenticated in order to prevent impersonation. This approach then requires that the requested device (*Active Router*) has information about *Sources* and permissions or it imposes a communication with an authorization server every time a *Source* sends an active packet. The second paradigm demands that every time *Source* sends a packets, a credential that proves the *Source's* permissions must be presented. Then the requested device (*Active Router*) only needs to verify the credential. However, credential generation and distribution may be more than a trivial task.
   The solution proposed in this paper will be designed based on the second paradigm, since we consider that it provides better scalability attributes. Note that since the access control list can not reside on every *Active Router*, because of the Zero user knowledge requirement, it must reside in a remote location, imposing a remote query every time a *Source* need to be authorized.

4.1.2. Public Key Cryptography Vs. Symmetric Key Cryptography
In order to allow the intended use, a credential must contain verifiable authorization information i.e. the permissions granted to the holder of the credential. Besides, it must be possible to verify that the issuer of the credential has the authority to grant these permissions, (it will be called a valid issuer). It is also critical to validate that the user that is presenting the credential is the same user that the credential was granted to.

In order to fulfill the above stated characteristics of a credential, public key encryption can be used. So, a credential containing the *Source*'s permission and the *Source*'s public key is signed with the private key of the valid issuer. Then, the *Active Router* must be capable of verifying the authenticity of the credential, using the valid issuer public key, and also it must be capable of verifying that the requesting user has the private key that corresponds to the public key included in the credential. Even though this mechanism provides all the required features, the usage of public key cryptography is very demanding in term of processing, specially when considering that for every active packet, two public key signatures must be verified.

In order to obtain a less demanding solution, symmetric key cryptography can be used. However, building a similar system using symmetric key would require the usage of two different symmetric keys (a first one shared by the valid issuer and the *Active Routers* and another key shared by *Source* and the *Active Routers*). This system would still demand for two signature verifications and it would present the additional problem of key distribution. So, in order to improve the scalability of the solution, we will next explore the possibility of using only one symmetric key, shared by the valid issuer, *Source* and the *Active Routers*.

The requirements imposed to this key are:

- Different keys for different *Sources*. (i.e. the key must be linked to a *Source*)
- Different keys for the same *Source* at different moments (i.e. the key must have a validity period)
- Different keys for different active codes by the same *Source* (i.e. the key must be linked to a service)

So, the key issued by the valid issuer is linked to a *Source*, a code and a validity period.

Then, if this key is used to sign (HMAC [4] signature) an active packet that request the execution of a particular active code, the active packet itself plays the role of a credential. Basically, an *Active Router* receives a signed active packet that includes the requested code identification, the *Source* and the time when the active packet was generated. Then if the *Active Router* has a valid key linked to the *Source* and the requested code, it can verify the authenticity of the active packet, without any further information. This mechanism imposes the usage of an *Authorization Server* (the valid issuer role), that generates the keys. So, in order to execute a code in the network, *Source* must obtain the correspondent key from the *Authorization Server* in a secure way. This is not a time critical task, since it is only performed when the service is requested and it is possible to execute it in advance . However, once the service is authorized and the key is generated, the *Authorization Server* must communicate it to all *Active Routers* in the network, so they are aware of the new authorization. This is does not seems to be the most scalable solution, because of the amount of communications needed between the *Authorization Server* and the *Active Routers*.

We will next present an improved solution that minimizes the required interaction between this elements. The basic idea is that the key can be almost autonomously generated in every *Active Router* when it is needed. In order to achieve this, we will associate a key to every Active Code that can be loaded in the *Active Routers*. The key associated with active code C_i is called K_{ci}. These keys are known by the *Code Server* and by the *Authorization Server*. Then, when a *Source* S requests authorization for the execution of code Ci at a moment T and for a period P, the *Authorization Server* generates the key K as the HMAC of the concatenation of Kci, S, T and P. The key K is then transmitted to the *Source*, so it can sign the active packet with it. If we analyze the characteristics of K we can see that: K is linked to an active code (Kci); K is linked to a *Source*(S); K has a validity period (T, T+P); K can not be generated by any *Source*, since they do not have Kci. However, if the *Code server* confidentially transmits the active code [COD] it also attaches Kci, then the *Active Routers* are capable of regenerating K without contacting the *Authorization Server* every time an active packet arrives or when a new *Source* requests an already downloaded code. The *Active Routers* have all the information needed to generate K, i.e. S, T, P and Ci are included in the service request and Kci is obtained when they download the code from de *Code Server*. Note that the solution is based on shared secret keys, so the security level of the solution can be defined by setting the number of parts share the keys Kci (authorized *Active Routers* for a given code Ci) and the frequency Kci are changed.

4.2. Code Downloading

Another key feature that must also be provided is a secure way to download code (and keys Kci) from the *Code Server* into the *Active Routers*. However, this is not as time critical as user authorization since it is only performed once, when the first packet arrives. The following packets will benefit from a cached copy of the code and the Kci. So, a protocol that allows a secure communication between two parts is needed. We will use TLS [5] since it provides all the needed features. However, the usage a new protocol specifically designed for this task would result in improved efficiency, since TLS is a generic protocol. Then the *Code Server* must have a digital certificate (public key cryptography is used), and a TLS session is established between the *Code Server* and the *Active Router*, before the code is downloaded. The *Active Router* does not need a digital certificate, since non repudiation features are not required. Its authentication can be performed using a user and password, transmitted through the TLS session.

4.3. Non Repudiation

Since charging is involved when the user requests a service non repudiation features must be provided. In order to assure non repudiation, public key cryptography must be used when the user request authorization to the *Authorization Server*, as it will be described in the following section. However, our solution will not use public key cryptography in the credentials because of performance issues. Summarizing, the proposed solution provides non repudiation features when the service is requested to the *Authorization Server*, but it does not provides them when active packets are processed by *Active Routers*; this is tradeoff between performance and security features that we consider acceptable for most scenarios.

4.4. The Security Solution: Step by Step

In this section we will describe the complete mechanism which is illustrated in figure 2.

First (step 1 in figure 2) , *Source* requests authorization (to the *Authorization Server*) to execute an active code Ci in the network. This request is done in a secure way, meaning that public key cryptography and digital certificates are used by both parts. So, *Source's* request is signed with the private key of *Source* and its digital certificate is also included. This request is encrypted with the public key of the *Authorization Server*. Then the *Authorization Server* after receiving and verifying the request, it generates K as the HMAC of the *Source's* identification (S), the key associated to the requested code (Kci), the moment of the request (T) and the validity period requested by Source (P). Then the *Authorization Server* sends a signed message containing K. The message is encrypted with the public key of *Source*. At this moment, the *Authorization Server* has all the needed information for charging, since it has the requested service, the user, authenticated in a non repudiable way, and the time that this service was requested.

Source decrypts the message and obtains K. Then (step 2 in figure 2), it generates active packets , that includes its own identification S, the moment of the request T, the validity period P and an identifier of the solicited active code Ci. This message is signed performing a HMAC of the message plus K.

When an *Active Router* receives active packets, it first verifies that the packet is not obsolete, i.e. it is within the validity period and then it verifies the solicited active code availability. In case the code (and Kci) is not locally available, it downloads it, using a secure (TLS) connection from the *Code Server* (step 3 in figure 2). Then the *Active Router* generates K, using S, T and P extracted from the packet and Kci obtained from the *Code Server* when the code was downloaded. If the HMAC signature is verified, it means that *Source* has been authorized to execute the requested code, so it processes the packet using the solicited code and forwards it to the next hop. The same procedure is repeated on every *Active Router* along the path until the packet reaches *Destination*. The following active packets of the flow will benefit from cached copies of the active code and Kci in every *Active Router*.

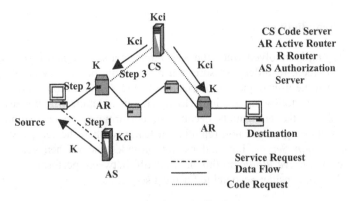

Figure 2. Key Distribution.

5. Related Work

There are many proposals for charging and accounting in a Multiservice Network [6], but few works have addressed the problem of the security mechanisms related to charging. One of these proposals has been done in the context of the CATI project [7].This IST project aims to investigate mechanisms of charging based on accounting QoS-based end-to-end communications. In CATI is proposed a security architecture for a scenario of multiple ISPs, an E-Commerce Service Provider, a customer, and a payment provider.

On the other hand, some proposals consider charging and accounting as an active application more. AIACE[8] and PEACH [9] explore the idea of using active services in order to enable charging and accounting system more scalable and flexible. In AIACE accounting active services are loaded on-the-fly and perform accounting tasks on behalf of accounting servers. PEACH is similar in many aspects to AIACE but focus more on the fast change in charging logic to reflects changes in business policies.

But to the best of our knowledge, FAIN [10] is the only active network platform previous to SARA that considers the charging for the use of active services. However, the scope of FAIN security architecture is limited mainly to protecting Active Network infrastructure from users and active code.

6. Conclusions

We have presented a security solution that provides authentication and authorization services for active networks based on the SARA platform. These features allow proper charging for active services. Futhermore, the solution performance is guaranteed by the usage of symmetric key cryptography. The scalability of the solution is assured by the authorization model, based on credentials, and the key distribution mechanism, that minimizes key exchanges by allowing key generation at every *Active Router* in an autonomous fashion. The security level of the solution is determined by the re-keying frequency i.e. how often Kci are changed. Then, we conclude that this solution enables the deployment charging mechanisms for SARA in a public testbed. Finally, it should be noted that it is possible to extend this architecture to other active network platforms as long as a central active code repository exists in the other platform.

References

1. David Wetherall, Ulana Legedza y John Guttag, "Introducing new Internet services: Why and How", IEEE Network Magazine [1998]
2. SARA home site. http://matrix.it.uc3m.es/~sara.
3. GCAP IST project home page. http://www.laas.fr/GCAP
4. Krawczyk H., M. Bellare, R. Canetti, HMAC: Keyed-Hashing for Message Authentication, RFC 2104, April 1997.

5. Alan O. Freier, Philip Karlton, Paul C. Kocher. TLS Working Group. The SSL protocol version 3.0. (November 18, 1996).
6. M. Falkner, M. Devetsikiotis, and I. Lambadaris, An Overview of Pricing Concepts for Broadband IP Networks, IEEE Communications Review, Vol. 3, No. 2, 2000.
7. B. Stiller, T. Braun, M. Günter, B. Plattner: "The CATI Project: Charging and Accounting Technology for the Internet", 5th European Conference on Multimedia Applications, Services, and Techniques (ECMAST'99), Madrid, Spain, May 26-28, 1999, LNCS, Springer Verlag, Heidelberg, Vol. 1629, pp 281-296.
8. Franco Travostino . "Towards an Active IP Accounting Infrastructure", Proc. 3rd IEEE Conference on Open Architectures and Network Programming, OPENARCH 2000, Tel Aviv, Israel, March 2000.
9. Lee, B. & O'Mahony,D., A Programmable Approach to Resource Charging in Multi-Service Networks, Proceeding of the IEEE International Conference on Software, Telecommunications and Computer Networks (SoftCoM 2000), October 10-14, 2000, Split, Croatia.
10. FAIN project (IST-1999-10561-FAIN) Deliverable Initial Active Network and Active Node Architecture, CEC Deliverable Nr: D2, editor: Spyros Denazism May 2001

Author Index

Lecture Notes in Computer Science

For information about Vols. 1–2419
please contact your bookseller or Springer-Verlag